Coastal Dunes
Form and Process

Edited by

Karl Nordstrom

Norbert Psuty

*Institute of Marine and Coastal Sciences, Rutgers University,
New Brunswick, New Jersey 08903, USA*

and

Bill Carter

*Department of Environmental Studies, University of Ulster,
Coleraine, Co. Londonderry, BT52 1SA, Northern Ireland*

JOHN WILEY & SONS

Chichester · New York · Brisbane · Toronto · Singapore

Other Wiley Editorial Offices

John Wiley & Sons, Inc., 605 Third Avenue,
New York, NY 10158-0012, USA

Jacaranda Wiley Ltd, G.P.O. Box 859, Brisbane,
Queensland 4001, Australia

John Wiley & Sons (Canada) Ltd, 22 Worcester Road,
Rexdale, Ontario M9W 1L1, Canada

John Wiley & Sons (SEA) Pte Ltd, 37 Jalan Pemimpin 05-04,
Block B, Union Industrial Building, Singapore 2057

Library of Congress Cataloging-in-Publication Data:

Coastal dunes : form and process / edited by Karl Nordstrom, Norbert
  Psuty and Bill Carter.
    p. cm.—(Coastal morphology and research)
  Includes bibliographical references.
  ISBN 0-471-91842-3
    1. Sand dunes. 2. Coast changes. I. Nordstrom, Karl.
  II. Psuty, Norbert P. III. Carter, Bill (R. W. G.) IV. Series.
  GB632.C63   1990
  551.3'75—dc20   90–34519
                                                           CIP

British Library Cataloguing in Publication Data:

Coastal dunes.
  1. coastal regions. Sand dunes
  I. Nordstrom, Karl   II. Psuty, Norbert   III. Carter, Bill
  IV. Series
  551.457

  ISBN 0-471-91842-3

Typeset by Dobbie Typesetting, Tavistock, Devon
Printed in Great Britain by Biddles Ltd, Guildford, Surrey

Coastal Dunes

Coastal Morphology and Research

Series Editor: Eric C. F. Bird

CORAL REEF GEOMORPHOLOGY
André Guilcher

COASTAL DUNES
Form and Process

Edited by

Karl Nordstrom, Norbert Psuty and Bill Carter

Contents

List of contributors

Dr Bernard O. Bauer, *Department of Geography, University of Southern California, Los Angeles, Ca. 90089-0663, USA*

Dr Ryszard K. Borówka, *Quaternary Research Institute, Adam Mickiewicz University, ul. Fredry 10, 61-701 Poznań, Poland*

Dr R. W. G. Carter, *Department of Environmental Studies, University of Ulster, Coleraine, Co. Derry, Northern Ireland BT52 1SA*

Dr Robin G. D. Davidson-Arnott, *Department of Geography, University of Guelph, Guelph, Ontario, N1G 2W1, Canada*

Dr Paul A. Gares, *Department of Geography, Colgate University, Hamilton, New York 13346, USA*

Mr Y. Gertner, *National Oceanographic Institute, Israel Oceanographic and Limnological Research Ltd, Haifa, Israel*

Professor Victor Goldsmith, *Department of Geology and Geography, Hunter College of the City University of New York, 695 Park Avenue, New York, NY 10021, USA*

Dr Patrick Hesp, *2 Fleming Street, Carlingford, NSW, Australia 2118*

Dr Shintaro Hotta, *Department of Civil Engineering, Nihon University, Chiyoda-Ku, Tokyo 101, Japan*

Mr Mark N. Law, *Department of Geography, University of Guelph, Guelph, Ontario, M1G 2W1, Canada*

Professor Anton McLachlan, *Department of Zoology, University of Port Elizabeth, Port Elizabeth, Republic of South Africa*

Associate Professor Karl F. Nordstrom, *Institute of Marine and Coastal Sciences, Doolittle Hall, Busch Campus, Rutgers University, New Brunswick, New Jersey 08903, USA*

Professor Antony R. Orme, *Department of Geography, University of California, Los Angeles, California 90024, USA*

Dr Shea Penland, *Louisiana Geological Survey, PO Box G, University Station, Baton Rouge, Louisiana 70893, USA*

Professor Norbert P. Psuty, *Institute of Marine and Coastal Sciences, Doolittle Hall, Busch Campus, Rutgers University, New Brunswick, New Jersey 08903, USA*

Dr Ken Pye, *Postgraduate Research Institute for Sedimentology, The University, Whiteknights, Reading, Berkshire RG6 2AB, England*

Professor William Ritchie, *Department of Geography, The University of Aberdeen, Elphinstone Road, Aberdeen AB9 2UK, Scotland*

Dr Maralyn J. Robertson-Rintoul, *ECOS, 52 Guild Street, Aberdeen AB1 2NB, Scotland*

Associate Professor Peter S. Rosen, *Department of Geology, Northeastern University, Boston, Massachusetts 02115, USA*

Associate Professor Douglas J. Sherman, *Department of Geography, University of Southern California, Los Angeles, Ca. 90089-0663, USA*

Professor Bruce Thom, *Department of Geography, University of Sydney, Sydney, NSW 2005, Australia*

Dr Peter Wilson, *Department of Environmental Studies, University of Ulster, Coleraine, Co. Derry, Northern Ireland BT52 1SA*

Foreword

During the past few decades there has been rapid growth in geomorphology. By the 1960s a number of textbooks on geomorphology had been published, most of them with a chapter on coastal geomorphology; in succeeding decades more specialised books on coastal geomorphology appeared; and now further specialisation is taking place. The Coastal Morphology and Research series is presenting books on topics such as coral reefs and cliffed coasts, which were covered in chapters of earlier works on coastal geomorphology.

The need for a specialised treatment of coastal dune geomorphology was apparent, but it was difficult to find a single author who felt he could encompass the whole of this field. Instead, two American dune specialists, Norbert Psuty and Karl Nordstrom from Rutgers University, and a British dune geomorphologist, Bill Carter from the University of Ulster, elected to produce a series of papers with the help of others active in dune research.

The aim was rather more ambitious than the conventional symposium, which usually produces a set of papers that are then simply edited and published. Professors Psuty, Nordstrom and Carter are distinguished scientists who have worked in this field over some decades, and they have spent three years working with the contributors on the papers originally submitted, in a process that I would call interactive editing. There has been much discussion of the research problems, requiring revision and extension of the original papers, and providing more general material that the three editors have used in the introductory and concluding chapters.

Coastal dunes show a great deal of variety in form, related to their sedimentology, their climatic setting (past and present), and ecological factors influencing their associated vegetation. An understanding of their morphodynamics requires studies from a variety of coastal environments around the world, and the 22 contributors to this volume have brought research from Australia, Britain, Ireland, Canada, Israel, Japan, Poland and South Africa, as well as the United States.

The result, we feel, is a collection of papers that represents the state-of-the-science in the broad field of coastal dune geomorphology, ranging from studies of aeolian processes and sand movement to the nature and sources of dune sediments, dune morphologies and their morphodynamics, and the role of soil and vegetation in the evolution of coastal dune landscapes. The collection will provide a scientific background to the study of dune geomorphology that will be of value not only to specialists, and students developing an interest in this field, but also to those concerned with management, planning, and development issues in coastal dune areas.

Eric Bird

Acknowledgements

During the three years' preparation time that this volume has consumed, we have been indebted to numerous individuals. In particular, we would like to thank the authors for their forebearance as well as their willingness to amend original manuscripts and provide additional material. We also acknowledge the help given by the various referees who found time to comment on the draft manuscripts. Readers should note that American and English spellings have been used consistently within the text of each chapter, according to the preference of the individual authors. Thus both 'eolian' and 'aeolian' are used, as well as 'meter' and 'metre', 'stabilize' and 'stabilise' and so on. For the sake of Anglo–American relationships the editors considered that this was the only way to resolve the arguments!

Technical assistance was provided by staff from both the Department of Environmental Studies at the University of Ulster and the Center for Coastal and Environmental Studies (now part of the Institute of Marine and Coastal Sciences) at Rutgers University. In particular, Mrs Mary McCamphill retyped much of the volume, and Mr Kilian McDaid drew or redrew several figures and mounted the photographs. Mr Nigel McDowell was responsible for turning a number of poor quality photographs into publishable images, Ms Jenny McArthur checked the references and Dr Clare Carter helped compile the index. However, as usual, responsibility for all deficiencies and errors rests with the editors.

Karl Nordstrom
Norbert Psuty
Bill Carter

February 1990

Chapter One

The study of coastal dunes

R. W. G. CARTER, K. F. NORDSTROM AND N. P. PSUTY

1. INTRODUCTION

This book is offered as a contribution to the understanding of active coastal dunes from a geomorphological perspective. There has been no long-standing tradition of geomorphological work on coastal dunes. Much of what has been accomplished has been led by management objectives, often in conjunction with engineering-orientated goals (van de Maarel, 1979), or as part of biological projects (Cowles, 1899; Olson, 1958; Ranwell, 1972).

Until very recently, there was a tendency to rely on the seminal contributions of Bagnold (particularly *The Physics of Blown Sand and Desert Dunes* first published in 1941) together with a small collection of widely cited papers—for example Zingg (1952), Landsberg (1956), Cooper (1958), Olson (1958). Many of the earlier investigators did not even attempt to identify the difference between coastal and interior dunes. In the last decade, process-based studies of coastal dunes have begun to revise and even question many long-held views (see, for example, Sherman and Hotta, this volume) and there has been a noticeable shift towards exploring the morphodynamic status of dunes (Short and Hesp, 1982). Simultaneously, it has become evident that coastal dunes face changing and increasing threats from human activities, resulting in widespread loss of geomorphic diversity, ecological productivity and amenity. Many geomorphic investigations have been a by-product of management studies through the initiatives of individual geomorphologists. Very few geomorphic data are required or expected by management funding agencies, reducing the likelihood that field information will be gathered and analysed. Despite these problems there has been a recent upsurge in interest in coastal dunes marked by several special thematic issues of journals (Psuty, 1988; Gimingham *et al.*, 1989; Bakker, Jungerius and Klijn, 1990) and conferences (including the formation in 1986 of the interdisciplinary *Eurodunes* group which organizes biennial symposia). A survey of GeoAbstracts (Parts A & E) reveals that the proportion of beach to

Coastal Dunes: Form and Process. Edited by K. F. Nordstrom, N. P. Psuty and R. W. G. Carter
©1990 John Wiley & Sons Ltd

dune geomorphological publications has remained about 3 to 1 since the early 1970s, but the total number of papers has increased. Currently about 20 peer-reviewed papers on coastal dune geomorphology appear each year. Recent reviews of dune geomorphology include those by Ranwell (1972), Goldsmith (1978), Pye (1983), Sarre (1987) and Carter (1988).

Recent advances in dune geomorphology include the linking of shoreline morphodynamics to the dune formation (Short and Hesp, 1982; Stewart and Davidson-Arnott, 1988), the development of storm-driven cyclic models of environmental change for low barrier dunes (Hosier and Cleary, 1977; Sorensen and McCreary, 1985), and the re-evaluation of many of the long-accepted transport formulae (Horikawa, Hotta and Kraus, 1986; Hotta, 1988). Despite these efforts, coastal dune research is still at a rather primitive stage. The role of wave action has not been placed in proper perspective relative to eolian processes and the significance of dune mobility to the long-term viability of the coastal ecosystem is poorly understood; many investigators still uncritically accept techniques and results from arid land dune studies. This book attempts to overcome some of these deficiencies, by establishing a broad theme of processes and morphology, but with a restriction to basic research issues, most of which are under-represented relative to applied research.

2. OCCURRENCES

Coastal dunes are found above the high water marks of sandy beaches. They occur on ocean, estuary and lake shorelines from the Arctic to the Equator (Fig. 1). The dunes discussed in this book are all formed by wind transport of predominately loose, sand-sized (2 to 0.2 mm) sediment of different origins. Coastal dunes find their greatest expression on windward coasts, especially where suitably textured sediment is abundant, for example on the Atlantic coasts of northwest Europe, the Pacific northwest of North America, southeast Australia or southwest Africa. (While all coastal dunes are eolian in character, readers should note that, confusingly, the term 'dunes' is used by geologists and engineers to refer to water-lain bedforms.) Coastal dunes comprise a variety of sediments, including quartz, calcareous particles (foraminfera, comminuted shell and coral), heavy minerals and volcanic dust. The basic requirement is for particles with a hydraulic equivalence approximating well-rounded, sand-sized (2 to 0.2 mm) quartz grains with a specific gravity in air of 2.65 g cm^{-3}. Dunes occur on many types of shoreline, from low, gently sloping surfaces to cliffed coasts. Dune formation is a function of sediment grain size, the characteristics of the beach profiles, and the wind regime. Dunes form most readily on low angle shorelines under dissipative (spilling) wave domains, where the foreshore dries out at low water or where swash bars weld regularly onto the upper beach (Short and Hesp, 1982; Carter, 1986). The rate of entrainment of sand by the wind depends both on the characteristics of the near-surface

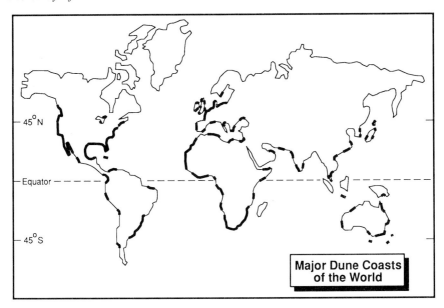

45°N

Equator

45°S

**Major Dune Coasts
of the World**

Fig. 1   Distribution of the major dune coasts of the world

airstream and on the nature of the surface. Both these factors may change rapidly in the course of a few hours. The deposition of sand is controlled by topography, the presence of obstructions (litter lines, tree trunks), obstacles (buildings, seawalls) and above all vegetation. The role of vegetation, in fixing, trapping and accumulating sediment, and the role of the waves in shaping the beach, replenishing sediment sources and reworking the foredunes distinguish coastal dunes from the majority of active dunes in arid regions. The growth of dunes is closely related to the delivery rate of sediment. Sedimentation is often restricted by limits in the density and growth rate of vegetation. On a worldwide basis (Fig. 1), sediment availability is greatest on paraglacial coasts (those coasts affected by glaciations); where past sea-level fluctuations have allowed shelf material to migrate onshore and where present coast erosion supplies sand-sized material. Dunes are also common around river mouths and inlets where sediment is available and easily transported, and along flat, 'trailing edge' coasts where barrier islands are common. However, dunes can form in almost any climatic zone. Even the humid tropics, at one time considered to lack coastal dunes, boast extensive systems (Swan, 1979; Pye, 1983). On tropical islands, often the only geomorphic feature above water level is formed of windblown material.

Dunes range in size and longevity from tiny, ephemeral forms to massive, persistent dunefields. Generally, preservation potential increases with volume. The smallest dunes are the most vulnerable to frequent wave overwash, whereas the largest forms can adjust to all but the most extreme climatic and tectonic

events. Dunes may be gaining or losing, or retaining a fairly constant volume.
They may also be mobile or fixed in position. The more stable forms are usually
fixed by vegetation. Sand-binding vegetation thrives in all but the most extreme
climatic or high sand flux conditions, including frequent drought and dessication
(even in temperate latitudes), nutrient limitation, exposure to high winds and
high salinity. Many dune plants have physiological, reproductive or behavioral
characteristics (Crawford, 1989) to overcome these stresses. Unstable dunes are
mobile, often migrating inland as parabolic, transverse or barchan forms,
sometimes as sand sheets (Hesp *et al.*, 1989). There is a fine line between stable
and mobile dunes, and many systems have switched from one form to the other,
often as a result of human activities (Higgins, 1933; Cooper, 1958. Garcia-Nova,
Ramirez-Diaz and Torres-Martinez, 1976). However, some dune systems fall
in the discontinuity between the two forms, often displaying complex mosaics
of fixed and free forms over relatively small areas as sand is accumulated, lost
or redistributed by wave or wind processes. These systems are characterized
by erosion scarps, blowouts, redeposition dunes, deflation hollows and
impersistent slip faces (Carter, Hesp and Nordstrom, this volume). It may be
that this geomorphic diversity is crucial to maintaining the quasi-stability of
these systems.

Many dunes are artificial in as much as they have been formed by humans,
both by design and by accident. Many of the European dunes were encouraged
by the planting of *Ammophila* from the early Middle Ages on, and the US dunes
have been extensively secured by planting and the construction of sand fences.
Some dunes have developed as by-products of environmental changes elsewhere.
For example the growth of dunes at tidal inlets is often associated with the
reclamation of adjacent lagoons, leading to a diminution of the tidal prism and
a subsequent shrinkage of the inlet bedforms (Carter, 1988). The impedance
of drift in front of jetties, piers and terminal groynes may lead to beach
accumulation, resulting in dune growth.

3. DUNE GEOMORPHOLOGY

Viewed in geologic time, coastal dunes are transient, mobile forms that stand
a relatively low chance of being preserved in the stratigraphic column. This is
not to decry the geologic role of dunes, as they play a major part in coastal
evolution, supplying shoreface deposits and sealing lagoons. Dunes react to
environmental change on a variety of time scales, especially to variations in
sediment supply and to sea-level change.

Morphodynamic interactions in dunes are not well understood, partly because
the chaotic relief of many coastal dunes defies simple description. The response
of a dune surface to a deep, turbulent fluid (air) flow is hard to predict beyond
the basic two- and three-dimensional dune shapes. Some of the flow-bedform
and structural index relationships established for desert dunes (Wilson, 1972;

Hunter, 1977) can be applied in a coastal context (e.g. Hesp, 1981), although this approach tends to be limited to depositional structures in the absence of vegetation cover. Air flow over complex, mobile relief is liable to be constantly adjusting, so that the relationship between dune topography and the overpassing fluid may be described as a strongly cybernetic one, resulting in a stochastic (or non-determinate) landscape. Although a classification of coastal dune types based on complex process-response linkages is not possible at this time, geomorphologically relevant groupings have been made on the basis of age (e.g. van Straaten, 1963; Hesp, 1987), structure (Bigarella, 1972; Goldsmith, 1978), position relative to the shoreline (Davies, 1972), morphology (King, 1972) or shoreline morphodynamics (Short and Hesp, 1982; Carter, 1988). Probably the most universally applied typology is a loose one based on an amalgam of positional, morphological, stability and age factors, embracing incipient (or yellow, or primary, or embryo) foredunes nearest the sea, established (or white, or secondary, or stabilized) secondary dunes farther inland, and a range of unstable, mobile forms, including barchan, parabolic, transverse and linear dunes (Fig. 2). Not all dune systems have these components. Most develop from the incipient dune stage to the controlled foredune stage. Incipient dunes may form and grow into a new foredune seaward of the former foredune, which then becomes part of the secondary dune. Many of the larger dunefields pass through a mobile stage, that may eventually be controlled either by vegetation (biologically arrested) or by morphology (topographically trapped).

Dunes develop within a broad framework of environmental controls, including tectonics, sea-level change, sediment availability and beach and nearshore conditions (Belknap and Kraft, 1981; Davis and Clifton, 1987). Locally, and over relatively short time scales (months, years, decades), sediment availability is probably the most important factor, although within a regional setting, over longer periods (centuries, millennia) sea level changes are crucial, as they redistribute coastal materials over wide areas, and ultimately determine local levels of wave attack. During periods of sea-level rise, large amounts of sediment are released by coast erosion, some of which is transported alongshore and accumulates as dunes. If the rate of sediment delivery to the dunes is high and the dunes are not fixed by vegetation, then extensive mobile dune sheets may form. However, if the dunes are stabilized and material is constantly reworked, the result is often a single, large foredune ridge, which migrates slowly inland as the sea level rises. In some instances, rising water level may lead to instability as the vegetation is overwhelmed by increased sand flux. It is likely that dunes have enhanced sensitivity to changes in sea level at certain times (Short, 1988). During a lowering of sea level, sediment is fed from the shelf within the falling wave base. On low-angle, high-energy coasts where sediment is abundant, this may result in a sustained sand supply to the shore over several thousand years, creating extensive beach and dune ridge plains (Roy, Thom and Wright, 1980). If the dunes are mobile, sand sheets or dunefields may result, but if vegetation

Fig. 2 Topographic variety in dune form. A: Linear and transverse ridges in southeast Australia (photograph Patrick Hesp); B: Foredunes and shadow dunes in the Donaña National Park, southern Spain. C: Dune deflation surfaces and embryo dunes on Spiekeroog, West Germany. D: Parabolic dunes on Cape Cod

is able to fix the dunes, it is likely that a prograded series of foredunes will develop. Slight fluctuations in sea level may result in combinations of these dune types. It is not necessary to invoke sea-level changes to explain these patterns; sediment supply variations can impart similar styles of dune development under stationary sea levels. For example, periodic cliff erosion or river flooding due to occasional storms may both result in a series of dune ridges. The extensively prograded Dutch dunes appear to have formed during a slow late-Holocene sea-level rise (van Straaten, 1963) and similar patterns are evident in Nova Scotia, Canada, under the present rapidly rising water levels (Carter *et al.*, 1989). Constant longshore feeding of sediment will lead to progradation of the shoreline. These changes are unlikely to continue for long, as they would lead to shoreline plan equilibrium and a cessation of transport, but it is conceivable that a constant supply from longshore transport could lead to sand sea development as in Namibia (Hesp *et al.*, 1989), or on a more local basis to headland bypassing, where sand from one littoral compartment is moved to another via a dune (Stapor, May and Barwis, 1983).

On a shorter time scale, many coastal dunes occupy a pivotal position in terms of coastal stability. The dunes supply, store and receive sand blown from and to the adjacent beaches. The periodic supply of sand to the beach is a major feedback loop lessening coast erosion by flattening the nearshore slope and reducing storm wave energy at the shoreline. In inter-storm periods sediment is returned to the beach and replaced in the dunes. This process was probably first described by Bremontier in 1833, and has been rediscovered at intervals ever since. The constant exchange of material between beach and dune is important, as it supplies fresh sand and nutrients to renew or sustain plant growth, and adds topographic variety to the system. Many foredunes grow within the context of these sediment exchanges, perhaps surviving long enough between storms to become established features.

4. HUMANS AND DUNES

Much of the world's coastal sand is stored subaerially as dunes, and it is this material that forms the front-line of defence against both short- and long-term fluctuations in the sea surface. Sand dunes undergo major changes as a result of the predicted sea level rises. One view (e.g. Komar, 1983) holds that dunes are vulnerable landforms eroding rapidly when attacked by waves. An alternative view suggests that dunes are adaptive, persistent and offer effective protection against wave attack (Ritchie and Penland, this volume). Perhaps the key to understanding the geomorphology of dunes is in recognizing that they are dynamic and opportunistic forms, developing rapidly where and when conditions are favorable. Dunes are especially active following storm erosion, because new sand surfaces form, releasing sediment for removal by the wind. The geomorphological characteristic of dunes that is of greatest human value is their ability to respond quickly to changing conditions and to buffer sudden shoreline alterations. Thus it is not surprising that much of the previous research on coastal dunes has focussed on their use as a form of shore protection.

Dunes should be viewed in a far broader context, as integral parts of larger coastal systems exchanging mass, energy, biota and information (through feedback) with abutting environments. Dunes play a strongly deterministic part in the ecosystem, as many biological gradients (shelter/exposure, water abundance/deficit, saline/fresh, nutrient rich/poor) are associated with physiographic variation (Fig. 3). These gradients exist at different scales both within and between dune systems. The beach/dune interface appears to be an important ecological front, supporting sustained levels of biological activity within the context of sediment flux and morphological change. On many coasts, the dune-beach-nearshore system operates homostatically, with numerous checks and balances controlling its behaviour. Thus the alteration or removal of one element of the system from the remainder is inadvisable and can lead to major adjustment.

Coastal dunes are often sensitive to environmental change, which in many cases may have been triggered by human activities, perhaps dating back several thousand years (e.g. Enright and Anderson, 1988; Wilson and Farrington, 1989). The relationship between coastal communities and dunes has altered significantly over the last two centuries. Initially dunes posed a hazard largely from wind drift (blowing sand). In its worst manifestation dunes engulfed houses, fields or even entire settlements (Higgins, 1933). Dunes were treated with respect, and care was often taken not to disturb them unduly. Nonetheless many areas had to be abandoned as sand shifted. Gradually, dunes have become controlled, mainly by planting sand-stabilizing vegetation (Carter, 1988), although in some places blowing sand is still considered a hazard (Sorensen and McCreary, 1985; Nordstrom, 1989). Dunes are now readily manipulated by humans, which is often a major attraction, particularly for developers. Many dune systems have been destroyed by competing uses, whereas others have been modified to attain a human utility value quite distinct from their intrinsic or ecological value (Nordstrom, 1990). The economic value of many dunes has increased rapidly in recent years. Dunescapes are sought after by developers because they provide ideal 'soft' locations for leisure and recreation activities. In many countries, including Spain, Mexico, the USA and Australia, coastal dunes are exploited for major tourist initiatives (Fig. 4), often with little or no consideration for the natural environment (Carter, 1988). The new dune-user groups bring with them changed perceptions of the dunes (Carter, 1985). Where the traditional view is of a dynamic, hazardous environment, the modern appreciation is of a static, unchanging landscape, eminently suitable for subjugation. In recent years, dunes are being threatened by humans, rather than vice versa.

Dune resources, including mineral, water and aesthetic, are exploited to varying degrees throughout the world. This exploitation often leads to conflict.

10

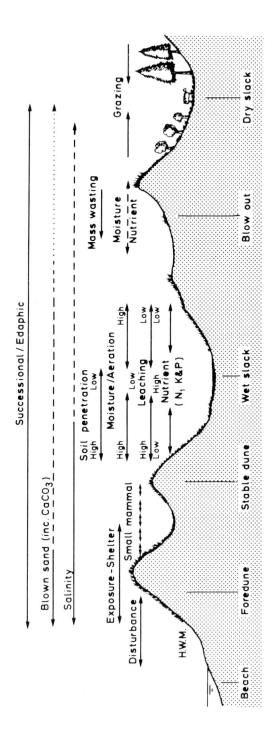

Fig. 3 Environmental gradients across coastal dunes (reproduced from Carter, 1988, by permission of Academic Press, London)

Fig. 4   The construction of an entirely new coastal city for 250 000 people at Matalascanas near Seville has been sited within the dune system. The foredunes have been removed to provide better views for the owners and renters of villas and condominiums

In the absence of effective planning strategies like zoning, dunescapes often become battlegrounds for competing and diverse uses. For example, Engel (1983) traces the turbulent history of the South Shore of Lake Michigan outside Chicago, where the dunes were heavily used for recreation by the urban population, and simultaneously coveted for industrial development. These conflicts require external resolution. The most commonly applied solution is to devise a series of activity zones with dunes becoming fragmented on the basis of land-use. Perhaps the most advanced model of dune zoning occurs in the Netherlands (van de Maarel, 1979), where dunes are zoned on a national basis to provide for shoreline protection, water supply, nature conservation, recreation and military training. Proper planning involves consideration of all the roles played by dunes in coastal systems, including geomorphological, ecological and human values.

5. PROSPECT

This volume is an attempt to address the need to define the geomorphological role of coastal dunes by examining many of the unresolved issues facing geomorphologists. The focus of the book is on basic research, but the results have application to management questions. We have attempted to provide a perspective on the variety of coastal dune forms that can occur, using examples from different parts of the world, representing dunes of different size, developmental stage, relationship to wave processes, sediment budgets and human alterations. The book is organized thematically into four sections as an overall structure for the individual chapters. The content of these sections corresponds to a traditional means of grouping coastal dune studies, but there is considerable thematic overlap between sections. The fundamental difficulty in categorizing individual chapters lies in the sheer range of ways by which dunes may be studied, a range that reflects the complex inter-relationships and feedbacks inherent in most dune environments. Such complexity underscores the need for research to identity specific linkages between processes and resultant landforms.

REFERENCES

Bagnold, R. A. (1941) *The Physics of Blown Sand and Desert Dunes*. Chapman & Hall, London, 265pp.

Bakker, T. W. M., Jungerius, P. D. and Klijn, J. A. (eds) (1990) European coastal dunes, *Catena Suppl.* **18**, 1–240.

Belknap, D. F. and Kraft, J. C. (1981) Preservation potential of transgressive coastal lithosomes on the US Atlantic Coast. *Mar. Geol.*, **42**, 429–442.

Bigarella, J. J. (1972) Eolian environments—their characteristics, recognition and importance. In Rigby, J. K. and Hamblin, W. K. (eds), *Recognition of Ancient Sedimentary Environments*. SEPM, Tulsa, Okla., pp. 12–62.

Carter, R. W. G. (1985) Approaches to sand dune conservation in Ireland. In Doody, P. (ed.), *Sand Dunes and their Management*. Nature Conservancy Council, Peterborough, pp. 29–41.

Carter, R. W. G. (1986) The morphodynamics of beach ridge development, Magilligan, Northern Ireland. *Mar. Geol.*, **73**, 191–214.

Carter, R. W. G. (1988) *Coastal Environments*. Academic Press, London, 617 pp.

Carter, R. W. G., Forbes, D. L., Jennings, S. C., Orford, J. D., Shaw, J. and Taylor, R. B. (1989) Barrier and lagoon coast evolution under differing sea-level regimes: examples from Ireland and Nova Scotia. *Mar. Geol.*, **88**, 221–242.

Cooper, W. S. (1958) *Coastal sand dunes of Oregon and Washington*. Geol. Soc. Amer. Mem., **72**, 169pp.

Cowles, H. C. (1899) The ecological relations of the vegetation on the sand dunes of Lake Michigan. *Bot. Gaz.*, **27**, 95–117, 167–202, 281–308, 361–391.

Crawford, R. M. (1989) *Studies in Plant Survival*. Blackwell Scientific Publications, Oxford, 296pp.

Davies, J. L. (1972) *Geographical Variation in Coastal Development*. Oliver & Boyd, Edinburgh, 204pp.

Davis, R. J. Jr and Clifton, H. E. (1987) Sea-level change and the preservation potential of wave-dominated and tide-dominated coastal sequences. In Nummedal, D., Pilkey, O. H. and Howard, J. D. (eds), *Sea-level Fluctuation and Coastal Evolution*. SEPM, Tulsa, Okla., pp. 167–178.

Engel, J. R. (1983) *Sacred Sands: The Struggle for Community in the Indiana Dunes*. Wesleyan Univ. Press, Middleton, Conn., 352pp.

Enright, N. J. and Anderson, M. J. (1988) Recent evolution of the Mangawhai Spit dunefield. *J. R. Soc. NZ*, **18**, 359–367.

Garcia-Nova, F., Ramirez-Diaz and Torres-Martinez, A. (1976) El sistema de dunas de Donaña. *Natural Hispan.*, 5, 1–56.

Gimingham, C. H., Ritchie, W., Willetts, B. B. and Willis, A. J. (eds) (1989) Coastal sand dunes. *Proc. R. Soc. Edinburgh* **B96**, 1–313.

Godfrey, P. J., Leatherman, S. P. and Zaremba, R. (1979) A geobotanical approach to classification of barrier systems. In Leatherman, S. P. (ed.) *Barrier Islands*. Academic Press, New York, pp. 99–126.

Goldsmith, V. (1978) Coastal dunes. In Davis, R. A. Jr (ed.), *Coastal Sedimentary Environments*. Springer-Verlag, Berlin, pp. 171–235.

Hesp, P. A. (1981) The formation of shadow dunes. *J. sediment. Petrol.*, **51**, 101–112.

Hesp, P. A. (1987) Morphology, dynamics and internal stratification of some established foredunes in southeast Australia. *Sed. Geol.*, **55**, 17–42.

Hesp, P. A., Illenberger, W., Rust, I., McLachlan, A. and Hyde, R. (1989) Some aspects of transgressive dunefield and transverse dune geomorphology and dynamics, south coast, South Africa. *Zeit. Geomorph. Suppl-Bd.*, **73**, 111–123.

Higgins, L. S. (1933) An investigation into the problems of the sand dune areas on the South Wales coast. *Arch. Cambriensis*, **88**, 26–67.

Horikawa, K., Hotta, S. and Kraus, N. (1986) Literature review of sand transport by wind on a dry sand surface. *Coast Eng.*, **9**, 503–526.

Hosier, P. E. and Cleary, W. J. (1977) Cyclic geomorphic patterns of washover on a barrier island in southeastern North Carolina. *Environm. Geol.*, **2**, 23–31.

Hotta, S. (1988) Sand transport by wind. In Horikawa, K. (ed.), *Nearshore Dynamics and Coastal Processes*. University Tokyo Press, pp. 218–238.

Hunter, R. E. (1977) Basic types of stratification in small eolian dunes. *Sedimentol.*, **24**, 361–387.

King, C. A. M. (1972) *Beaches and Coasts* (2nd edn) Arnold, London.

Komar, P. D. (1983) The erosion of Siletz Spit, Oregon. In Komar, P. (ed.), *Handbook of Coastal Processes and Erosion*, CRC Press, Boca Raton, Florda, 65–76.

Landsberg, S. Y. (1956) The orientation of dunes in relation to winds. *Geogr. J.*, **122**, 176–189.

Maarel, van de E. (1979) Environmental management of coastal dunes in the Netherlands. In Jefferies, R. L. and Davy, A. J. (eds), *Ecological Processes in Coastal Environments*. Blackwell Scientific Publications, Oxford, pp. 543–570.

Nordstrom, K. F. (1989) Dune grading along the Oregon coast, USA: a changing environmental policy. *Appl. Geogr.*, **8**, 101–116.

Nordstrom, K. F. (1990) The intrinsic characteristics of depositional coastal landforms. *Geogr. Rev.*, **80**, 68–81.

Olson, J. (1958) Lake Michigan dune development I. *J. Geol.*, **66**, 254–263.

Psuty, N. P. (ed.) (1988) Dune/Beach Interaction, *J. Coast Res. Special Issue No.*, **3**, 1–136. CERF, Charlottesville, Virginia.

Pye, K. (1983) Formation and history of Queensland coastal dunes. *Zeit., Geomorph. Suppl.-Bd.*, **45**, 175–204.

Ranwell, D. S. (1972) *The Ecology of Salt Marshes and Sand Dunes*. Chapman & Hall, London, 258pp.

Roberts, H. H., Ritchie, W. and Mather, A. S. (1973) Cementation in high latitude dunes. *Coast. Stud. Bull., Louisiana State Univ.*, **7**, 95–112.

Roy, P. S., Thom, B. G. and Wright, L. W. (1980) Holocene sequences on an embayed, high energy coast; an evolutionary model. *Sediment. Geol.*, **26**, 1–19.

Sarre, R. (1987) Aeolian sand transport. *Prog. Phys. Geogr.*, **11**, 155–182.

Short, A. D. (1988) The south Australia coast and Holocene sea-level transgression. *Geogr. Rev.*, **78**, 119–136.

Short, A. D. and Hesp, P. A. (1982) Wave, beach and dune interaction in southeast Australia. *Mar. Geol.*, **48**, 259–284.

Sorensen, J. C. and McCreary, S. T. (1985) Institutional arrangements for coastal resource management in developed and developing countries. *Proc. Coast. Zone, '85, ASCE*, 1–25.

Stapor, F. W., May, J. P. and Barwis, J. (1983) Eolian shape-sorting and aerodynamic traction equivalence in the coastal dunes of Hout Bay, Republic of South Africa. In Brookfield, M. E. and Ahlbrandt, T. S. (eds), *Eolian Sediments and Processes*. Elsevier, Amsterdam, pp. 149–164.

Stewart, C. J. and Davidson-Arnott, R. G. D. (1988) Morphology, formation and migration of longshore sandwaves: Long Point, Lake Erie, Canada. *Mar. Geol.*, **81**, 63–77.

Straaten, van L. M. J. U. (1963) Aspects of Holocene sedimentation in the Netherlands. *Verh. Kon. Ned. Geol. Mijnb. Gen. Geol. Series*, **21**, 149–172.

Swan, B. (1979) Sand dunes in the humid tropics: Sri Lanka. *Zeit. Geomorph. NF*, **23**, 152–171.

Wilson, I. (1972) Aeolian bedforms—their development and origins. *Sedimentol.*, **19**, 173–210.

Wilson, P. and Farrington, O. (1989) Radiocarbon dating of the Holocene evolution of Magilligan Foreland, Co. Londonderry. *Proc. R. Ir. Acad.*, **89B**, 1–23.

Zingg, A. W. (1952) Wind tunnel studies of the movement of sedimentary material. *Proc. 5th Hydraul. Conf. Iowa University*, **34**, 111–135.

Section I

EOLIAN PROCESSES AND SAND TRANSPORT

An understanding of dune development inevitably must be linked to the mechanics of air flow across the beach and dunes and the pathways and rates of sediment transport associated with eolian processes. Many previous investigations of these processes in coastal dunes have relied on data and models derived from investigations in wind tunnels and from field sites in inland arid regions. The four chapters presented in this section provide perspective on the application of much of this earlier work to the study of coastal dunes. Chapter Two, by Sherman and Hotta, begins with a critical assessment of the theory and measurement of sediment transport. Their chapter functions both as a primer on eolian transport and an appraisal of unresolved research issues. One of the principal issues they identify is the need to improve the means of identifying surface shear stresses through an analysis of velocity profiles. Sherman joins forces with Bauer and others in Chapter Three to determine the nature of the vertical velocity profile as the wind adjusts to new boundary conditions in passing from the water to the beach. Their work complements an earlier study by Hsu (1987) that focussed on alterations in flow within the dunefield. These studies of meso-scale alterations are complemented by Robertson-Rintoul's detailed analysis of wind flow patterns over isolated dunes within a dunefield, presented in Chapter Four. The results of her study reveal the complex nature of small-scale flow patterns, and they indicate that flow patterns can be dramatically different from flows over ideal surfaces.

Field investigations invariably reveal differences between predicted transport rates determined in controlled environments and values determined in more complex real-world settings. The coastal zone incorporates many limitations to achieving full transport, as indicated in Bauer *et al*. These differences are treated more explicitly in Chapter Five by Goldsmith, Rosen and Gertner, who use data from sediment traps and meteorological stations to analyze the spatial variability of sediment transfers.

REFERENCE

Hsu, S. A. (1987) Structure of airflow over sand dunes and its effect on eolian sand transport in coastal regions. *Proc. Coast. Sed. '87, ASCE*, 188–201.

Chapter Two

Aeolian sediment transport: theory and measurement

DOUGLAS J. SHERMAN
Department of Geography, University of Southern California

AND

SHINTARO HOTTA
Department of Civil Engineering, Nihon University

1. INTRODUCTION

The formation of coastal sand dune systems is a response to a conceptually simple, but physically complex series of processes associated with sediment transport by fluids. Much of our understanding remains rudimentary because of the difficulty in both developing deterministic, physically-based models and measuring the appropriate parameters in the field. Despite the decades of research on the basic processes associated with aeolian transport and dune formation, Hotta (1985, p. 5.45) notes correctly that 'a systematic and well-controlled field measurement of the sand transport rate has not yet been carried out,' and Horikawa, Hotta and Kraus (1986, p. 523) conclude a review with the statement that, 'more effort should be directed toward refining measurement techniques for the sand transport rate by wind.'

This chapter reviews the critical concepts in aeolian sediment transport and discusses methods for field experiments designed to elucidate transport processes. Comprehension of these processes is a prerequisite for understanding the dynamics of coastal dune formation and behaviour, and is therefore fundamental to knowing these landforms. The discussion is in two parts: (i) general concepts of dune formation; and (ii) principles of aeolian sediment transport.

Coastal Dunes: Form and Process. Edited by K. F. Nordstrom, N. P. Psuty and R. W. G. Carter
©1990 John Wiley & Sons Ltd

2. DUNE FORMATION

Conceptually we can recognize that the formation of coastal dunes is dependent upon there being a source of sediments, a wind fast enough to move the sediment, and a location, at least partly removed from wave activity, for the preferential deposition of sand. Vegetation, where present, will enhance deposition of grains in transit and anchor local deposits.

These basic notions cannot be taken for granted. Pye (1983, p. 531), for example, states that 'there are two basic requirements for the formation of coastal dunes: (i) availability of adequate supplies of well-sorted beach sands; and (ii) onshore winds capable of moving sand for at least part of the year.' However, neither of these conditions is basic for dune formation. Although coastal dunes are usually formed by the onshore movement of beach sands there are locations where the movement of sediments into dunes is alongshore or offshore (e.g. Cooper, 1967, p. 54; Psuty, 1969; Leatherman, 1976; Rosen, 1979). The other notion, that well-sorted sands are required for dune development, runs contrary to accepted concepts of aeolian sediment transport. The early work of Bagnold (1941, p. 67), for example, showed transport rates are increased across surfaces with 'sand with a very wide r of grain sizes.'

If we assume onshore transport, the outline of dune formation begins the movement of wind over the sand surface. If wind velocity exceeds a cri value then sediments also will begin to move. The sand will continue to r landward until the net transport rate decreases. Decreased transport will re for example, from an increase in surface slope, or a decrease in wind velc The first shoreward encounter between wind and vegetation can cause an at gradient in surface shear stress and moving sand will be deposited quickly. local deposition will increase the beach slope and make it harder for sar continue to move inland. This focusses deposition and results in the sim form of dune generation.

The propriety of the general conditions for dune formation given above be demonstrated by a consideration of the processes associated with ae sedimentation. Again, the underlying concepts are straightforward, understanding remains difficult.

3. AEOLIAN PROCESSES

The movement of sand by wind results from momentum transfer from the air to the sediment. This basic concept is at least implicit in all models of aeolian sediment transport, although formal expressions of the relationship vary substantially. Because none of the existing approaches works well under field conditions, the emphasis here will be on the difficulties in applying models in prototype environments.

3.1. Airflow over a surface

Air flowing across a surface will impart a shear stress to that surface as a result of a net downward flow of momentum. The shear stress (τ) is the force the wind exerts on the sand surface:

$$\tau = \rho u_*^2 \tag{1}$$

where ρ is air density (approximately 1.22 kg m^{-3}), and u_* is shear velocity. For turbulent flow, momentum transfer depends on the aerodynamic roughness of the surface, and is proportional to the change in velocity with elevation. Assuming that velocity increases logarithmically with height, this relationship is described by the general wind profile equation (von Kármán, 1934):

$$u_z = (u_*/\varkappa) \ln[(z-h)/z_0] \tag{2}$$

where u_z is wind velocity at an elevation z above the sand surface, $\varkappa$ is von Kármán's Constant (approximately 0.40), z_0 is the roughness length (corresponding to the elevation at which wind speed goes to zero), and h is the displacement height (the aerodynamic boundary, usually assumed to be at the sand surface, $h = 0$ m). For unvegetated sand surfaces, the displacement height is small relative to the roughness length and can be disregarded. Equation (2) predicts shear velocity based on wind measurements and a knowledge of the roughness length and thereby links wind velocity with shear stress. This is one of the most fundamental relationships in understanding sediment transport.

3.2. Initiation of sediment movement

For cohesionless or moveable beds composed of sedimentary particles, individual grains will begin motion when the shear stress exceeds a threshold value. The forces opposing motion are due to gravity and any partially cohesive agent (e.g. surface moisture). Ignoring cohesive forces, the fluid motion must overcome gravity only.

The fluid forces of lift and drag operate to move the grain upward and downwind, respectively. For a sand grain resting on the surface, the lift force arises due to the uneven velocity distribution over the particle. Relatively high velocity flow over the top of the grain will cause an upward directed pressure gradient and thus a net lifting force. The drag force results from the force of the air against exposed areas of the grain, and is a function of wind speed and the aerodynamic roughness of the surface. Therefore the drag is reflected in the slope of the vertical velocity gradient, and a critical shear velocity, u_{*t}, for the initiation of motion can be derived:

$$u_{*t} = A[gd(\rho_s - \rho)/\rho]^{0.5} \tag{3}$$

where A is the square root of the Shields Function (e.g. Miller, McCave and Komar, 1977), g is gravitational acceleration, d is mean grain diameter, and

ρ_s is sediment density. Estimates of A vary from about 0.1 (Bagnold, 1941) to 0.2 (Lyles and Woodruff, 1972). Iverson *et al.* (1987) attribute the variation in A through a similar range to differences in the density ratio between sediments and the transporting fluid. The analysis of Hotta (1985) suggests that equation (3) is valid for sands larger than 0.1 mm diameter. More detailed discussions of the physics of the initiation of motion can be found in Middleton and Southard (1978), Allen (1982) and Sarre (1987).

As the shear velocity exceeds the threshold for motion, sediment transport begins, with the sand moving as suspension, saltation, or traction (creep) load. Because true suspension is rare for sand-size particles it will not be considered further. Movement by saltation or creep is quite common on beaches and coastal dunes.

The saltation load represents the greatest fraction of aeolian sand transport (Bagnold, 1941; Wilson and Cooke, 1980). This is due partly to feedback between the saltation load and the flow structure of the wind near the surface. It has been demonstrated repeatedly that saltation load distorts velocity profiles in the vicinity of the boundary, with an apparent increase in the surface roughness (Bagnold, 1941; Zingg, 1952; Horikawa and Shen, 1960; Belly, 1962; Hsu, 1973; Hotta, 1985). Bagnold (1941) noted that as a result of the saltation load, velocity profiles associated with a range of shear velocities seemed to converge on what he termed a focal point. He interpreted this phenomenon as an indication of the threshold velocity and of the height of local ripples. He suggests (p. 59): 'no matter how strongly the wind is made to blow . . . the wind velocity at a height of about 3 mm (the focal point) remains almost the same.' From these results, the velocity gradient equation must be rewritten as:

$$U_z = (u_*/\varkappa) \ln (z/z') + u' \tag{4}$$

where z' and u' represent the coordinates of the focal point. Bagnold (1941) also notes that during active saltation the value of A in equation (3) falls to about 0.08, as the result of increased momentum transfer by impacting grains. Chepil (1945) found A to be about 0.085 for the impact threshold.

3.3. Sediment transport over ideal surfaces

Ideal surfaces, with regard to predicting wind-blown sand transport, are horizontal, dry, unobstructed and unvegetated. The saltation population is assumed to be in equilibrium with the local flow field. For these circumstances, there are a large number of equations, based upon estimates of shear velocity, designed to relate airflow conditions and sediment characteristics for the prediction of sand transport rates. These are reviewed by Horikawa, Hotta and Kubota (1986) and Sarre (1987). The primary approaches are reflected in the models of Bagnold (1941), Kawamura (1951), Horikawa and Shen (1960), Kadib (1965), and Hsu (1973). The basic models are discussed briefly here, with emphasis on the environmental parameters needed to solve the equations. Special attention is paid to the complications arising from applying these models under the non-ideal conditions found in beach-dune environments (see also Bauer *et al.*, 1990; and Davidson-Arnott and Law, 1990). Detailed guidelines for engineering applications are presented by Hotta (1988).

An early equation for the sand transport rate was derived empirically by O'Brien and Rindlaub (1936) based upon the wind speed cubed and incorporating a threshold velocity. Horikawa and Shen (1960) modified this equation to:

$$q = 9.96 \times 10^{-7} \, (u_* + 10.8)^3 \tag{5}$$

where q is the transport rate in gf s^{-1} cm, and $u_* > 0.20$ m s^{-1}. This equation appeals in its simplicity. If values of the focal point coordinates are known or can be estimated, then (5) can be solved using (4) with measurement of wind speed at one elevation only.

Bagnold (1941) based his model on laboratory observations of saltation and basic principles of physics. Assuming that the movement of sand is the result of momentum transfer directly from the wind, he obtained:

$$q = C(d/D)^{0.5} (\rho/g) \, u_*^3 \tag{6}$$

where C is an empirical coefficient related to the sediment size distribution, and D is a reference grain size of 0.25 mm. Bagnold (1941, pp. 67, 72) assigned values for C ranging from 1.5 for nearly uniform sand, to 1.8 for typical dune sands, to 2.8 for moderately to poorly sorted sands, to a maximum of 3.5 over hard, relatively immobile surfaces (e.g. pebbles, or lag deposits in swales or blowouts). This model requires measurement of local grain size and at least an estimate of sorting. It is also implicit that the constant does not change abruptly, as indicated, but rather grades with changes in sorting.

For typical dune sands, Bagnold (1941, p. 106) reduced his equations to obtain:

$$q = 1.5 \times 10^{-9} \, (u - u_t)^3 \tag{7}$$

where q is in cgs units, u is measured at 1 m, and u_t is a threshold wind velocity equal to u'.

Lettau and Lettau (1977) provided a similar formula, expressed in terms of shear velocity:

$$q = C'(d/D)^{0.5} \, (u_* - u_{*t}) \, u_*^2 \, \rho/q \tag{8}$$

where C' equals 4.2. The use of either (7) or (8) requires that a threshold condition also must be measured or predicted.

Kadib (1964) assumed that part of the momentum transfer was due to stress caused by the impact of saltating grains, and derived a similar model from considerations of the physics of grain motion:

$$q = K\rho/g(u_* + u_{*t})^2(u_* - u_{*t}) \tag{9}$$

where K is a coefficient with values varying from an average of about 1.0 from laboratory work, to an average of 2.7 from field experiments. The higher value is attributed to flow unsteadiness in natural conditions (Horikawa, Hotta and Kraus, 1986). This model has the same data requirements as (7) or (8).

Kadib (1964, 1966) used Einstein's (1950) bed load transport model to modify the Bagnold approach, suggesting that:

$$q = \phi g [\rho_s/(\rho_s - \rho)]^{0.5} [1/(gd^3)]^{0.5} \tag{10}$$

where ϕ is an index of the intensity of transport, depending on flow intensity (see Kadib, 1966, or Sarre, 1987 for details).

From dimensional considerations, Hsu (1974, 1987) related the transport rate to a Froude Number:

$$q = H[u_*/(gd)^{0.5}]^3 \tag{11}$$

where H has the dimensions of q and is related empirically to grain size:

$$\ln H = (-0.42 + 4.91 \, d) \times 10^{-4} \tag{12}$$

Most studies that attempt to measure aeolian sediment transport in the laboratory or in the field use several of the above models for purposes of comparison, usually with mixed to poor results (e.g. Svasek and Terwindt, 1974; Berg, 1983; Sarre, 1984 (as cited in Sarre, 1987) Hotta, 1985; McCluskey, 1987; and Bauer *et al.*, this volume). Each of the models presented above has been preferred by at least one subsequent study.

There are several general criticisms of these approaches, including (i) the O'Brien–Rindlaub/Howikawa–Shen approach is too general and ignores several fundamental transport parameters (e.g. sediment size) and is therefore not appealing on deterministic grounds; and (ii) the Kadib and Hsu models can be modified by the use of constants to approximate the Bagnold/Kawamura results and are thus redundant. The Kadib model is especially troublesome because of the reliance on graphical solutions, although it is notable that both Berg (1983) and Sarre (1987) report that this model worked best in their field applications (however, Sarre, 1988, contradicts his own finding).

3.4. Sediment transport over non-ideal surfaces

All of these models assume that transport occurs over an ideal, or quasi-ideal, surface. Where this assumption is not tenable, local transport rates can deviate substantially from predicted rates. Accurate modelling of aeolian sedimentation, therefore, requires terms to correct for these variations.

3.4.1. Slope effects

For movement over non-horizontal surfaces, Howard *et al.* (1978) propose a correction factor for predicted transport rates:

$$q_c = q/\cos \theta' \ (\tan \alpha + \tan \theta') \tag{13}$$

where q_c is the slope-corrected estimate, θ' is the slope in the direction of transport, and α is the dynamic friction angle (assumed to be approximated by an angle of repose, 33 degrees). The transport slope differs from the local slope, θ, as $\tan \theta' = \tan \theta \cos \chi$, where χ is the angle between the wind vector and surface gradient (Howard, 1977). The value of including a slope correction factor is certain. Although (13) has also been used or cited by Lancaster (1985) and Sarre (1987), it is not a useful approximation. For a horizontal surface, the corrected and uncorrected transport rates should be identical, $q_c = q$. However, substituting $0°$ slope into (13) results in $q_c = 1.6q$, a substantial error. It is likely that a $\tan \chi$ term has been omitted from the numerator of (13).

Hardisty and Whitehouse (1988) have suggested that the slope effect is best estimated by $q_c = B_q$, where the slope correction, B_1 is found from:

$$B_1 = [\tan \alpha/(\tan \alpha - \tan \theta)]^7 \tag{14}$$

This empirical relationship results from more than 450 field observations, and fits their data well.

3.4.2. Sediment cohesion

Several researchers have attempted to quantify the effect of the bonding of surface sediments on the local transport rate. These models tend to address the role played by either surface moisture or salt crusting. The principal role played by either agent is to increase the threshold of motion.

There have been a number of attempts to formalize the relationship between the water content of the surface and near-surface sands and the threshold shear velocity (e.g. Belly, 1964; Johnson, 1965; Svasek and Terwindt, 1974; Logie, 1982). Based upon consideration of existing literature, Hotta *et al.* (1984) designed a series of laboratory and field experiments to assess the influence of water content on sand movement. They found that the standard threshold equation (3) can be modified to:

$$u_{*tw} = u_{*t} + 7.5w \tag{15}$$

where w is the per cent water content of the upper 5 mm of sand. Equation (15) is valid for prototype conditions of $0\% < w < 8\%$, and $0.2\,mm < d < 0.8\,mm$ (Hotta *et al.*, 1984). A water content of 8% is approximately that found in the upper portions of the swash zone (Iwagaki, 1950).

Based upon laboratory work, Belly (1964) found the moisture effects on entrainment thresholds to be described by:

$$u_{*tw} = u_{*t} (1.8 + 0.6 \log_{10} w) \tag{16}$$

This relationship was found for 0.44 mm sands and over a moisture content range of 0 to 4%.

There have also been attempts to link directly the general effect of moisture content on sediment transport (e.g. Iwagaki, 1950; Woodruff and Siddoway, 1963; Sarre, 1988). Hotta (1985), again based upon experimentation, substituted a water-compensated threshold shear velocity, based upon equation (15), that includes a dimensionless 'evaporation factor', I_w:

$$u_{*tw} = u_{*t} + 7.5 w \tag{17}$$

where I_w varies, depending on the evaporation rate, from 0.0 to 1.0 when $u_{*t} < u_* < u_{*tw}$, and is 1.0 for $u_* > u_{*tw}$.

In a contradictory finding based on field measurements, Sarre (1988) has concluded that moisture contents up to 14% have no effect on transport rates. This conclusion was based on the quality of agreement between his observed transport rates and those predicted by models that did not include moisture terms. This suggests that the issue of moisture effects is not yet well resolved.

The formation of crusts due to evaporative concentration of salts will also increase the threshold shear stress. Nickling (1978, 1984) has pioneered this line of research. Nickling and Ecclestone (1981) have proposed a modified, threshold shear stress equation to account for the cohesive effects of NaCl:

$$u_{*ts} = u_{*t} (0.97^{0.103 \, s/c}) \tag{18}$$

where s/c is mg salt g^{-1} soil.

It has also been noted that biological agents can bind sediments and thereby reduce dramatically aeolian transport rates. The role of algae in crust formation is described in Fletcher and Martin (1948), Campbell (1979), and Forster and Nicholson (1981), among others. Van der Ancker, Jungerius and Mur (1985) used a laboratory experiment to measure transport from encrusted and non-encrusted surfaces. They found that transport from a well-developed crust was less than 1% of that from an unprotected surface.

3.4.3. Vegetation effects

Finally, it is apparent that the presence of vegetation can play a central role in controlling local sediment transport. The primary influence of vegetation is to modify the near-surface velocity field in a manner that reduces the bed shear stress (Hesp, 1981). Olson (1958), Bressolier and Thomas (1977), and Pye (1983) attribute the reduction of τ to an increased roughness length related to the characteristics of the vegetation. Olson (1958) reported a 30-fold increase in z_0 after planting *Ammophila*. Bressolier and Thomas (1977) show that the roughness length increases with both vegetation height and density, especially with the latter. According to Pye (1983, p. 537), vegetation reduces the wind shear near the surface by 'raising the roughness height . . . and decreasing the slope of the wind velocity profile,' thus enhancing sedimentation rates. Although Pye's Fig. 4 (after Olson, 1958) appears to show a reduced velocity profile slope, the impression is an artifact of plotting the independent variable elevation on the dependent variable, y, axis. The real effect of increasing the roughness length is to increase the velocity gradient. Inspection of equations (1) and (2), above, reveals that for a given wind velocity, an increase in the surface roughness length indeed will increase the boundary shear stress and should increase transport over vegetated surfaces. The confusion arises from the failure to recognize the role played by the displacement height.

The displacement height, h, is the average elevation of the surface roughness elements. It is also the elevation at which the average surface shear stress acts (Jackson, 1981). Because vegetation is a roughness element, it will thus increase the average surface elevation as a function of the height and density of the plants. Where vegetation is relatively low and sparsely distributed, or absent entirely, the sand surface will be the primary control on h.

For dense vegetation, h is about two-thirds mean plant height, H_v (Oke, 1978; or $0.7H_v$, Jackson, 1981). The role of vegetation, therefore, is to increase the elevation where the mean shear stress occurs. If this statistical elevation is lifted above the highest elements of the sediment surface, then transport can be sporadic at best (if occasional gusts penetrate to the sand). The high roughness length values reported in the articles above must include an unspecified displacement elevation, above which occurs the true roughness length.

Consideration of the models presented above leads to a recognition of the appropriate environmental parameters that must be measured or derived in order to specify the transport conditions. These parameters can be classified as sedimentologic, geomorphologic, atmospheric, or biologic factors. Experimental design for measuring these factors is presented below.

4. SEDIMENTOLOGIC PARAMETERS

4.1. Grain size characteristics

The sedimentologic factors to be measured include grain size characteristics and quantities of sand moved. Grain size characteristics that are fundamental to using the equations above are mean or median diameter and some measure of sorting is needed. These values can be derived through the analysis of sieve, settling tube, or other size distribution data, and there is a large literature describing the pertinent procedures (e.g. Folk, 1966, 1974; Goudie, 1981; or CERC, 1984). It has also been shown that grain shape is an important influence on transport potential where sediments are markedly non-spherical (MacCarthy, 1935; MacCarthy and Huddle, 1938; Mattox, 1955; Torobin and Gauvin, 1960; Williams, 1964; Jensen and Sorensen, 1986).

The design of a sampling programme is also crucial to a well-run field experiment. Because of the complex dynamics of the nearshore zone, the characteristics of beach sands exhibit large spatial and temporal variability (King, 1972). This implies that a description of source-area sediments may require multiple samples in both time (Bagnold, 1941; Cooke and Warren, 1973; Carter, 1976) and space (Krumbein and Slack, 1956; Greenwood, 1978).

4.2. Measuring transport rates

The accurate measurement of sediment transport rates in field experiments is difficult. There are three conceptually distinct methods of estimating transport: (i) detailed measurement of volumetric changes in source/sink areas; (ii) sediment trapping; and (iii) tracer experiments.

Where source and sink areas can be clearly defined, net sediment transport will be reflected in topographic changes, and can be approximated through field mapping. For example, Bagnold (1941), and Inman, Ewing and Corliss (1966), measured the migration of barchan dunes, and the latter study related dune height and the migration rate to sediment transport using the relationship:

$$L = qT/\gamma_s H_d \qquad\qquad (19)$$

where L is the distance travelled in time T, H_d is dune height, and γ_s is the specific weight of the sand. Illenberger and Rust (1988) have also used this approach to model a sediment budget for the Alexandria dunefield in South Africa based on aerial photography and field measurements of dune geometry.

By assuming that foredune vegetation is an effective trap, transport rates have been measured by micro-profiling to find elevation changes in sample grids (Hesp, 1983; Davidson-Arnott and Law, this volume). Horikawa *et al.* (1984) used an inverted form of this general approach with great success, by excavating large trenches in a beach and detailing the rates of infilling due to deposition of aeolian sediments.

Most attempts at prototype measurement of sediment transport rely on small-scale trapping. There have been few major advances in trap technology since the review by Horikawa and Shen (1960), and their Type V-5 trap may be a near optimum design (Horikawa and Shen, 1960, Fig. 34 and pp. 47–48). Some notable innovations in trap design include the development of a continuous weighing trap (Lee, 1987), a Venturi-compensating trap (Illenberger and Rust, 1986), large (1.5 m elevation), vertical, total traps (Hotta, 1985), and varieties of horizontal, total traps (Belly, 1964; Walker and Southard, 1984). For experiments where deployment of a large number of traps is desirable, especially in experiments where they will be unattended, the cylindrical traps described by Leatherman (1978) or Rosen (1978) are an excellent alternative to more complex and expensive designs. This is a viable option, and of great practical utility, when the efficiency of the cylindrical traps can be measured or estimated.

There has also been some field experimentation using tracers to estimate transport rates. Perhaps the best designed and most useful study was that of Berg (1983). Using fluorescent tracer, Berg modelled the migration of the concentration centroid across an almost flat sand surface. Tracer was prepared and analysed using the methods described in MacArthur (1980), and Berg tagged discrete sediment size classes. When he deployed his tracer, Berg (1983, p. 103) diluted his samples and placed them in a trench 5 mm deep. Perhaps the primary criticism of this approach is that under conditions of equilibrium transport or accretion, the release of tracer from this shallow inset may be inhibited or prevented, and thus not truly reflect transport rates.

5. GEOMORPHOLOGIC PARAMETERS

The geomorphologic characteristics to be measured include the local slope and variability of adjacent surfaces, and in some instances it is desirable to measure bedform geometry. Most of this work can be accomplished using standard field survey and mapping methods (Goudie, 1981). Repetitive, detailed mapping, using either break-in-slope or grid methods, is required for longitudinal studies of net changes across a study area. The grid method requires precise and accurate reoccupation of former survey points in order to minimize measurement error. This problem can be mitigated by installing a series of secure semi-permanent benchmarks tied to reference baselines and elevations.

For shorter period investigations, a survey area can be networked with a series of pins to denote location. Elevation changes can be measured against the exposed length of the pin (this is a small-scale version of pipe-profiling, for example Horikawa *et al.*, 1984). For some applications it may be desirable to use washers with the pins to produce a depth-of-activity grid (Greenwood and Mittler, 1984). This method will provide additional information concerning change in elevation when observers cannot be present.

These data will also provide the information needed to perform slope corrections on predicted transport rates, and can also be useful in demonstrating topographical influences on the local wind field. As a field expedient, the geomorphogic attributes of a site can be approximated using an inclinometer and a tape measure. This is also valuable for quick, one-person surveys.

6. ATMOSPHERIC PARAMETERS

The atmospheric characteristics influencing sediment transport are fluid density, wind speed and direction, temperature, precipitation, surface moisture and, to a lesser extent, relative humidity and solar radiation. All of these attributes can be measured or derived using standard meteorological methods.

6.1. Air density

When applying the transport equations, it is usually assumed that air density is a constant, about 1.2 kg m^{-3}. However, this value varies with temperature, humidity, and elevation. Chepil (1945) has suggested that under typical conditions the total variation in density will seldom exceed 3%, and he dismisses this variation as insignificant. However, this assumption may not always be tenable under prototype conditions.

The density of air decreases with rising temperatures and increasing humidity. The density of cold, dry air (at $-5°C$) is about 1.32 kg m^{-3}, and the density of warm, moist air (at 40°C) is about 1.10 kg m^{-3}. Because air density enters linearly into most initiation of motion and transport equations, this variation can translate into an error term as large as 20% if all other parameters are perfectly specified. This source of error can be eliminated by correcting for temperature changes (e.g. table in Monteith, 1973, p. 221). The variation with humidity is slight and usually can be disregarded.

Air density also varies with barometric pressure. However, over the range of typical sea-level air pressures, density changes only about 3%. Thus Chepil's (1945) dismissal of barometric pressure as an important parameter is appropriate for marine environments. However, important variations in pressure may occur when evaluating aeolian transport at higher elevations. This has potential impacts for the comparison of field experiments performed in lacustrine or desert environments. Near sea level, mean pressure drops at a rate of about 1 mb per 10 m increase in elevation (Barry and Chorley, 1982). This means offsets of more than 100 mb (10%) for the Great Salt Lake region in Utah, for example.

6.2. Wind speed and direction

There are many techniques developed for the measurement of wind speed and direction. Wind speed is usually monitored with anemometers using propellors or cups (Goudie, 1981, p. 227), or hot-wires or films (Shaw, Kidd and Thurtell, 1973; Lang and Leuning, 1981) or acoustics (Shaw, Kidd and Thurtell, 1973; Hotta, 1985). For most field experiments, Gill-type 3 cup anemometers combine the requisite attributes of simplicity, reliability, and accuracy, with a threshold speed of about 0.4 m s^{-1}, an accuracy of $\pm 1/2\%$ over the operating range (up to 50 m s^{-1}), and a life expectancy of 2 to 4 years of continual operation. Gill anemometers produce a signal by two means, either by directly generating a DC voltage, or by conditioning (using a photo-chopper) an input AC supply, and thus require some form of recording device. There are also a variety of hand-held wind instruments that are especially practical for reconnaissance or use in remote locations.

Wind direction is usually measured with a wind vane. These can be either hand-held instruments that measure a relative azimuth, or electronic devices, generally using a potentiometer to condition an input voltage. Viable, but less conventional, field expedients include the measurement of the strike of scour-hollows downwind of obstructions (Howard, 1977), or the strike of the depositional pattern of hand-dropped sands.

For measuring velocity profiles (to derive u_*) it is necessary to deploy vertical arrays of anemometers and employ some form of recording device. Bauer *et al.* (this volume) measured velocity gradients across a beach with two towers, each with six, Gill three-cup, DC-generating, anemometers (at elevations of 0.3, 0.5, 1.0, 2.0, 4.0, and 8.0 m), and topped with a Gill microvane. The instruments were controlled by a micro-computer and a data acquisition unit that sampled the instruments at 1-second intervals for ten-minute periods.

An example of a resulting velocity profile is presented in Fig. 1. For the ideal logarithmic profile predicted by equation (2), all of the points should fall on a line (drawn here as a least squares line). Small variations from the line can indicate calibration errors in the instruments, an external source of regular variation, or unsteadiness in the flow field. The variation in Fig. 1 is probably associated with local topographic effects (Bauer *et al.*, this volume).

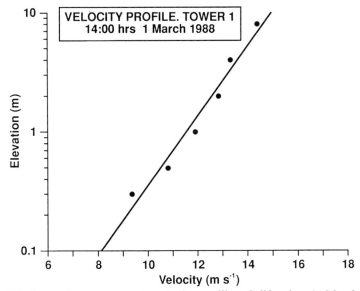

Fig. 1 Wind speed measurements at Castroville, California, 1 March 1988. The regression line ($R^2 = 0.97$) suggests a best estimate of $u_* = 0.59$ m s^{-1}, although considerable variability exists

It is frequently suggested that wind measurements be recorded for 30-minute periods (Greeley and Iverson, 1985, p. 43; Monteith, 1973, p. 88). However, it is important to note that velocity fields are frequently unsteady over that time span and shorter sample durations are appropriate. For the 10-minute averages coefficients of variability at each elevation are about 10% to 15%, suggesting a stable wind field.

6.3. Temperature

For detailed experimentation the measurement of temperature is important for two reasons: (i) it has an important influence on air density (see above); and (ii) steep temperature gradients will alter the velocity profile (Ming, Panovsky and Ball, 1983). There are many methods available for temperature measurement, ranging from hand-held thermometry to the use of thermisters or thermocouples (Omega, 1988).

For the correction of air density, an accuracy of $2°$ is sufficient to reduce most error from this source. Thus a common mercury thermometer is adequate. Accurate measurement of thermal gradients requires a sampling scheme similar to that used for wind speed. An ideal system would employ a series of temperature sensors (e.g. thermocouples) co-arrayed with anemometers and linked to a recording system. Berg (1983) used a pair of sensors to record temperature at two elevations, making readings off the instrument dial. In a scaling experiment, Bauer *et al.* (this volume) deployed vertical arrays of four pocket thermometers in shelters on anemometer towers. The results, and those of Berg (1983), suggest that errors in u_* resulting from ignoring the thermal gradient seldom exceed 10%. In particular, Berg (1983, p. 107) showed a maximum distortion of the velocity gradient of about 6%. In some coastal environments, however, the effect can be larger, especially for wind speeds and shear velocities close to the threshold for the initiation of motion.

6.4. Precipitation and surface moisture

Some indication of the occurrence of precipitation is important for two reasons: (i) intense, wind-driven rain can transport sediments by splash processes; and of greater importance (ii) the residual moisture can greatly reduce aeolian transport by increasing the threshold shear stress.

Sediment transport due to rain splash is of secondary importance on most beaches, with net transport being small relative to aeolian transport. However, splash processes can move sediments in conditions where no motion is predicted by the aeolian equations, and it may also be important during severe storms. Key concepts and methods for measuring splash transport can be found in Ekern and Muckenhirn (1947), Mutchler (1967), Morgan (1978), Poesen and Savat (1981), and Savat and Poesen (1981).

To determine the residual moisture effects, an estimate of the water content of the surface and near-surface sediments must be obtained. At present no one has devised a viable method for determining the wetness of only the surface sediment. Although Svasek and Terwindt (1974) have attempted to measure moisture content near the surface (0.20 m) with a neutron probe, with mixed results, most techniques involve sampling the sediments and returning the sand to a laboratory for drying. The near-surface sand is collected (usually taking the top 5 mm) in airtight containers. In the laboratory, the sample is weighed, dried, and then reweighed. The change in weight thus represents the moisture content lost through evaporation. The value of w in equation (15) is derived as $w = \{(W_t - W_s)/W_s\} \times 100$, where W_t is total sample weight and W_s is the dry weight of the sand.

6.5. Relative humidity and solar radiation

Relative humidity and incoming solar radiation can be important parameters in aeolian sediment transport because of their potential control on the moisture content of the sand surface. Relative humidity can either increase or decrease moisture content. High relative humidity over cold surfaces, at night for example, can increase moisture content through the formation of dew on the sand grains (this has been assumed by Belly, 1962; and Johnson, 1965). Low relative humidity can expedite the drying process and increase transport rates. Although the measurement of relative humidity is simple, the correlation of the measurements with changes in moisture content has not yet been established and will require careful experimentation.

Solar radiation will also influence moisture content through controlling evaporation rates. Hyde and Wasson (1983) have demonstrated that the incidence of short-wave radiation on sand surfaces does influence transport processes. Again, however, the quantitative linkages have not yet been established. The surface radiation balance can be measured with a net radiometer.

7. BIOLOGIC PARAMETERS

Biological influences are manifested through increased threshold velocities for sediment entrainment and through increased displacement heights. The extent of these effects will depend upon the species present in an area, and their densities. These factors can be estimated using quadrat sampling.

The use of quadrat sampling is discussed in detail in Ripley (1981). The concept involves the counting (or other measurement) of appropriate organisms found within a randomly selected area, or over a complete study site (e.g. a foredune).

For the measurement of sand-binding by algae, their presence and density must be determined, and some measure of the strength with which they hold sediments. For the estimation of displacement height associated with plant forms, representative heights, widths, and densities must be determined. These descriptors may include the variable geometry of plants in high winds. Plant geometry and density can be used to estimate displacement height. Where vegetation is sparse, the relationship is:

$$h/H_v = \lambda c_D c_m \tag{20}$$

where c_D is a drag coefficient, c_m a moment coefficient, and λ is vegetation density (Jackson, 1981). For dense vegetation, c_D and λ approach unity, and c_m equals h/h_v.

Counihan (1971) and Lee and Soliman (1977) have investigated the relationship of h/H_v and λ for intermediate roughness densities. Their data are co-plotted in Jackson (1981, Fig. 2). Regression analysis of points digitized from Jackson suggests the approximate relationship, $h/H_v = 2.6 \lambda$ for vegetation densities up to about 0.3 ($R^2 = 88\%$, 0.0001 significance).

8. DATA ANALYSIS

For the most part, analysis of field data to satisfy the requirements of the transport equations is straightforward, and many of the techniques have been discussed. The analysis of velocity profiles, however, deserves special consideration. Because it is difficult to measure surface shear stresses directly (Hayashi, 1984), these values are usually derived from estimates of the wind shear velocity and the surface roughness length. These values, in turn, are derived from the gradient in wind speed.

The line drawn to the data in Fig. 1 is based on linear regression. The y intercept, -7.8, indicates a roughness length of about 0.4 mm. This value can be used to solve equation (2) for u_*. However, because the line is not a perfect fit to the data ($R^2 = 0.97$), u_* will depend upon the wind speed measurement selected. For the Fig. 1 data, u_* estimates range from 0.57 m s^{-1} to 0.67 m s^{-1}, and average 0.61 m s^{-1}. The shear velocity can also be estimated from:

$$u_* = \varkappa/s \tag{21}$$

where s is the regression slope. For the Fig. 1 data, $s = 0.676$, and u_* becomes 0.59 m s^{-1}. These differences seem small, yet because most transport equations contain terms of u_*^3, the resulting error can be substantial. For example (ignoring threshold requirements), if the true shear velocity is 0.59 m s^{-1} and 0.61 m s^{-1} is used instead, the difference in predicted transport will exceed 10%. For the range in u_* reported above, the maximum potential error exceeds 60%.

This analysis suggests three items: (i) it is very difficult to derive accurate estimates of u_*; (ii) great care must be taken in the design and implementation of experiments to measure the wind field; and (iii) except for very high values of R^2 for regression data, it is advisable to calculate a range of results rather than a single value. These results also suggest that the interpretation of results of other experiments requires careful consideration of the methods used to derive u_*.

9. CONCLUSIONS

Our understanding of the aeolian processes associated with sediment transport indicates a complex interaction between wind and sand. Although a great deal of background research has been published, the complete specification of environmental controls on transport rates under prototype conditions still requires detailed and extensive field experiments. The detailed studies will be needed to elucidate and quantify specific aspects of the transport process, for example, the effect of relative humidity on transport across a beach berm. The extensive experiments will require substantial effort and equipment. These projects should aim at both the integration of existing knowledge into experimental design and the testing of specific components of the transport models. This includes the development of new equipment and methods for measuring both the fluid processes and sediment transport. The constraints of time and money make much of this work difficult to execute. There is no question, however, that our understanding has progressed to a point where a next round of advances is most likely to derive from field experiments.

ACKNOWLEDGEMENTS

Part of the research for this chapter stems from work sponsored by the California Department of Boating and Waterways, through Contract No. 85–42–175–45, to DJS. Responsibility for all statements rests with the authors.

REFERENCES

Allen, J. R. L. (1982) *Sedimentary Structures: Their Character and Physical Basis*, vol. 1. Elsevier, Amsterdam.

Bagnold, R. A. (1941) *The Physics of Blown Sands and Desert Dunes.* Methuen, London.

Barry, R. G. and Chorley R. J. (1982) *Atmosphere, Weather and Climate*, 4th edn. Methuen, London.

Bauer, B. O., Sherman, D. J., Nordstrom, K. F. and Gares, P. A. (1990) Aeolian transport measurement and prediction across a beach and dune at Castroville, California. This volume, pp. 39–55.

Belly P.-Y. (1962) Sand movement by wind. Unpublished M. S. thesis, University of California.

Belly, P.-Y. (1964) *Sand movement by wind.* US Army Corps of Engineers, Coastal Engineering Research Center Technical Memo. No. 1, Washington, DC.

Berg, N. (1983) Field evaluation of some sand transport models. *Earth Surf. Proc. Landf.*, **8**, 101–114.

Bressolier, C. F. and Thomas, Y. (1977) Studies on wind and plant interactions on French Atlantic coastal dunes. *J sediment. Petrol.*, **47**, 331–338.

Campbell, S. (1979) Soil stabilization by a prokaryotic desert crust: implications for Pre-Cambrian land biota. In *Origins of Life.* Reidel, Holland, pp. 335–348.

Carter, R. W. G. (1976) Formation, maintenance and geomorphological significance of an eolian shell pavement. *J. sediment. Petrol.*, **46**, 418–429.

Chepil, W. (1945) Dynamics of wind erosion, II: Initiation of soil movement. *Soil Sci.*, **6**, 397–411.

Coastal Engineering Research Center (1984) *Shore protection manual*. US Army Corps of Engineers, Coastal Engineering Research Center, Washington, DC.

Cooke, R. U. and Warren A. (1973) *Geomorphology in Deserts*. University of California Press, Berkeley.

Cooper, W. S. (1967) *Coastal dunes of California*. Geological Society of America, Memoir 104, 131 pp.

Counihan, J. (1971) Wind tunnel determination of the roughness length as a function of the roughness density of three dimensional elements. *Atmos. Env.*, **5**, 637–642.

Davidson-Arnott, R. G. D. and Law, M. N. (1990) Seasonal patterns and controls on sediment supply to coastal foredunes, Long Point, Lake Erie. This volume, pp. 177–200.

Einstein, H. E. (1950) *The bed-load function for sediment transportation in open channel flows*. US department of Agriculture, Technical Bulletin 1026, Washington, DC.

Ekern, P. and Muckenhirn, R. (1947) Waterdrop impact as a force in transporting sand. *Soil Sci. Soc. Proc.*, **12**, 441–444.

Fletcher, J. and Martin, W. (1948) Some effects of algae and moulds in the rain crusts of desert soils. *Ecol.*, **29**, 95.

Folk, R. L. (1966) A review of grain-size parameters. *Sedimentol.*, **6**, 73–93.

Folk R. L. (1974) *Petrology of Sedimentary Rocks*. Hemphill, Austin, Texas.

Forster, S. and Nicholson, T. (1981) Microbial aggregation of sand in a maritime dune succession. *Soil Biol. Biochem.*, **13**, 205–208.

Goudie, A. (ed.) (1981) *Geomorphological Techniques*. Allen & Unwin, London.

Greeley, R. and Iverson, J. (1985) *Wind as a Geological Agent*. Cambridge University Press, Cambridge.

Greenwood, B. (1978) Spatial variability of texture over a beach-dune complex, North Devon, England. *Sediment. Geol.*, **21**, 27–44.

Greenwood, B. and Mittler, P. W. (1984) Sediment flux and equilibrium slopes in a barred nearshore. *Mar. Geol.*, **60**, 79–98.

Hardisty, J. and Whitehouse, R. J. S. (1988) Evidence for a new sand transport process from experiments on Saharan dunes. *Nature*, **322**, 532–534.

Hayashi, Y. (1984) Direct measurement of friction velocity using drag plate in a wind tunnel. *Ann. Rep. Instit. Geosci., University of Tsukuba*, **10**, 44–48.

Hesp, P. A. (1981) The formation of shadow dunes. *J. sediment. Petrol.*, **51**, 101–112.

Hesp, P. A. (1983) Morphodynamics of incipient foredunes in N.S.W. Australia. In Brookfield, M. and Ahlbrandt, T. (eds), *Eolian Processes and Sediments*. Elsevier, Amsterdam, pp. 325–342.

Horikawa, K., Hotta, S., Kubota, S. and Katori, K. (1984) Field measurement of blown sand transport rate by trench trap. *Coast. Eng. Japan*, **27**, 213–232.

Horikawa K., Hotta, S. and Kraus, N. (1986) Literature review of sand transport by wind on a dry sand surface. *Coast. Eng.*, **9**, 503–526.

Horikawa, K., Hotta, S. and Kubota, S. (1986) Field measurements of vertical distribution of wind speed with moving sand on a beach. *Coast. Eng. Japan*, **29**, 163–178.

Horikawa, K. and Shen, W. (1960) *Sand movement by wind action (on the characteristics of sand traps)*. US Army Corps of Engineers, BEB Technical Memo. No. 119.

Hotta, S. (1985) Wind blown sand on beaches. Unpublished PhD dissertation, University of Tokyo.

Hotta, S. (1988) Sand transport by wind; and Instruments and procedures for wind-blown sand. In Horikawa, K. (ed.), *Nearshore Dynamics and Coastal Processes.* University of Tokyo Press, pp. 218–238, 475–482.
Hotta, S., Kubota, S., Katori, S. and Horikawa, K. (1984) Sand transport by wind on a wet sand beach. *Proc. 19th Coast. Eng. Conf., ASCE*, 1265–1281.
Howard, A. (1977) Effect of slope on threshold of motion and its application to orientation of wind ripples. *Bull. Geol. Soc. Am.*, **88**, 853–856.
Howard, A., Morton, J., Gad-El-Hak, M. and Pierce, D. (1978) Sand transport model of barchan dune equilibrium. *Sedimentol.* **25**, 307–338.
Hsu, S. (1973) Computing eolian sand transport from shear velocity measurements. *J. Geol.*, **81**, 739–743.
Hsu, S. (1974) Computing eolian transport from routine weather data. *Proc. 14th Conf. Coast. Eng., ASCE*, 1619–1626.
Hsu, S. (1987) Structure of airflow over sand dunes and its effect on eolian sand transport in coastal regions. *Coastal Sediments '87, ASCE*, New York, 188–201.
Hyde, R. and Wasson, R. J. (1983) Radiative and meteorological control on the movement of sand at Lake Mungo, New South Wales, Australia. In Brookfield, M. and Ahlbrandt, T. (eds) *Eolian Sediments and Processes.* Elsevier, Amsterdam, pp. 311–323.
Illenberger, W. K. and Rust, I. C. (1986) Venturi-compensated eolian sand trap for field use. *J. sediment. Petrol.*, **56**, 541–543.
Illenberger, W. K. and Rust, I. C. (1988) Sand budget for the Alexandria coastal dune field, South Africa. *Sedimentol*, **35**, 513–521.
Inman, D., Ewing, G. and Corliss, J. (1966) Coastal sand dunes of Guerrero Negro, Baja California, Mexico. *Bull. Geol. Soc. Am.*, **77**, 787–802.
Iversen, J., Greeley, R., Marshall, J. and Pollack, J. (1987) Aeolian saltation threshold: the effects of density ratio. *Sedimentol.*, **34**, 699–706.
Iwagaki, Y. (1950) On the effect of the sand drift on the coast by wind for the filling up with sand in Ajiro-Harbour. *J. Japan. Soc. Coast. Eng.*, **6**, 265–271 (in Japanese).
Jackson, P. (1981) On the displacement height in the logarithmic velocity profile. *J. Fluid Mech.*, **111**, 15–25.
Jensen, J. and Sorensen, M. (1986) Estimation of some aeolian saltation transport parameters: a re-analysis of Williams' data. *Sedimentol.*, **33**, 547–558.
Johnson, J. W. (1965) Sand movement on coastal dunes. *Fed. Inter-Agency Sediment. Conf. Proc.*, US Department of Agriculture Misc. Pub. No. 970, Washington, DC, pp. 747–755.
Kadib, A. A. (1964) *Addendum II: Sand transport by wind, studies with sand (c.0.145 mm diameter).* US Army Corps of Engineers, Coastal Engineering Research Center Technical Memo. 1, II–1–II–10.
Kadib, A. A. (1965) *A function for sand movement by wind.* University of California Hydraulics Engineering Laboratory Report HEL–2–8, Berkeley.
Kadib, A. A. (1966) Mechanism of sand movement on coastal dunes. *J. Waterw. Harb. Div.*, *ASCE*, **92**, 27–44.
Kawamura, R. (1951) *Study of sand movement by wind.* Translated as University of California Hydraulics Engineering Laboratory Report HEL 2–8, Berkeley.
King, C. A. M. (1972) (2nd edn) *Beaches and Coasts.* Edward Arnold, London.
Krumbein, W. E. and Slack, H. H. (1956) *Relative efficiency of beach sampling methods.* US Army Corps of Engineers, Beach Erosion Board Technical Memo. No. 90.
Lancaster, N. (1985) Variations in wind velocity and sand transport on the windward flanks of desert sand dunes. *Sedimentol.*, **32**, 581–593.
Lang, A. and Leuning, R. (1981) New omnidirectional anemometer with no moving parts. *Bound. Layer Meteorol.*, **20**, 445–457.

Leatherman, S. P. (1976) Barrier island dynamics: overwash processes and aeolian transport. *Proc. 15th Coastal Eng. Conf.*, *ASCE*, New York, 1958–1974.

Leatherman, S. P. (1978) A new eolian sand trap design. *Sedimentol.*, **25**, 303–306.

Lee, J. (1987) A field experiment on the role of small scale wind gustiness in aeolian sand transport. *Earth Surf. Proc. Landf.*, **12**, 331–335.

Lee, B. and Soliman, B. (1977) An investigation of the forces on three-dimensional bluff bodies in rough wall turbulent boundary layers. *J. Fluid Eng.*, **99**, 503–510.

Lettau, K. and Lettau, H. (1977) Experimental and micrometorological field studies of dune migration. In Lettau, K. and Lettau H. (eds), *Exploring the World's Driest Climate*. University of Wisconsin-Madison, IES Report 101, pp. 110–147.

Logie, M. (1982) Influence of roughness elements and soil moisture on the resistance of sand to wind erosion. *Catena Suppl.*, **1**, 161–173.

Lyles, L. and Woodruff, N. (1972) Boundary-layer flow structure: effects on detachment of noncohesive particles. In Shen, H. (ed.), *Sedimentation*. Department of Civil Engineering, Colorado State University, pp. 2.1–2.16.

MacCarthy, G. (1935) Eolian sands: a comparison. *Am. J. Sci..*, **30**, 81–95.

MacCarthy, G. and Huddle, J. (1938) Shape-sorting of sand grains by wind action. *Am. J. Sci.*, **35**, 64–73.

Mattox, R. (1955) Eolian shape-sorting. *J. sediment. Petrol.*, **25**, 111–114.

McArthur, D. (1980) Fluorescent sand tracer methodology for coastal research. *Mélanges 14*, Louisiana State University Museum of Geosciences, Baton Rouge.

McCluskey, J. (1987) The role and magnitude of eolian processes in the Barrier Island environment. Unpublished PhD dissertation, Rutgers University.

Middleton, G. and Southard, J. (1978) *Mechanics of Sediment Movement*, Society of Economic Paleontologists and Mineralogists, Short Course No. 3, Tulsa, Okla.

Miller, M., McCave, I. N. and Komar, P. D. (1977) Threshold of sediment motion under unidirectional currents. *Sedimentol.*, **24**, 504–527.

Ming, Z., Panofsky, H. H. and Ball, R. (1983) Wind profiles over complex terrain. *Bound. Layer Meteorol.*, **25**, 211–228.

Monteith, J. H. (1973) *Principles of Environmental Physics*. Edward Arnold, London.

Morgan, R. P. C. (1978) Field studies of rainsplash erosion. *Earth Surf. Proc.*, **3**, 295–299.

Mutchler, C. (1967) Parameters for describing raindrop splash. *J. Soil Water Conserv.*, **22**, 91–94.

Nickling, W. G. (1978) Eolian sediment transport during dust storms. *Can. J. Earth Sci.*, **15**, 1069–1084.

Nickling, W. G. (1984) The stabilising role of bonding agents on the entrainment of sediment by wind. *Sedimentol.*, **31**, 111–117.

Nickling, W. G. and Ecclestone, M. (1981) The effects of soluble salts on the threshold shear velocity of fine sand. *Sedimentol.*, **28**, 505–510.

O'Brien, M. and Rindlaub, B. (1936) The transportation of sand by wind. *Civ. Eng.*, **6**, 325–327.

Oke, T. (1978) *Boundary Layer Climates*. Methuen, New York.

Olson, J. (1958) Lake Michigan dune development; 1. Wind velocity profiles. *J. Geol*, **66**, 254–263.

Omega (1988) *Temperature Measurement Handbook and Encyclopedia*. Omega Engineering, Stamford, Connecticut.

Poesen, J. and Savat, J. (1981) Detachment and transportation of loose sediments by raindrop splash, Part 2. *Catena*, **8**, 19–41.

Psuty, N. P. (1969) Beach nourishment by eolian processes, Paracas, Peru. *Proc. 10th Meeting, New York-New Jersey Division, Association of American Geogr.*, **3**, 117–123.

Pye, K. (1983) Coastal dunes. *Progr. Phys. Geogr.*, **7**, 531–557.

Ripley, B. (1981) *Spatial Statistics*. Wiley, New York.

Rosen, P. (1978) An efficient, low cost aeolian sampling system. *Current Res. A, Geol. Surv. Canada*, **78–1A**, 531–532.

Rosen, P. (1979) Aeolian dynamics of a barrier island system. In Leatherman, S. P. (ed.), *Barrier Islands*, Academic Press, New York, pp. 81–98.

Sarre, R. (1984) Sand movement from the intertidal zone and within coastal dune systems by Aeolian processes. Unpublished PhD dissertation, University of Oxford.

Sarre, R. (1987) Aeolian sand transport. *Progr. Phys. Geogr.*, **11**, 155–182.

Sarre, R. (1988) Evaluation of aeolian sand transport equations using intertidal zone measurements, Saunton Sands, England. *Sedimentol.*, **35**, 671–679.

Savat, J. and Poesen, J. (1981) Detachment and transportation of loose sediments by raindrop splash, Part 1. *Catena*, **8**, 1–17.

Shaw, R., Kidd, G. and Thurtell, G. (1973) A miniature three-dimensional anemometer for use within and above plant canopies. *Bound. Layer Meteorol.*, **3**, 359–380.

Svasek, J. and Terwindt, J. H. T. (1974) Measurements of sand transport by wind on a natural beach. *Sedimentol.*, **21**, 311–322.

Torobin, L. and Gauvin, W. (1960) Fundamental aspects of solids-gas flow; Part IV: The effects of particle rotation, roughness and shape. *Can. J. Chem. Eng.*, **38**, 142–153.

van der Ancker, J., Jungerius, P. and Mur, L. (1985) The role of algae in the stabilization of coastal dune blowouts. *Earth Surf. Proc. Landf.*, **10**, 189–192.

von Kármán, T. (1934). Turbulence and skin friction. *J. Aeronaut. Sci.*, **1**, 1–20.

Walker, J. and Southard, J. (1984) A sticky-surface trap for sampling eolian saltation load. *J. sediment. Petrol.*, **54**, 652–654.

Williams, G. W. (1964). Some aspects of the eolian saltation load. *Sedimentol.*, **3**, 257–287.

Wilson, S. and Cooke, R. U. (1980) Wind erosion. In Kirkby, M. J. and Morgan, R. C. P. (eds), *Soil Erosion*. Wiley, New York, pp. 217–251.

Woodruff, N. and Siddoway, F. (1963) A wind erosion equation. *Proc. Soil Sci. Am.*, **29**, 602–608.

Zingg, A. (1952). Wind-tunnel studies of the movement of sedimentary material. *Proc. 5th Hydraul. Conf.*, Iowa University, 111–135.

Chapter Three

Aeolian transport measurement and prediction across a beach and dune at Castroville, California

BERNARD O. BAUER AND DOUGLAS J. SHERMAN
Department of Geography, University of Southern California

KARL F. NORDSTROM
Institute for Marine and Coastal Sciences, Rutgers University

AND

PAUL A. GARES
Department of Geography, Colgate University

1. INTRODUCTION

Accurate prediction of aeolian sedimentary processes is fundamental to the development of physically-based models of coastal dune evolution. Under ideal conditions, the movement of sand-sized materials by wind is reasonably well predicted by transport equations derived from theoretical or laboratory work. The assumptions associated with these derivations include a steady, uniform flow field, with a logarithmic velocity profile, over a flat, extensive surface composed of dry, uniform sand. Rarely are these conditions encountered in nature. On beaches and dunes the departures from ideal conditions can limit severely the applicability of existing sediment transport models (Sherman and Hotta, this volume).

In this chapter we describe the results of a field study designed to investigate sediment transport processes across a beach and dune system through the measurement of coincident wind and sediment transport parameters. Measured transport rates are compared to rates predicted by the equations of Bagnold

Coastal Dunes: Form and Process. Edited by K. F. Nordstrom, N. P. Psuty and R. W. G. Carter
© 1990 John Wiley & Sons Ltd

(1941) and Kawamura (1964) and the discrepancies discussed. Particular attention is given to potentially erroneous interpretations of flow data when ideal conditions are assumed or when instruments are deployed without appreciation for topographical modulation of the flow field.

2. THEORETICAL BACKGROUND

The Bagnold (1941) and Kawamura (1964) equations were used in this study because they are recognized as the benchmark models for predicting sand transport rates (see review in Horikawa, 1988). According to Bagnold (1941), the discharge of sediment, q, can be predicted by:

$$q = C\left(\frac{d}{D}\right)^{0.5}\frac{\rho}{g} u_*^3 \tag{1}$$

where C is a constant (varying from 1.5 to 3.5), d is mean sediment diameter, D is a reference grain size of 0.25 mm, g is gravitational acceleration, ρ is air density, and u_* is shear velocity over a saltation layer (Bagnold uses u_g).

Kawamura (1964) presented an alternative expression:

$$q = K\frac{\rho}{g}(u_* + u_{*c})^2(u_* - u_{*c}) \tag{2}$$

where K is a constant (2.78 according to laboratory findings), and u_{*c} is the threshold shear velocity for the initiation of sand movement. The threshold shear velocity can be calculated using Bagnold's (1941) relationship:

$$u_{*c} = A \left[\left(\frac{\rho_s - \rho}{\rho}\right)gd\right]^{0.5} \tag{3}$$

where the constant, A, has an assumed value of 0.1 (Sarre, 1987), g is gravitational acceleration, and ρ_s is sediment density. Table 1 shows the range in u_{*c} for common values of d and ρ.

The primary differences between the two approaches involve assumptions concerning the mode of momentum transfer and the use of a threshold term

Table 1   Representative values of threshold shear velocity (m s^{-1}) (after Bagnold, 1941)

		Grain diameter (mm)			
		0.3	0.4	0.5	0.6
Air density	1.1	0.266	0.307	0.344	0.376
(kg m^{-3})	1.2	0.255	0.294	0.329	0.360
	1.3	0.245	0.283	0.316	0.346

to describe the initiation of motion. The discussion by Sherman and Hotta (this volume) provides more details about these differences. Empirical testing of the transport models requires measurement of shear velocity, sand discharge, and grain size characteristics. The purpose of the field experiment discussed herein was to determine the values for these variables across a beach and dune system.

3. STUDY LOCATION AND EXPERIMENTAL DESIGN

The study area is located along the eastern shore of Monterey Bay, immediately north of the Salinas River, near Castroville, California (Fig. 1). The field site is on a sandy barrier fronting a former outlet of the Salinas River. Most of the barrier is wave-formed, but aeolian processes interacting with the vegetation and driftwood have modified the barrier by building a low foredune and hummocky back-barrier surface (Fig. 2). The dune and its colonizing vegetation are low in elevation; the dune crest averages about 0.7 m above the back beach. Small, funnel-shaped, unvegetated blowouts, created by westerly (onshore) winds, extend inland across the dune crest. These blowouts are generally less than 2 m wide. Vegetation includes many mid-dune species because the barrier is eroding. The principal species on the dune crest is the ice plant, *Mesembryanthemum chilense*, growing to a height of about 0.1 m above the sand surface. Shrubs, such as coast buckwheat (*Eriogonum parvitolium*) are located landward of the dune crest. Most shrubs are lower than 0.4 m high.

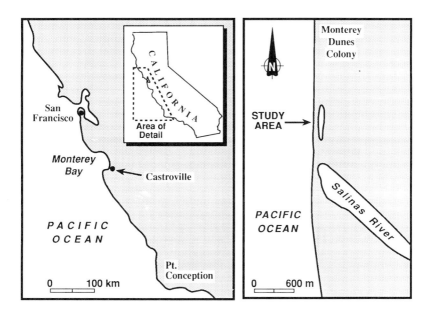

Fig. 1   Location of study site

Small clumps of sea rocket (*Cakile maritima*) are found seaward of the dune crest. The density of dune vegetation is less than 50% of the ground cover, with numerous patches less than 20%.

Wind speed was measured using DC-generating, Gill-type, 3-cup anemometers, with a threshold speed of about $0.4 \, \text{m s}^{-1}$, and an accuracy of $\pm 0.5\%$ over the operating range (up to $50 \, \text{m s}^{-1}$). A total of 14 anemometers were deployed across the beach-dune system. Wind direction was measured at four locations with Gill-type microvanes. All instruments were cable-linked to a Hewlett-Packard 3852A Data Acquisition and Control Unit that was, in turn, controlled with an HP 9000, series 320 micro-computer. The sensing instruments were sampled near-simultaneously at 1-second intervals for ten-minute periods.

Two 8-metre towers were erected on the beach at the locations depicted in Fig. 3. Each tower was topped with a wind vane and held six anemometers placed at elevations of 0.3, 0.5, 1.0, 2.0, 4.0, and 8.0 m. Thermometers were also deployed at three elevations on each tower to estimate atmospheric stability conditions. Tower 1 was located on the nearly horizontal back beach, approximately 10 m landward of the berm crest, and 10 m from the foredune. Tower 2 was 3 m landward of the berm crest. Vertical velocity profiles from these towers were used to derive estimates of u_*. Additional reference stations, consisting of an anemometer and wind vane pair, were located at the foredune crest and on the backslope of the dune (Fig. 3).

Fig. 2   Site characteristics

Blowing sands were trapped using cylindrical traps (Leatherman, 1978; Rosen, 1978) and a Helley–Smith Bedload Sampler (Helley and Smith, 1971). The mouths of the cylindrical traps were 0.05 m wide and 0.40 m high. The overall trapping efficiency of this design has been rated as about 70% (Marston, 1986). The Helley–Smith trap was designed for bedload trapping in fluvial systems and has a 0.1 m by 0.1 m orifice. The efficiency of this trap in air is unknown; however, visual observations during the experiments suggest that disturbance of the transport field was minimal at the mouth.

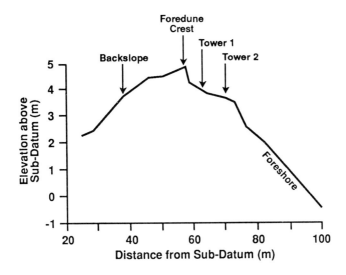

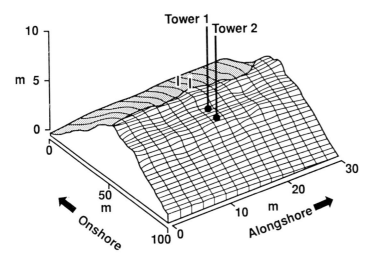

Fig. 3 Topography and instrument deployment configuration

Sets of the cylindrical traps were deployed near the anemometer reference stations across the dune, at the entrance of a blowout in the dune crest line, and near the base of Tower 1 (Fig. 3). The traps at the blowout and Tower 1 filled rapidly and were emptied as needed through the experiment. The Helley–Smith was used near the base of Tower 2. Samples were taken with this trap from 10-minute periods every 30 minutes for the duration of the experiment.

4. MEASUREMENT RESULTS

4.1. Vertical velocity profiles

Average wind velocities for the period 0400 to 1900 hours, 1 March 1988, at the top of the two anemometer towers are shown in Fig. 4. The greatest velocities occurred between 1200 and 1800 hours, which corresponds to the period when sediment transport was active. The close correspondence between the two 8 m anemometer traces suggests that the roughness transition from the aerodynamically smooth water surface (where spatially uniform flow conditions exist) to the rougher beach surface was insufficient to cause vertical development of an internal boundary layer to the top of either tower. The 8 m anemometers were high enough to be either in the free stream flow or within the fully developed boundary layer of the oceanic flow field.

The development of an internal boundary layer across the beach was apparent at lower elevations (Fig. 5). At 0430 hours, average velocities at all anemometer elevations were small and nearly uniform except close to the surface. Thus, the internal boundary layer at this time was less than 0.5 m in depth. In contrast, at 1330 hours, the absolute velocities were large and the distortion of the vertical

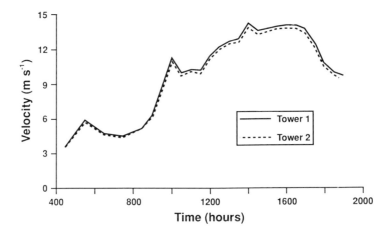

Fig. 4   Wind velocity record from 8 m anemometers at towers 1 and 2 (1 March 1988)

profiles was much more pronounced. The shape of the velocity profile at Tower 1 was markedly different from that at Tower 2, and this disparity shows that the depth of the internal boundary layer increased quickly in the downwind direction. Excepting the 8 m anemometers, flow velocities were consistently smaller at the downwind location (Tower 1), possibly because of flow reduction in the downwind boundary layer or because of overspeed (Hsu, 1987) at the upwind location (Tower 2). Overspeed is due to vertical convergence of streamlines when a uniform flow field encounters a pronounced rise in topography, such as a berm crest or dune (Lancaster, 1985; Livingston, 1986).

4.2. Horizontal distribution of wind speed and direction

Temporal fluctuations in wind speed and direction at different locations across the beach and dune system were similar, even in the lee of the dune, as demonstrated by the 1 m anemometer traces shown in Fig. 6. However, the absolute magnitude of flow velocity varied spatially. Tower 2 velocities were consistently greater than at Tower 1 because of boundary layer development in the downwind direction. Flow velocities at the dune crest were fastest, probably because of the overspeed effect. Flow velocities in the lee of the dune were smallest, as expected for flows over bluff bodies. Although these trends are not surprising, it is interesting to note that the critical threshold value for sediment entrainment of 4 to 5 m s^{-1} reported by Bagnold (1941) and Chepil

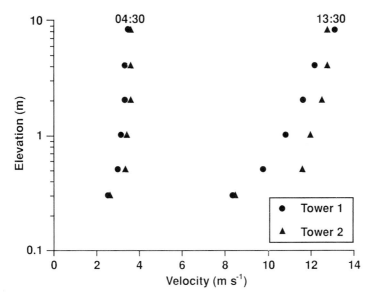

Fig. 5   Selected vertical velocity profiles (1 March 1988)

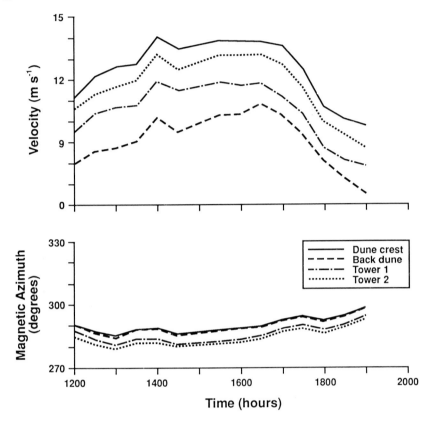

Fig. 6   Wind speed (measured at 1 m elevations) and wind direction across the beach and dune (1 March 1988). Magnetic azimuth of 270° is cross-shore

(1945) was always exceeded at all anemometer locations during this time period, even though sediment was not being moved continuously.

The wind vane records in Fig. 6 demonstrate that the flow field was fairly constant in its approach. Thus, any differences in measured sediment transport cannot be attributed to a spatially varying directional flow field. The 5-degree offset between the tower vanes and the others is likely due to orientation inaccuracies. The dune may have caused some deflection of the flow lines. However, it seems unlikely that the degree of deflection would remain constant across a broad range of velocities. In either case, the offset is minimal and probably has little influence on sediment transport conditions. Furthermore, the standard deviation of the direction records was always less than 5 degrees during this event, demonstrating the consistent nature of the flow field.

Table 2 Sediment transport rates

Location	Grain size (mm)	kg m^{-1} storm^{-1}	kg m^{-1} hr^{-1}
Tower 2	0.573	36.25	6.39
Tower 1	0.419	45.74	8.07
Blowout	0.408	151.79	26.77
Backdune	0.370	6.09	1.07

4.3. Sediment transport

Measured sediment transport rates across the beach and dune are presented in Table 2. The values are averages for the duration of the transport event (i.e. from 1200 to 1730 hours) adjusted to compensate for the different geometries of the traps. Helley–Smith samples were taken at 30-minute intervals to coincide with the wind records. The sampling duration was 10 minutes, and these samples were considered to be representative of the entire half-hour interval. The 30-minute transport rates were summed and normalized by the duration of the event. The cylindrical traps were employed as integrating traps. The trap in the backdune did not fill up throughout the entire event, whereas at the blowout location, the trap filled and was emptied five times. The average transport rates presented in Table 2 for the cylindrical traps are the total weight of collected material normalized by the duration of the event. The average grain sizes, however, were weighted according to the percentage of material transported within each time interval (i.e. each full tube or each Helley–Smith sample).

The smallest transport rates were in the lee of the dune, well down the back slope. Transport rates were substantially greater at the blowout location, but the greatest transport rates were measured at the dune crest. The cylindrical trap at the dune crest filled by 1340 hours, much earlier than neighbouring traps. Unfortunately, this location could not be monitored subsequently because of trap failure. The cylindrical traps located beside Tower 1 filled very slowly. Less than one-third the amount of sediment trapped at the blowout location was trapped at Tower 1 (Table 2), and even smaller rates of sediment transport were measured closer to the berm at Tower 2.

5. ANALYSIS

Most models of sediment transport incorporate shear velocity (u_*) instead of the absolute velocity (u) in the parameterization of the transport process (Sherman and Hotta, this volume). In most laboratory studies and over extensive flat surfaces, the vertical velocity profile is logarithmic down to within millimetres of the bed. For these conditions, shear velocity is proportional to the slope of the vertical velocity gradient and any two points on the profile within the boundary layer will yield the same shear velocity. However, for shore-normal flows across a beach and dune system there can be considerable variability in the form of the vertical velocity gradient. Portions of the velocity profiles presented in Fig. 5 are indeed logarithmic, falling along a straight line on log-elevation plots, but they are not logarithmic throughout the entire depth of the boundary layer. In order to estimate shear velocity, then, one must make a choice of which pair of anemometers to use in the calculations. Often eye-fit lines are drawn through the data to approximate the average slope, but in either case, the procedure is subjective. This has profound implications for the prediction of sediment transport rates since even slight deviations from the true shear velocity will yield large inaccuracies because of the cubic shear velocity term in equations (1) and (2).

In this study, linear regression was used to determine the slope and intercept of the velocity gradient by the method of least-squares. All of the anemometers in each vertical array were included in the regression. The calculations produced a range in shear velocity between 0.48 and 0.70 m s^{-1} at Tower 1, and between 0.57 and 0.80 m s^{-1} at Tower 2, for the period between 1200 and 1800 hours when sediments were being moved. These values are greater than the calculated threshold shear velocities presented in Table 1, as expected for periods of active transport. Roughness length values from these regressions also correspond closely with values reported in the literature. At Tower 1, for example, calculated roughness lengths varied from 0.17 mm at the beginning of the period to about 3.62 mm at the end when the surface was covered with coarse-grained ripples. Bagnold (1941) gives an average value of around 0.3 mm for many dune sand surfaces, and when ripples are present, values of around 3.0 mm are expected (Hsu, 1971).

Despite the good agreement between the magnitude of shear velocities and roughness lengths calculated from the experimental data with those reported in the literature, some uncertainties remain. Shear velocities at Tower 2 were consistently greater than at Tower 1, implying that the ability of the flow field to entrain and transport sediment decreased in the landward direction. Although this spatial trend corresponds with the absolute velocity (*u*) traces presented in Fig. 6, the sediment transport data in Table 2 show an increase in sediment transport from Tower 2 to Tower 1. Also, the magnitude of shear velocity continued to increase with time and peaked around 1700 hours when sediment transport was decreasing. A second series of regressions was performed on subsets of the velocity data in order to determine what effect instrument deployment might have on the interpretation of aeolian processes. Fig. 5 showed that a logarithmic model applied only to the upper portions of the vertical velocity profiles so the 0.3 m anemometers were excluded from the second regression. Indeed, instruments are rarely placed at very low elevations during field experiments because of fouling by saltating sediments and because there is often a need to deploy anemometers at a number of locations in order to characterize spatial trends.

The R^2 improved to greater than 0.98 for almost all cases when the 0.3 m anemometers were excluded from the regression. Tower 1 shear velocities were only slightly smaller than before. However, in contrast to the first regression for which the shear velocities fluctuated considerably throughout the measurement period, the values from the second regression remained essentially constant. For the Tower 2 data, shear velocities from the second regression were much smaller than from the first regression, falling between 0.20 and 0.35 m s^{-1}. These magnitudes are about one-third of those obtained from the full array, and more importantly, are often below the threshold of entrainment.

6. DISCUSSION

It is apparent from the two regression exercises that the number of instruments and their placement can be of considerable importance to the interpretation and prediction of sediment transport processes. For the first regressions, the values of shear velocity and roughness length corresponded closely with those reported in the literature, but the logarithmic assumption was not realistic. Conversely, when the 0.3 m instruments were excluded from the regressions, the logarithmic model was approximately correct, but the values of shear velocity and roughness length were unreasonable. Calculated roughness lengths for Tower 1 were of the order of nanometers (10^{-9} m), which is much smoother than expected. In addition, shear velocities at Tower 2 were much less than those at Tower 1, which contradicts the results of the first regressions. Note that both approaches are equally valid, even though the implications for sediment transport are rather different.

Many interpretations of a single vertical velocity profile are possible depending on the number and position of anemometers used in the regression. Nevertheless, there should be only one true value of shear velocity, and it is desirable to determine the optimal instrument deployment configuration that minimizes the total number of instruments in the profile, but yields the most accurate estimate of the true shear velocity from the regression. Unfortunately, the true shear velocity cannot be measured directly using anemometers and therefore a comparison becomes problematic. An alternative approach is to back-calculate shear velocities using the sediment trap data in the empirical equations, and to adopt these as a surrogate for true shear velocities. Fig. 7 shows the results of such calculations using Bagnold's relationship with the empirical coefficient, C, fixed conservatively at 1.5. The rationale for presenting these back-calculated shear velocities rather than attempting to predict rates of sediment transport from the anemometer data is that variations in grain size through the wind event (Table 2) can be incorporated into the calculations explicitly, thereby yielding a more accurate representation of the natural processes.

Also plotted on Fig. 7 are estimates of shear velocity derived using the 2 m wind speeds in the following logarithmic velocity profile relationship:

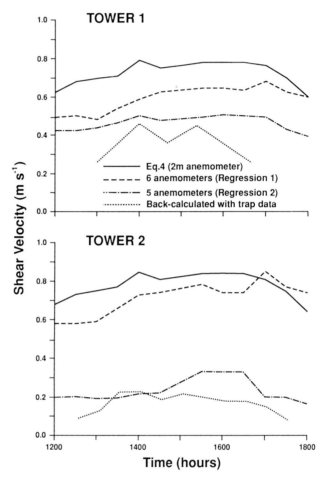

Fig. 7   Shear velocities for the sediment transport period using regression procedures, logarithmic law of equation 4, and sediment trap data

$$u_z = \frac{u_*}{\varkappa} \ln\left(\frac{z}{z_0}\right) \tag{4}$$

where u_z is wind speed, $\varkappa$ is Von Karman's constant (0.4), and z is anemometer elevation. For $z = 2$ m and assuming a roughness length of $z_0 = 3$ mm, equation (4) simplifies to $u_* = u_z/16.25$. The rationale for including this trace in Fig. 7 is that in many field studies only single anemometers are deployed at standard elevations. Logarithmic boundary conditions are assumed and the results used to estimate shear velocity.

Fig. 7 shows the variability between the shear velocities calculated with the various methods. There are large discrepancies between the calculated shear velocities using the anemometer data and those back-calculated from the sediment trap data; the latter are consistently smaller. Assuming that sediment transport is proportional to the cube of shear velocity, there could be a maximum disparity of 740% between predicted and measured transport at Tower 1, and of 4700% at Tower 2. The poorest correspondence was with equation (4) and it appears unlikely that the placement of single anemometers at 2 m elevations would produce realistic estimates of sediment transport rates across a beach similar to the one at Castroville. Incorporating a threshold velocity in equation (4), as suggested by Bagnold (1941), improves the fit slightly.

The closest correspondence between the calculated and back-calculated shear velocities was with the second regression when the 0.3 m anemometers were excluded. Even then, the trends are not consistent. In some cases, the calculated shear velocities increase when the back-calculated shear velocities decrease. Furthermore, if a threshold shear velocity were to be included into the calculations explicitly (as with Kawamura's relationship), then the second regression would predict active transport at Tower 2 between 1530 and 1630 hours only, which was not the case.

There are many possible explanations for the lack of correspondence between the shear velocity traces presented in Fig. 7. The presence of binding salts, pore moisture, or steep surface slopes may affect the magnitude of sediment transport. If the sand is wet, for example, surface tension and decreased surface roughness (due to water in the surface interstices) will increase the critical shear velocity for the initiation of motion. A water content of 4% (by weight), for example, would increase the threshold shear velocity to about 0.60 m s^{-1} for sediment sizes similar to those at Castroville (Horikawa, 1988). This factor could not be evaluated easily in the field. However, rainfall occurred prior to the measurement period, and even though this might explain differences between observed and predicted sediment transport, it does not explain the variability in the temporal trends evident in Fig. 7. In addition, because of the uniform moisture conditions across the beach and dune, spatial trends apparent in the data are probably not a consequence of spatially varying moisture conditions.

The effect of surface slope on transport rate is believed to be minimal. If the model of Hardisty and Whitehouse (1988) is used with a berm slope of 5° (a maximum for Castroville), the maximum error from assuming a flat surface is about 12%. For most berm locations, the error is much less than 10%, and is therefore omitted from the calculations. Similarly, measurements of temperature profiles at the tower locations demonstrated that buoyancy effects were negligible and should have little bearing on the shape of the vertical velocity profiles. Thus, the obvious physical factors associated with field conditions cannot be invoked to explain the discrepancy between calculated and back-calculated shear velocities.

A fundamental reason for the differences between observed and predicted sediment transport involves the derivation of the empirical equations.

The empirical formulae are based on the assumption of a uniform and steady flow field and the existence of an equilibrium sediment transport system (i.e. the wind is carrying as much sediment as it can at all times). However, boundary layer development across the beach and dune system during this study was non-uniform and non-stationary, as must have been the sediment transport system. Note that for Tower 2 in Fig. 7, the back-calculated shear velocities were always less than the expected critical threshold shear velocities for all but the finest, dry sands (Table 1). It is also important to stress that a value of $C = 1.5$ was used in the Bagnold relationship to derive these back-calculated shear velocities, and if a larger value for C had been used, the values of shear velocity would have been smaller yet. It would appear, then, that in the absence of equilibrium conditions in the sediment transport system, back-calculated shear velocities are not good representations of the true shear velocity.

Equilibrium sediment transport conditions are attained, if at all, only on extensive, flat reaches of beach, with shore-parallel winds or across overwash flats. For cross-shore flows, the variations in topography cause variations in boundary layer development which affect the ability of the wind field to transport sediments (e.g. Svasek and Terwindt, 1974; Ming, Panofsky and Ball, 1983). As the oceanic wind field reaches the foreshore, it begins to entrain sediments if the critical conditions are exceeded. Few sediments are carried by the flow in this region because there is no sediment supply upwind. Downwind, the transport system becomes more fully developed (i.e. more sediments are carried in the flow) because sediments derived from the foreshore are already in transport and additional sediments are being entrained locally. At Castroville, the sediment transport system was most fully developed close to the dune crest where a sufficient reach of beach had been traversed to entrain a substantial amount of sediment. In addition, convergence of stream lines over the dune line caused increased velocities so that the flow field was even more competent. This explains why the quantity of sediment trapped increased in the direction of the dune crest. However, the restricted fetch length from the foreshore to the dune crest never allowed the sediment transport system to attain full development, even at the dune crest, and therefore the shear velocities derived from the trap data were consistently smaller than values obtained from the wind profiles. The foreshore and lower beach are transition zones, both in terms of increased aerodynamic roughness, and in terms of sediment availability, which is zero over the water.

The issue of which of the regressions presented above provides the better estimate of true shear velocity is most appropriately addressed in the context of a non-equilibrium sediment transport system. When the 0.3 m anemometers were excluded, the vertical velocity profiles were approximately logarithmic and the values of shear velocity were of the same magnitude as those derived from the sediment transport data. In addition, an increasing horizontal shear stress gradient towards Tower 1 was observed, as suggested by the transport data. However, this correspondence may have been fortuituous because the shear velocities were unrealistically small (below the critical threshold of entrainment in the case of Tower 2), as were the estimates of surface roughness. When all anemometers were used in the regression, estimates of surface roughness and shear velocity were larger than the back-calculated values, but corresponded better with those reported in the literature. Despite the strong deviation from a logarithmic profile, inclusion of the lowermost anemometer appears to be critical to estimating the true shear velocity at the surface, probably because these anemometers were always in the bottom boundary layer where the shear stress gradients were most pronounced. Although the latter regressions predict that shear velocities were greater at the berm crest and decreased toward the foredune (in contrast to the sediment trap data), this may not be unrealistic because the sediment transport system is not in equilibrium with the wind field. More sediments could potentially be carried if introduced to the flow upwind.

7. CONCLUSIONS

A close correspondence between the magnitude of predicted and measured transport rates is often used as the primary criterion for judging the validity of predictive models. This study demonstrates that such correspondence may occur even though the assumptions underlying the model are violated. More importantly, adopting a slightly different, but equally valid, analysis procedure yielded a contrasting interpretation of process at the study site. Indeed, the lack of correspondence between back-calculated shear velocities (based on quantities of trapped sediments) and calculated shear velocities (based on regressions of the vertical velocity profiles) suggested that the sediment transport system across the beach and foredune was not in equilibrium with the non-uniform flow field.

The implications of this study are important to the prediction of sediment transport across beaches. If the flow field is steady and uniform, if the vertical velocity profile is logarithmic, and if the sand surface is flat and extensive, an equilibrium condition develops between the energy available and the amount of sediment being transported. It matters little where instruments are deployed because such a system is well-behaved everywhere and easily modelled. However, on most beaches the flow field is unsteady and non-uniform, and the resulting horizontal and vertical velocity gradients cannot be characterized adequately through deployment of single anemometers at fixed elevations. Spatially varying conditions necessitate spatial approaches. Unfortunately, it is not evident from the present analysis that increasing the number of instruments would enhance the ability to predict sediment transport; it merely improves the ability to provide potential explanations for discrepancies between measured and predicted values. This unsatisfactory conclusion arises because, in transition zones, the amount of sediment in transport is not necessarily proportional to shear stress, as the empirical models suggest. Sediment traps installed in the foreshore or lower beach would not be expected to yield a great amount of sediment despite the large shear stresses because there is no source of sediment upwind. Thus, empirical relationships, such as those of Bagnold and Kawamura, would overpredict the rates of sediment transport because they were derived for equilibrium systems. That there is a functional relationship between the fluid motion and the sediment motion is not open to debate. The nature of this relationship under less than ideal conditions is far from obvious, requiring caution against the unqualified interpretation of sediment transport processes on the basis of assumed logarithmic profiles and shear velocities.

ACKNOWLEDGEMENTS

Funding for this study was provided by the California Department of Boating and Waterways (better known to us as George Armstrong and Ron Flick) and further assistance came from the University of Southern California. The support and encouragement of the editors, for this and many other projects, is greatly appreciated. Kilian McDaid redrafted the figures from some dodgy originals. Thanks to Bob Blair and Andy Hincenbergs for helping to lay-out and reel-up cable and more, to the weekend horses for not stepping on too many cables, and to the black, legless, dune lizards for generally keeping out of our way.

REFERENCES

Bagnold, R. (1941) *The Physics of Blown Sand and Desert Dunes*. Chapman & Hall, London 265 pp.
Chepil, W. S. (1945) Dynamics of wind erosion, II: Initiation of soil movement. *Soil Science*, **6**, 397–411.
Hardisty, J. and Whitehouse, R. J. S. (1988) Evidence for a new sand transport process from experiments on Saharan dunes. *Nature*, **332**, 532–534.
Helley, E. J. and Smith, W. (1971) *Development and calibration of a pressure-difference bedload sampler*. US Geological Survey Open File Report, 18 pp.
Horikawa, K. (ed.) (1988) *Nearshore Dynamics and Coastal Processes*. University of Tokyo Press, Tokyo, 522 pp.
Hsu, S. (1971) Measurement of shear stress and roughness length on a beach. *Journal of Geophysical Research*, **76**, 2880–2885.
Hsu, S. (1987) Structure of airflow over sand dunes and its effect on eolian sand transport in coastal regions. *Coastal Sediments '87, Vol. I*, New York, pp. 188–201.
Kawamura, R. (1964) *Study of Sand Movement by Wind*. University of California Hydraulics Engineering Laboratory Report HEL 2–8, Berkeley, 38 pp.
Lancaster, N. (1985) Variations in wind velocity and sand transport on the windward flanks of desert sand dunes. *Sedimentology*, **32**, 581–593.
Leatherman, S. (1978) A new eolian sand trap design. *Sedimentology*, **25**, 303–306.

Livingston, I. (1986) Geomorphological significance of wind flow patterns over a Namib linear dune. In Nickling, W. (ed.) *Aeolian Geomorphology*, Allen & Unwin, Boston, pp. 97–112.

Marston, R. A. (1986) Maneuver-caused wind erosion on impacts, South Central New Mexico. In Nickling, W. (ed.) *Aeolian Geomorphology*, Allen & Unwin, Boston, pp. 273–290.

Ming, Z., Panofsky, H. A. and Ball, R. (1983) Wind profiles over complex terrain. *Boundary-Layer Meteorology*, **25**, 221–228.

Rosen, P. (1978) An efficient, low cost aeolian sampling system. *Current Research Part A*, Geological Survey of Canada, Paper 78-1A, 531–532.

Sarre, R. (1987) Aeolian sand transport. *Progress in Physical Geography*, **11**, 155–182.

Sherman, D. J. and Hotta, S. (1990) Aeolian sediment transport: theory and measurement. This volume, pp. 17–37.

Svasek, J. and Terwindt, J. (1974) Measurements of sand transport by wind on a natural beach. *Sedimentology*, **21**, 311–322.

Chapter Four

A quantitative analysis of the near-surface wind flow pattern over coastal parabolic dunes

MARALYN J. ROBERTSON-RINTOUL
ECOS, Aberdeen

1. INTRODUCTION

Parabolic dunes have been described both from coastal environments and from desert locations (e.g. Landsberg, 1955; McKee, 1979; Pye, 1982). They are a class of dune in which the development of form is controlled in part at least by stabilisation from vegetation. The dunes develop a 'U' or 'V' shape as the arms of the dune are anchored and the noses migrate in the direction of the dominant sand-carrying wind. Typically, coastal parabolic dunes consist of a steep vegetated leeward slope and a windward slope of bare sand. The trailing arms of parabolic dunes are often anchored by vegetation.

Several major fields of parabolic dunes occur around the coastline of Scotland. They include the Barry Sands, Morrich More and North Forvie dune fields. The variation in morphology of the Scottish coastal parabolic dunes typifies the variation in form that has been described from both coasts and deserts. The Barry Sands parabolas, located on the northern side of the Tay Estuary, are elongate, relatively simple 'V'-shaped dunes. An index used frequently to describe the planform of parabolic dunes is a length to width, or form, ratio. The Barry Sands dunes have an average form ratio of 3.3, similar to the simple 'V'-shaped desert parabolas of the White Sands National Monument, New Mexico (McKee, 1979) and the elongate 'V'-shaped parabolic dunes described by Pye (1982) from the North Queensland coast. The dunes of Morrich More on the Dornoch First coast are compound 'V'-shaped dunes characterised by the presence of several trailing arms with a form ratio of 2.0. The North Forvie dunes, located north of Aberdeen, are 'U' and rake-shaped dunes with a mean

---

Coastal Dunes: Form and Process. Edited by K. F. Nordstrom, N. P. Psuty and R. W. G. Carter

ratio of 1.2, close to the mean for the parabolas of the Thar Desert which are 'U'-and rake-shaped in gross planimetric form (Tsoar, 1978). However, where the parabolic dunes of the Thar Desert are characterised by the presence of numerous trailing arms, those of the Forvie dunes possess secondary blowouts in the dune rims.

As complex dune forms grow and migrate downwind a pattern of airflow evolves around the bedform. Wilson (1972) found that for aeolian bedforms, flow structures are the product of the interaction between the dune and the airflow into which it projects. He also suggested that it was important to distinguish between the pattern of secondary flows at the initiation of a bedform and the flow pattern that became established over the mature dune form. The development of these flow structures is important to the transport of sand over the dune and therefore the sedimentary processes which govern its continued growth as a mature dune form.

Although the morphology of parabolic dunes has been considered in the literature (e.g. McKee, 1966; McKee, 1979; Pye, 1982), and there has been some attention directed to the history of development of certain parabolic dune fields (Landsberg, 1955; Pye, 1982), there has been little detailed field study of either the characteristics of the near-surface airflow over the three-dimensional form of parabolic dunes or the evolution of flow structures over the dune. As an understanding of the near-surface wind flow is crucial both to sand transport processes and to the resultant morphology of the dunes, this chapter considers the dynamics and interactions of the near-surface airflow over a coastal parabolic dune. The results should be compared with those of Bauer *et al.* (this volume) who worked on flatter aeolian surfaces.

Appreciation of the interrelationships between the near-surface airflow over the dune, landform morphology and the flow structure requires measurement and visualisation of airflow characteristics. Three parameters of the flow have been selected to characterise the airflow over the dune and two morphometric variables are used to represent dune form. A combination of Principal Components Analysis (PCA) and Cluster Analysis has been used to examine systematic trends in the data. This quantitative method of analysis, used previously in dune studies by Bressolier and Thomas (1977) to assess factors influencing surface roughness over dunes, is useful for evaluating trends which require the simultaneous assessment of a large number of measurements on a number of variables. The results from the statistical analyses are examined in relation to the flows identified during experiments.

2. EXPERIMENTAL SITE

The experimental site is located in the North Forvie dune field which forms part of the more extensive Sands of Forvie. The Sands are located approximately 19 km north of Aberdeen in northeastern Scotland (Fig. 1) with the North Forvie dunes sited on a rock plateau covered by glaciofluvial deposits. These Pleistocene deposits are overlain by Holocene aeolian sand which has been moulded into

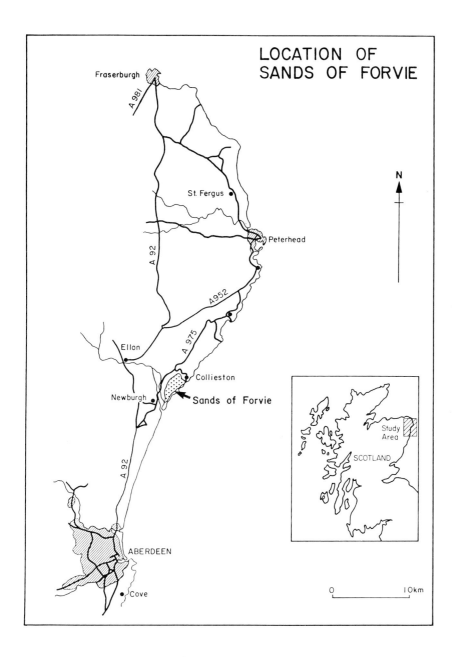

Fig. 1   The location of the Sands of Forvie

the discrete sand waves, sandhills and parabolic dune areas that are characteristic of the present day topography of North Forvie. The evolution of the Sands of Forvie dune system has been discussed by Landsberg (1955) and Ritchie, Rose and Smith, (1978).

The parabolic dunes of North Forvie are grouped into a series of bands with the experimental dune sited at the southeastern end of the middle band of dunes. In plan, the parabolic dunes are basically 'U'-shaped although some of the dunes experience distortion from a smooth outline due to the development of secondary blowouts within the dune rims. The planform of the experimental dune is shown in Fig. 2 together with relative heights of the dune crest above the adjacent deflation surface. The dune is broadly 'U'-shaped in plan, is about 200 m wide and 215 m in length. The dune is partially vegetated with an *Empetrum nigrum*–lichen community on its windward face and *Carex arenaria–Ammophila* sp. grassland on its crestal and leeward slopes.

The preferred orientation of the North Forvie parabolic dunes, including the experimental dune, is 247° representing the direction of regional sand drift. The resultant vector of the effective sand-moving winds is 257°, while the vector resultant for the unweighted regional winds is 240° with the greatest percentage occurring between southwest and northwest. Winds from 240° to 270° form the acceptance sector for velocity measurements with the main dune forming winds for the North Forvie parabolic dunes falling within this 30° sector. Wind data reported in this paper are for winds from the preferred orientation only.

The array of anemometer positions for the velocity readings is shown in Fig. 2. This arrangement sampled half of the flow field over the dune with the anemometers positioned in transects perpendicular to the dune crest. Each transect had monitoring stations located at the windward base, the crest and the leeside base of the dune. Transects were sited parallel to the long axis of the dune (Stations, 1, 2, and 3), through the zone of maximum curvature or turning point of the dune (Stations 4, 5, and 6) and across the trailing arm (Stations 7, 8, and 9). A control point or reference station in the uniform stream of flow (Station 10) was established on the flat deflation surface some distance windward of the dune.

The field experiments were conducted using lightweight expanded polystyrene cup anemometers attached to a central body machined out of round aluminium bar. The anemometer cups were linked to magnetic reed switches which sent DC-pulses to a data logger. Each anemometer was mounted on Dexion with the cups 1.6 m above the sand surface. Measurements of wind direction were made by means of directional vanes set up at the anemometer sampling points. Flow over the parabola was visualised using candles that generated a dense white smoke. Time lapse photography with an interval of 10 seconds provided sequential photographs of the flow. The 10 stations were monitored on several occasions to ensure reproducibility of results.

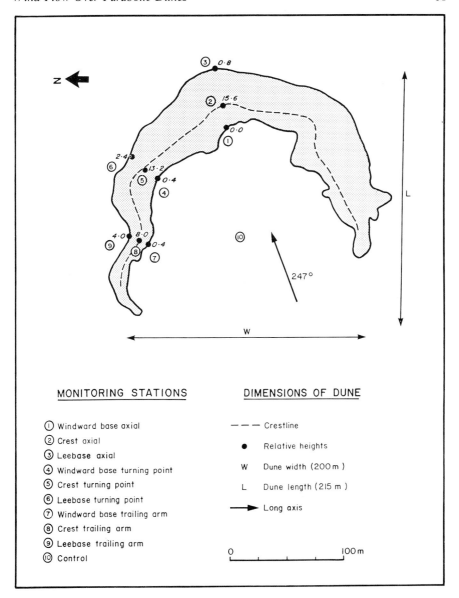

Fig. 2   Dune morphology and location of monitoring stations

3. FLOW PARAMETERS

Simultaneous velocity readings for each of the 10 dune monitoring stations were recorded on the data logger at 5-second intervals during a 10-hour period in November 1982. Examples of traces for part of the period are given in Fig. 3,

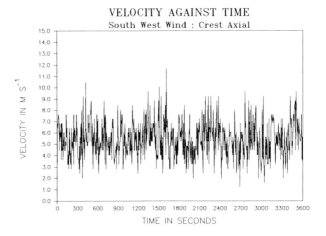

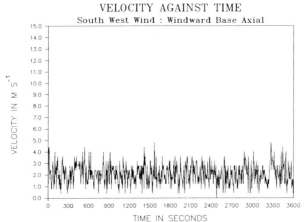

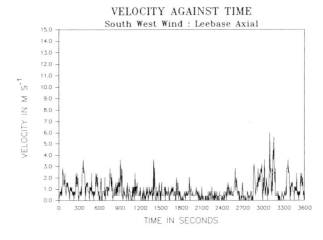

showing velocity readings at 5-second intervals from the axial transect of the dune including windward, crestal and leeside positions (Stations 1, 2, and 3).

Flow visualisation experiments were also carried out during the same period. A large volume of data was generated during this period and therefore, following Hsu (1971), the velocity data were reduced to 15-minute mean values in order to facilitate the subsequent statistical analyses. Wind data reduction, analysis procedures and graphical output are presented in detail elsewhere (Robertson-Rintoul, 1985).

Three parameters were selected to represent the flow characteristics in the statistical analysis. These were relative velocity, longitudinal turbulence intensity and turbulent energy. Relative velocities were calculated as:

$$U = u_{dune}/u_{control} \qquad (1)$$

where U = relative velocity

u = absolute velocity in m s^{-1}

The strength or intensity of the longitudinal component of turbulent velocity may be expressed as the root mean square of the velocity (Knott, 1978);

$$\sigma_{n-1}u = \sqrt{u'^2} \qquad (2)$$

where σ_{n-1} = standard deviation

u' = the fluctuating component of velocity in the x direction.

Total turbulent kinetic energy (E) has been found to be proportional to the mean square of the longitudinal component of the velocity fluctuations, that is, $E \propto u'^2$ (Bradshaw, 1975).

The 15-minute mean relative velocity patterns showed that the passage of the air across the parabolic dune was associated with the development of velocity gradients both parallel and normal to the flow (Fig. 4). Retarded flows occurred at the windward and lee dune bases, while accelerated flows were found at the dune crests. Absolute velocities at the control station were a maximum of 7.65 m s^{-1} and had an average value of 4.01 m s^{-1}. Relative velocities at the windward station ranged from 0.45 times the control station at the axial transect to 0.81 times the control station at the trailing arm. Crestal velocities ranged from 1.21 times the control station at the crest turning point. Leeside values exhibited high relative values of 1.28 times the control station at the turning point and low values of 0.16 times the control station at the leeside of the axial transect. Maximum relative velocities were recorded at the turning point of the dune.

Turbulence intensity and turbulent energy values at the control station were on average about 1.17 m s^{-1} and 1.36 m s^{-1} s^{-2}. A general feature of both turbulence measures over the dune was an overall increase in values from those measured at the control. High levels of turbulence are a feature of the

Fig. 3 Velocity traces for the axial transect of the study dune. Traces include the windward, crestal and leeside stations and show a part of the monitoring period only

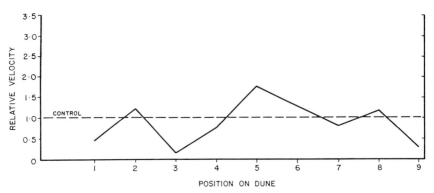

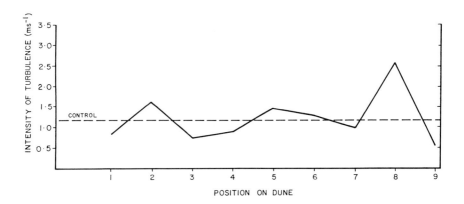

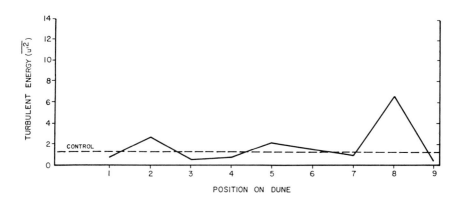

Fig. 4   Mean relative velocity, turbulent intensity and turbulent energy values for the dune monitoring stations. The plot represents average values calculated from the 15-minute means

readjustment of a sheared flow to passage across bedform roughened surfaces. Turbulence measures also varied with position over the dune showing both downwind and crosswind gradients (Fig. 4).

A crosswind variation in turbulence was superimposed on a cross-profile trend of increasing intensity from windward base to crest and decreasing intensity from crest to leeward base. Low turbulence intensity values were recorded at all the windward base stations (Fig. 4). Only a small range of values was recorded for these stations with typical velocities around $0.85 \, \mathrm{m \, s^{-1}}$ to $0.99 \, \mathrm{m \, s^{-1}}$. Velocity traces for these stations were characterised by relatively frequent weaker gusts. Peak turbulence intensities were generally recorded at the crest together with accelerated mean velocities. The velocity traces for crests were characterised by strong, rapidly occurring gusts (Fig. 4). Turbulence intensity showed minimum values at the axial and trailing arm positions where crests showed peak turbulent fluctuations. Higher values of $1.28 \, \mathrm{m \, s^{-1}}$ were recorded at the leeside turning point where the velocity traces were characterised by stronger winds with frequent gusts of moderate strength. Values for the other two leeside stations ranged from $0.75 \, \mathrm{m \, s^{-1}}$ at Station 3 to $0.55 \, \mathrm{m \, s^{-1}}$ at Station 9, exhibiting marked decreases of turbulence from the control station. Velocity traces for these two stations showed only short-lived gusts of moderate amplitude (Fig. 3).

Turbulent energy exhibited a generally similar pattern to that measured for turbulence intensity with marked downwind and crosswind gradients (Fig. 4). A small range of values was found for the windward stations around the dune base. Values ranged from about $0.6 \, \mathrm{m^{-1} s^{-2}}$ to $0.98 \, \mathrm{m^{-1} s^{-2}}$. Areas of peak turbulent energy were associated with the axial and arm zones of the dune with peak values occurring at the crest and minimum values at the leebase stations. Values for the leeside stations varied from $0.2 \, \mathrm{m^{-1} s^{-2}}$ at the trailing arm to $1.2 \, \mathrm{m^{-1} s^{-2}}$ at the turning point, where they attained their highest values. Peak turbulent energy values were recorded at the trailing arm crest. Here values reached $6.2 \, \mathrm{m^{-1} s^{-2}}$ whereas those at the axial crest averaged $2.8 \, \mathrm{m^{-1} s^{-2}}$ and those at the turning point about $1.8 \, \mathrm{m^{-1} s^{-2}}$. Minimum values of turbulent energy coincide with the zone of maximum curvature (turning point) of the dune. Maximum values are recorded where local orientation of the dune parallels the long axis, along the trailing arm of the dune.

4. MORPHOMETRIC PARAMETERS

Two parameters were measured to represent variation in morphology between the dune monitoring stations. Relative height of the dune station above the deflation surface indicates the topographic barrier effect of the dune to the incident airflow. These vary from zero at the windward axial dune station to 15.6 at the crest axial dune station. The second variable used was the angle of incidence of the airflow as the angle between the local wind direction and the trendline of the dune. This parameter varies with the planform curvature of the dune.

5. NUMERICAL ANALYSIS

The wind data demonstrate significant variation in flow velocities and turbulence characteristics between the dune monitoring stations. Flow variation over the dune is apparent in two senses: first there is a variation in the value of the flow parameters between the windward, crestal and leeside stations within transect; second there is a variation between transects in relative velocity, turbulence intensity and turbulent energy as location around the 3-dimensional form of the dune changes.

The data for flow and morphometry were analysed statistically to explore any systematic variations or trends.

5.1. Principal Components Analysis

Principal Components Analysis was used to analyse the data matrix of flow and form variables. Principal components are eigenvectors or latent roots of the basic variance–covariance data matrix and summarise the major relationships between the variables, revealing patterns of covariance. PCA maximises the variance of the first principal component from the basic variance–covariance matrix, with successive components extracting the maximum amount of residual variance. The first component may then be used as an index of the variables in the initial grouping, so that systematic variation in the original point distribution may be represented on a single scale by calculating component scores.

Results from the PCA of the raw data indicate there are two significant components (with eigenvalues > 1.0) (Kaiser, 1960) and that these components together explain 86.1% of the total variance in the original data set. These two components or eigenvectors, and the eigenvalues from the analysis, are given in Table 1.

Table 1   The eigenvalues and loading matrix for the principal components

Eigenvalues and cumulative proportion of total variance for two principal components

1	2.9214	0.5891
2	1.3702	0.8610

Principal components loading martrix

	1	2
1. Mean velocity	0.8282	−0.1598
2. Turbulence	0.9486	−0.0924
3. Turbulent energy	0.9482	−0.0897
4. Angle of incidence	−0.1332	0.9469
5. Height	0.6533	0.9459

The loadings on the first component show a strong positive correlation with relative velocity, turbulence and turbulent energy. There is also a weaker, but positive correlation with the relative height variable. These positive loadings on the flow and height variables indicate that as height increases over the dune so the flow parameters also increase. The highest points on the dune should experience high relative velocities and turbulence. The magnitude and direction of the loadings on the first component indicate that this new dimension represents a compounded index of the variables which have been used to characterise the airflow over the dune and the variable representing the topographic barrier which is responsible for the local vertical distortion of the airflow. Therefore it may be interpreted as a general process or flow component accounting for nearly 60% of the total variance in the data set.

For the second component high positive loadings are exhibited by the two morphological variables which quantify the height and planform effects of the dune providing an index of the morphological variables which may be regarded as a form component. This component accounts for an additional 27% of the variance in the original data set. A sufficiently high proportion of the residual variance represents a significant secondary trend superimposed on the dominant flow trend indicated by the first component.

The multivariate statistical analysis used here transforms the original five inter-correlated variables into two components without substantial loss of information. While this reduction in the number of variables is useful, the value of the technique lies in the analysis of the component scores. The component scores calculated from the first principal component were therefore plotted onto the axis of the first component (Fig. 5), thus projecting the scores into one dimension. The plot shows a continuum of standardised component scores which range in value from about $+2$ to -1.75. Variation in the magnitude and direction of the scores is coincident with position on the dune. Those points possessing high positive scores have been derived from monitoring data from the dune crest locations. Those points with high negative scores are associated with the leeside monitoring sites. Moderate values, that is those between about $+1$ and -1, are windward stations. The exception to this general pattern is the station at the zone of maximum curvature of the dune, the leeside turning point which occupies a plotting position between the crestal locations and the leeside stations indicating relatively high values on this first component.

The principal components in PCA are mutually orthogonal and therefore uncorrelated. Any analysis of the pattern or structure presented by the scores of the second (or subsequent significant components) may yield secondary trends superimposed on the dominant trend represented by the first component. The relationship between the major trend in the North Forvie flow data set and the superimposed secondary trend is evident when the scores for the flow component and the form component are plotted onto the plane defined by the orthogonal

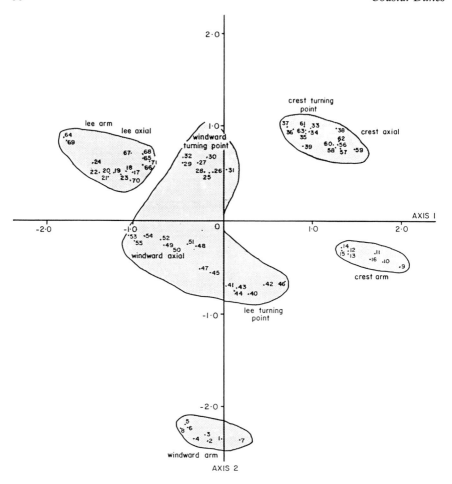

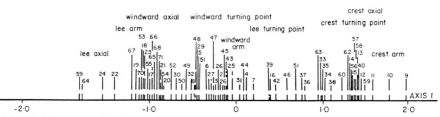

Fig. 5   The upper plot shows the Component 1 and Component 2 scores for each station plotted onto the principal plane. The boundaries of the groups are determined from Cluster Analysis. The lower plot is a projection of the Component 1 scores for each station onto the axis of the first Component

axes of the two principal components (Fig. 5) which reveals a clustering in the distribution of component scores over the principal plane.

The plot of the component scores along the first principal axis enabled the trend in the flow variables to be examined while holding the variables incorporated into the second component constant. In the principal plane plot, the planform and height effects of the dune have been allowed to vary and this has resulted in the grouping of data points into clusters over the principal plane. This plot therefore represents the temporal variation of the flow parameters over the dune as they vary with the spatial characteristics of the dune form. Therefore the principal plane plot may be regarded as a process-form plane that enables the statistical and graphical identification of the two-way interaction between flow and form over the coastal parabolic dune.

Before continuing to analyse the significance of the groups shown in the principal component plot (Fig. 5) the component scores from the two principal components were clustered statistically.

5.2. Cluster Analysis

Ward's (1963) hierarchical grouping algorithm, minimising an error sum of squares (ESS) objective function, is probably the best option for finding tight minimum variance clusters (Wishart, 1978) and therefore was used here.

Defining cut-off points for grouping procedures is generally acknowledged to be subjective and usually made after a visual appraisal of the dendrogram (Harbor, 1986). Ward (1963) proposed a method for determining the cut-off point in a clustering procedure, and this has been used in a number of studies (Wishart, 1970; Miles and Norcliffe, 1984). By this method the similarity level at which to accept a group may be suggested by plotting within group variance as a proportion of total variance against number of groups at that similarity level. The point at which there is a marked discontinuity in the plot suggests a cut-off for the grouping. In this analysis there was a marked discontinuity from the five to the four cluster cut-off point, indicating a large increase in the ESS. According to this method the level which best minimises within group variance is the five cluster grouping. Further, the convergence of the cluster analysis grouping scheme with that evident on the plot in Fig. 5 suggests that the groups produced from the cluster analysis provide a satisfactory grouping of the scores from the two principal components.

The arrangement and composition of the clusters on the principal plane suggest that certain morphological zones on the dune exhibit a typical process response to the airflow. Cluster one comprises points from the dune crests of the turning point and axial transects, while cluster two is made up of points from the leeside positions of the trailing arm and axial transects and cluster three is formed of points from the windward stations of the turning point and axial transects. The leeside turning point is also included in this cluster. The windward arm station and the dune crest of the arm transect form discrete clusters.

6. FLOW VISUALISATION

The flow visualisation experiments conducted over the dune at the same time as the monitoring of the wind flow suggest that process-form relationships are associated with a distinctive evolution of unsteady flow structures, within which boundaries were subject to considerable flux. The flow was viewed from the nine dune monitoring stations and therefore will be discussed relative to these stations.

6.1. The windward dune face

Some flow element variation around the dune was apparent within this morphological zone. At the windward slopes of Stations 1 and 4 the visible flow structures were similar and characterised by stalling, flow separation and reverse rotational flow as closed windward eddies (Fig. 6). As the decelerated flow approached the dune base, stalling of the airflow sometimes occurred so that a pronounced recirculatory motion was set up with the separation bubble at the slope almost isolated from the main flow. However, this effect was intermittent, as the reverse flow sequence was periodically replaced by either an upslope flow or a deflected upslope flow towards the dune crest, coincident with stronger gusts of wind.

The windward base flow structure seen at Station 7 is illustrated in Fig. 6. The dominant flow pattern for this station is a spiral flow moving along the windward face of the arm as a trailing arm vortex. Comparison with the velocity trace showed that this flow element occurred when higher wind speeds were monitored at the control station. Reductions in wind speed coincided with a change in flow pattern as the flow moved upslope and became incorporated into the fast-moving airstream that crossed the dune crest.

6.2. The dune crest

Flow patterns at the three dune crest locations are shown in Fig. 6. A feature of the flow—that was steady throughout the period of observation—was the separation of the stream from the surface. The separated flow in the crestal area consisted of the detachment point, shown on Fig. 6, and the free layer, so that as the flow crossed the dune crests, the smoke particles were confined to a concentrated stream or jet. At the downstream side of the crestal area the flow detached from the surface with the separated or free layer moving downwind. This point at the crestal locations coincided with the abrupt change in surface inclination between the crestal area and lee slope.

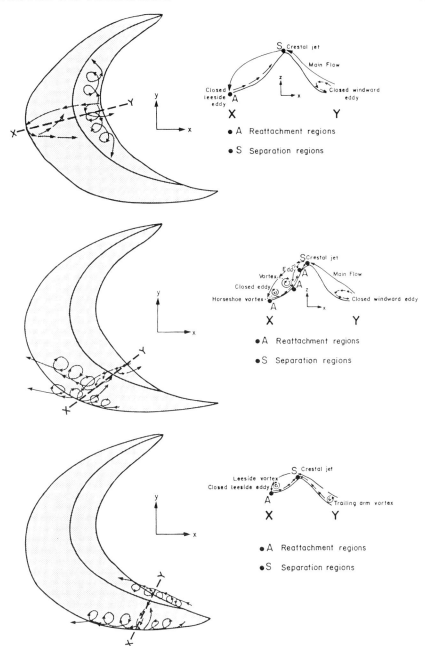

Fig. 6   The flow structures developed within the near-surface wind flow over the axial, turning point and trailing arm transects of the dune as revealed by flow visualisation. The flow structures are illustrated both in cross-profile and in planform

6.3. The leeward dune slope

The leeside of the axial transect exhibited a back flow into the cavity that was formed below the free layer of the separated crestal flow. An upper and lower level in the leeside flow was therefore distinguishable. The upper level comprised the separated boundary layer moving in a downstream direction and a lower layer formed of the ground level return flow.

A strong flow that followed the basal contours of the leeslope was apparent at the turning point of the dune. Flow trajectories had a clockwise rotation around an axis parallel with the local flow direction following a vortex towards the axial area of the dune. When this vortex flow reached the point of maximum curvature between the turning point and axial sections of the dune, detachment of the flow from the surface was noticeable (Fig. 6).

Air movements on the leeside of the trailing arm transect showed the return of a separated leeside flow, alternating occasionally with a spiral vortex, although reverse flow was probably the more persistent.

A short-lived reverse flow was also present at the leeside turning point of the dune. The upslope return flow at this point was clearly the ground level return flow of a transverse leeside eddy, probably associated with a large amplitude eddy formed by the detachment crestal flow (Fig. 6).

A composite picture of the flow patterns and structures in the near-surface flow over the dune has been determined from individual dune stations (Fig. 7). The axial transect probably exhibits the simplest pattern with flow elements dominated by diverging, decelerating and separating flow at the windward foot of the dune, a crestal jet and a leeside eddy. Such flow features are typical of flow over two-dimensional obstacles and they have been reported in dunes (Landsberg, 1942; Olson, 1958; Svasek and Terwindt, 1974; Hsu, 1977; Knott, 1978; Tsoar, 1978). At the turning point transect, an overall two-dimensional flow pattern could be observed, with the development of a closed windward eddy and crestal jet. However, this section of the dune exhibits pronounced bed curvature leading to development of marked leeside spiral vortices moving parallel to the dune base. Here the basic two-dimensional flow pattern was combined with flow parallel to the dune to give rise to a three-dimensional flow field. Knott (1978) and Hesp (1981) have described similar spiral vortex flows as the airstream is distorted around the dune base of barchan dunes and shadow dunes.

Flow elements at the trailing arm exhibited both the two-dimensional flow pattern observed at the axial transect, with a crestal jet and leeside eddy, and also the three-dimensional pattern of diverted and distorted flow patterns as spiral vortices developed on both the windward and leeward flanks.

7. PARABOLIC DUNE FORM, FLOW AND EVOLVED STRUCTURES

Field experiments have shown that four main types of flow structure could be distinguished within the near-surface airflow over the North Forvie coastal

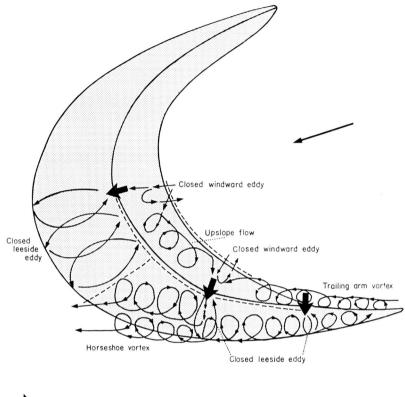

Closed windward eddy

Upslope flow

Closed windward eddy

Closed
leeside
eddy

Trailing arm vortex

Horseshoe vortex

Closed leeside eddy

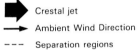 Crestal jet
→ Ambient Wind Direction
- - - Separation regions

Fig. 7 A composite diagram representing the flow structure evolution within the near-surface wind flow over the experimental site

parabolic dune. These were the jets at all crestal locations, the closed windward eddies at the axial and turning point transects, the leeside eddies at the axial and trailing transects and the spiral vortex flows at the windward arm and leeside turning point stations.

The relationship between the flow and form variables represented by the process-form plane and the observed flow structures at the various dune stations is shown on Fig. 5. The plot shows the sectors of the plane in which the flow structures evolve over the North Forvie parabolic dune, suggesting that there is a range of values for the flow and form variables at which crestal jets, eddy flow and vortex flow are likely to occur near the surface.

The first cluster of points shown on Fig. 5 represents the data points from the crestal locations of the turning point and axial dune stations, characterised by high to very high positive values for the flow component, exhibiting large increases of velocity with increased dune height. The stations also have high positive scores on the form component, with high angles of incidence and high elevations. The combination of the flow-form data and flow visualisation for these locations demonstrated markedly accelerated velocities and high turbulence values associated with clearly defined crestal jets. Knott (1978) recorded similar results for a desert barchan dune showing that convergence of streamlines and acceleration of flow gave rise to a crestal jet with velocity and turbulence maxima. Landsberg (1942), Landsberg and Riley (1943) and Olson (1958) also reported increased velocities in the crestal area of the dune and commented on the existence of the crestal zones as an area of sand erosion.

The second cluster of points contains the data from the leeside arm and axial stations, characterised by high negative scores for the process variable and moderate to high values for the form variable. These two stations were notable for their extremely low values for relative velocity, turbulent intensity and turbulent energy (Fig. 4) and both stations had similar, steep angles of incidence, 78° for the axial station and 70° for the arm station. During the flow visualisation experiments large-scale leeside eddies developed at these stations. The eddy at the arm station was seen to fluctuate with a spiral vortex flow as the angle of incidence narrowed, although the eddy flow appeared to be the dominant pattern.

The existence of the leeside eddy for two-dimensional flow over dunes has been demonstrated both through velocity measurements in the field and by flow visualisation experiments. Landsberg (1942) and Landsberg and Riley (1943) demonstrated a well-developed eddy to the leeward of a blowout in a Lake Michigan sand dune. Inman, Ewing and Corliss (1966) and Knott (1978) both showed a flow reversal at the slipface of dunes and Hsu (1977) measured a fall in wind speed and reversal of wind direction in the lee of coastal sand dunes on Padre Island, Texas. Together these studies illustrate the effect of the dune form on the velocity characteristics of the airflow and the flow structures which are produced from the interaction at the fluid solid boundary. The importance of the eddy is in the creation of a cavity or wake of relatively stagnant air which then becomes a depositional zone in which slipface formation can occur.

The points making up the third cluster plot around the origin of the principal plane and consist of values from the windward turning point, windward axial and leeside turning point stations. These stations are characterised by moderate values for the process component varying from about $+0.8$ to -1.0, and moderate values for the form component, ranging from about $+0.8$ to -0.8. At the windward turning point and axial stations flow visualisation demonstrated that the flow was dominated by stalling and production of closed eddies. Landsberg (1942), Landsberg and Riley (1943) and Olson (1958) all present diagrams of isovels over coastal dunes in Lake Michigan displaying a pronounced divergence of isovels at the foot of the dune. Olson also noted that the result of this was a stalling of air at the dune base. A similar shelter effect at the dune foot has been recorded by Carter (1977) and Sarre (1984). Sarre observed the effect of this flow expansion, demonstrating a lifting of smoke from the surface, and showed that these wind effects resulted in reduced sand transport rates in this area of the dune. Landsberg (1942) also commented that the distortion of the flow at the foot of the dune would prevent sand transport from that zone.

However, flow elements at the leeside turning point were more complex with vortex flow alternating with closed eddy flow. The relative displacement of the leeside turning point down the y axis within the cluster is a result of the influence of the form variable, and in particular, the angle of wind incidence, which for the windward turning point is *c*. 90°, for the windward axial point is *c*. 70° and for the lee turning point is *c*. 38°. As the angle of incidence closes, the points are depressed further along the y axis of the principal plane. Therefore this dune station is intermediate between those which are dominated by leeside eddies and those with a predominant spiral vortex pattern, such as are found at the windward arm station. The alternation of the leeside eddy and spiral vortex flow at this station appears to be associated with variation in wind speed and orientation, and hence a combination of both form and flow effects. Weaker winds saw the development of a leeside eddy whereas stronger gusts of wind coincided with the development of a spiral vortex. These findings are similar to those of Knott (1978) who demonstrated that a rise in wind speed was associated with the spiral vortex flow and a marked drop in speed with the leeside eddy. He also found that when the airflow encountered the crestline of the experimental dune at an angle between 0 and 60° the edges of the leeside eddy rolled up and formed a trailing spiral vortex. With angles varying between 60° and 90° a closed leeside eddy was formed. Tsoar (1978) found a similar change in leeside flow pattern and wind speed with variation in the angle at which flow crossed the crestline of the dune. He suggested, however, that it was difficult to determine empirically the range of angles within which the different flow patterns would develop. Although it is possible that the critical angle may be 40° with angles below this leading to the formation of a leeside spiral vortex.

The fourth cluster of points is made up of points from one dune station only, the windward arm station notable for its extreme position on the form axis. The cluster of points has moderately high values for the flow component as did that for the windward stations in cluster 3, but had extreme negative values for the form component. This result is due to the very small angle of wind incidence of this station, on average only 18°. The flow structure developed at this station was characterised by a diverging, trailing arm vortex. The occurrence of the spiral vortex at the leeside of the dune in the trailing arm dune sector may reflect the angle at which the flow crossed the crestline and therefore whether separation occurred in two or three dimensions.

The crestal position of the trailing arm also occupies a single cluster. Here the more acute angle of incidence experienced by this station plus the lower height relative to the two other crestal positions lowers the position of the cluster along the form or vertical axis, although it possesses high values for the flow component. Here a weaker crestal jet was seen in the smoke trajectories.

On the parabolic dune vortex flow may occur on either the windward or leeside of the dune, a response to planform curvature of the dune and the resultant flow distortion effects. Whether the station is associated with a flow structure dominated by a closed separation bubble or vortex flow appears to be determined by the combination of the wind speed and the degree of air stream distortion forced by the curving trendline of the dune. Stations with a higher angle of incidence tend to develop separation bubbles while those at acute angles appear to become dominated increasingly by vortices. Thus, the more curvilinear the crest the more likely the dune is to experience a combination of two-dimensional and three-dimensional flow.

8. CONCLUSION

The field experiments on the North Forvie parabolic dune suggest that at least four main types of flow structure develop over the dune, each occurring at particular morphological zones around the dune. Jets developed at all crestal locations; closed windward eddies were generated on the windward slopes, while at the axial and turning point transects leeside eddies were found. Spiral vortex flow occurred at the windward arm and leeside turning points. The results from the multivariate analysis and the flow visualisation experiments indicate that there could be an environmental range within which crestal jets, closed eddies and spiral vortices might be expected to occur near the surface across the strongly curved form of coastal parabolic dunes.

REFERENCES

Bagnold, R. A. (1941) *The Physics of Blown Sand and Desert Dunes*. Methuen, London.
Bradshaw, P. (1975) *An Introduction to Turbulence and its Measurement*. Pergamon Press, Oxford.
Bressolier, C. and Thomas, Y. -F. (1977) Studies on wind and plant interactions on French Atlantic coastal dunes. *J sediment. Petrol.*, **47**, 333–338.
Carter, R. W. G. (1977) The rate and pattern of sediment interchange between beach and dune. In *Coastal Sedimentology*, Tanner, W. F. (ed.), Florida State University and Coastal Research, Tallahassee, Florida, pp. 3–34.
Cooper, W. S. (1958) Coastal dunes of Oregon and Washington. *Geol. Soc. Am. Mem.*, No. 72.
Cooper, W. S. (1967) Coastal dunes of California. *Mem. Geol. Soc. Am.*, No. 104.

Cornish, V. (1897) On the formation of sand dunes. *Geogr. J.*, **9**, 298–309.

Davis, J. C. (1973) *Statistics and Data Analysis in Geology.* Wiley, New York.

Harbor, J. M. (1986) A comment on certain multivariate techniques used in the analysis of late Quaternary relative age data. *Prog. Phys. Geogr.*, **10**, 215–225.

Hesp, P. A. (1981) On the formation of shadow dunes. *J. sediment. Petrol.*, **51** (1), 101–111.

Hoyt, J. H. (1966) Air and sand movement to the lee of dunes. *Sedimentol.*, **7**, 137–143.

Hsu, S. A. (1971) Measurement of shear stress and roughness length on a beach. *J. Geophy. Res.*, **76**, 2880–2885.

Hsu, S. A. (1977) Boundary layer meteorological research in the coastal zone. *Geosc. Man.*, **18**, 99–111.

Inman, D. L., Ewing, G. C. and Corliss, J. B. (1966) Coastal sand dunes of Guerrero Negro, Baja California, Mexico. *Bull. Geol. Soc. Am.*, **77**, 787–802.

Kaiser, H. F. (1960) The application of electronic computers to factor analysis. *Educ. Psychol. Meas.*, **20**, 141–151.

Knott, P. (1978) The structure and pattern of dune-forming winds. PhD thesis, Dept. of Geography, University College, London.

Lancaster, N. (1985) Variations in wind velocity and sand transport on the windward flanks of desert sand dunes. *Sedimentol.*, **32**, 581–593.

Landsberg, H. (1942) The structure of wind over a sand dune. *Trans. Am. geophys. Un.*, **23**, 237–239.

Landsberg, H. and Riley, J. (1943) Wind influences on the transportation of sand over a Michigan sand dune. *Proc. 2nd Hydraulic Conf. Iowa Stud. Eng Bull.*, **27**, 342–353.

Landsberg, S. V. (1955) The morphology and vegetation of the Sands of Forvie, PhD thesis, University of Aberdeen.

McKee, E. D. (1966) Structures of dunes at White Sand National Monument, New Mexico. *Sedimentol.*, **7**, 1–69.

McKee, E. D. (ed.) (1979) A study of global sand seas. *Geol. Surv. Prof. Paper*, 1052, US Govt. Printing Office, Washington, DC.

Miles, N. and Norcliffe, G. (1984) An economic typology of small scale rural enterprises in Central Province, Kenya. *Trans. Inst. Brit. Geogr. NS*, **9**, 163–187.

Olson, J. S. (1958) Lake Michigan dune development 1: Wind velocity profiles. *J. Geol.*, **66**, 254–263.

Pye, K. (1982) Morphological development of coastal dunes in a humid tropical environment; Cape Bedford and Cape Flattery, North Queensland. *Geogrf. Annaler*, **64A**, 213–227.

Ranwell, D. S. (1958) Movement of vegetated dunes at Newborough Warren, Anglesey. *J. Ecol.*, **46**, 83–100.

Ritchie, W. and Mather, A. S. (1984) *Beaches of Scotland.* Report for The Countryside Commission for Scotland, Dept. Geography, University of Aberdeen.

Ritchie, W., Rose, N. and Smith, J. S. (1978) *Beaches of North East Scotland.* Report for The Countryside Commission for Scotland, Dept. Geography, University of Aberdeen.

Robertson-Rintoul, M. J. (1985) The morphology and dynamics of parabolic dunes within the context of the coastal dune systems of mainland Scotland, unpublished DPhil. thesis, University of Oxford.

Sarre, R. D. (1984) Sand movement from the intertidal zone and within coastal dune systems by aeolian processes, unpublished DPhil. thesis, University of Oxford.

Sharp, R. P. (1966) Kelso Dunes, Mojave Desert, California. *Bull. Geol. Soc. Am.*, **77**, 1045–1074.

Svasek, J. N. and Terwindt, J. H. J. (1974) Measurement of sand transport by the wind on a natural beach. *Sedimentol.*, **21**, 311–322.

Tsoar, H. (1978) *The dynamics of longitudinal dunes.* European Research Office, US Army Report No. DA-ERO. 76-G-072.

Ward, J. H. (1963) Hierarchical grouping to optimise an objective function. *J. American Statist. Ass.*, **58**, 236–244.

Wilson, I.G. (1972) Aeolian bedforms—their development and origins. *Sedimentol.*, **19**, 173–210.

Wishart, D. (1970) Some problems in the theory and application of numerical taxonomy, PhD thesis, University of St Andrews.

Wishart, D. (1978) *Clustan User Manual*, 3rd ed., University of St Andrews.

Chapter Five

Eolian transport measurements, winds, and comparison with theoretical transport in Israeli coastal dunes

VICTOR GOLDSMITH,
Department of Geology and Geography, Hunter College, City University of New York

PETER ROSEN,
Department of Geology, Northeastern University, Boston

AND

YARON GERTNER
National Oceanographic Institute, Israel Oceanographic and Limnological Research

1. INTRODUCTION AND LITERATURE REVIEW

In the last decade, significant advances have been made in describing and quantifying the multivariate processes acting on the Israeli coastal zone. Most of this effort has been focused on the dynamics of waves and the role of waves in transporting and depositing sand in the nearshore zone (Goldsmith and Golik, 1980). Wind transport of sand between the beach and dune system interacts directly with the nearshore transport by waves, but little is known about the magnitude of these wind transfers relative to wave transfers, and the net transport resulting from wind. Eolian transport must be assessed quantitatively, in order to understand fully the Israeli coastal sediment budget.

While theoretical equations exist to predict sand transport by wind under ideal conditions, little work has been done to define actual transfers in the field setting, encompassing such complicating factors as moisture and vegetation. Sand

Coastal Dunes: Form and Process. Edited by K. F. Nordstrom, N. P. Psuty and R. W. G. Carter
© 1990 John Wiley & Sons Ltd

transfers by wind on and off the Israeli beaches were determined by field measurements along both the north and south coasts of Israel over 15 months of monitoring, encompassing two winters.

The coast of Israel is a suitable field setting for delineating wind and wave sediment transport because the system lacks complications found on many other coasts. Firstly, the coastal geomorphology is not complex. Secondly, the Mediterranean climate consists of fairly predictable weather cycles, both short term and seasonal. Transfer trends can be reasonably well correlated with these cycles. The seasonal lack of precipitation greatly simplifies the model. While theory predicts that little transport takes place during rain, observations show that this is not true.

A regional eolian sediment budget has been developed by trapping sand transfers in four directions. The net movement of sand by wind for the system has been derived. Wind records concurrent to monitoring provide a data base both for hindcasting and forecasting transport in the region.

Sediment transport by wind was first quantitatively described by Bagnold (1954). Chepil's results (1945) were consistent with Bagnold's model. Most field-based studies of wind transport (i.e. Belly, 1964; Kadib,1964; Johnson and Kadib, 1965; Hsu, 1971, 1973, 1974; Gutman, 1977; Tsoar, 1983) have been based on the use of, or modification of Bagnold's equation. Swart (1987) has recently studied the prediction of eolian transport. Kutiel, Danin and Orshan (1979/1980) studied the effect of sand mobility, substrate, and soil content on dune plant communities and succession in the vicinity of Caesarea. Danin and Yaalon (1982) studied the interaction between soil processes and vegetation under differing rainfall conditions in the north and south of Israel. Wasson and Nanninga (1986) tried to estimate the vegetation effect on blown sand by theoretical equations. Moreno-Casasola (1986) studied the effects of sand on plant community distributions by monitoring sand movement and plant succession.

The theoretical development of the Bagnold equation and application to measurements of eolian sediment transport are succinctly detailed in Bagnold (1954), supported by Horikawa, Hotta and Kraus (1986), and reviewed by Sherman and Hotta (this volume). The application to eolian transport in coastal sand dunes is thoroughly detailed in Goldsmith (1985). The prediction of transport in an unvegetated and moisture-free setting, or over a short time interval, has been reasonably successful. The volume transport (Q) is directly proportional to the cube of the wind shear velocity (U_*), as follows:

$$Q = C[(d/D)^{0.5}(\rho/g)U_*{}^3]^{0.5} \qquad \text{(Goldsmith, 1985, equation 5)}$$

where C = sorting constant
d = mean grain diameter
D = grain diameter of standard 0.25 mm sand
ρ = fluid density (wind)
g = gravity

or, substituting for the Israeli coast, $Q = 1.8 [0.31/0.25] [1.25 \times 10^{-6}] [U_*{}^3]$.

Pye (1983) and Goldsmith (1985) have reviewed the literature on coastal sand dunes, as well as the applicable eolian sediment transport models. The importance of the eolian contribution to the total coastal sediment budget has also been shown by Bowen and Inman (1966), Pierce (1969), and Vallianos (1970). Gares (1987) studied the eolian sediment transport and dune formation on undeveloped and developed shorelines in New Jersey. (See Chapter 16.)

It is important to distinguish between vegetated coastal dunes and desert dunes which occur along a coast, such as those studied in the Sinai by Tsoar (1974, 1978, 1983, 1986). The vegetation acts as a 'baffle', decreasing shear stress, increasing surface roughness and the thickness of the calm wind layer, thereby increasing deposition, causing the dune to grow vertically in place rather than through slipface deposition and migration (Bagnold, 1954; Olson, 1958; Goldsmith, 1973). This results in a different process and morphology from desert dunes. The definitive works on coastal dunes are cited in Goldsmith (1985) and Pye (1983). Attempts have been made to incorporate the vegetation effect on surface roughness into the Bagnold equation (e.g. Olson, 1958; Bressolier and Thomas, 1977).

Another factor complicating the direct application of the Bagnold equation to eolian prediction in vegetated coastal dunes is the role of moisture. Belly's (1964) experiments in wind tunnels showed that even small amounts of moisture in the sand (2 to 3%) can significantly increase threshold shear values and decrease transport. However, field observations in North Carolina, New England, New York and eastern Canada by Rosen show that actual transport during severe precipitation events can range from none to very high values (e.g. Rosen, 1979). A means of describing soil moisture relative to the surface grain of sand does not exist, and renders the predicative use of field measurements low, since in many areas major winds occur during precipitation events. One reason is that the wind exerting shear stress is also causing evaporation, making moisture content measurements difficult. In a recent work, Hotta *et al.* (1984) attempted to develop an hypothesis for eolian transport on a wet sand surface, using field data, and met with moderate success for low water content only. Israel is comparatively rain-free, allowing the moisture problem to be eliminated during most of the year.

Direct measurement of sand transport by wind in the field has been attempted with numerous types of samplers (i.e. Horikawa and Shen, 1960; Leatherman, 1978; Rosen, 1978; Jones and Willetts, 1979; Pye and Tsoar, in press). Most studies with these samplers were oriented towards the dynamics of dune and beach deposition. Rosen (1979) utilized eolian sand trap field data to define long-term transfers of sand by wind on a small barrier beach.

2. TRAP CONSTRUCTION AND MEASUREMENT TECHNIQUES

Sixty-one eolian sediment traps were deployed in various configurations at five locations along the coast of Israel (Fig. 1) in November and December 1985.

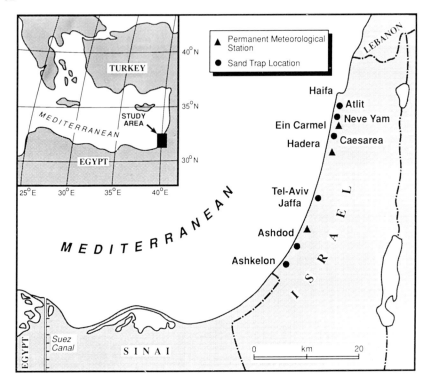

Fig. 1   Eolian sand trap location on the coast of Israel

The trap was based on a model deployed by Rosen (1978). The traps were constructed at the Oceanographic Institute at a cost of approximately $35 each plus labor (Fig. 2). The traps were placed in groups of four, facing SW, NW, NE and SE. The orientation of the traps was determined by inspection of the dune topography, which showed a marked orientation of wind shadows and other features to the southwest winds. This orientation means that the traps are oriented obliquely to the shoreline, complicating calculations of onshore/offshore transport. Nevertheless, this orientation was chosen in order to measure the maximum transport which was hypothesized to come from the southwest, the main storm wind direction along the Israeli coast.

The relation of the traps to the dune topography, distance from the sea, etc., is shown on the profiles measured at the beginning of the study (Fig. 3). Vegetation begins at the most seaward trap, which was placed at the first rise in slope. The Neve Yam location, with a wide flat beach and low dunes, is different from the Ashdod and Ashqelon profiles which have narrow, steep beaches backed by hills. These hills are composed of a core of quartz eolianite (locally called Kurkar) with mobile eolian sand superimposed on top of the rock.

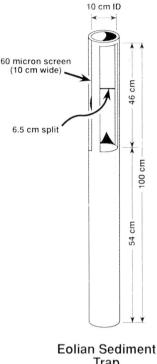

**Eolian Sediment
Trap**

Fig. 2   The PVC eolian sediment trap

Profile locations were chosen in areas where the rock does not show, and does not appear to directly influence eolian transport.

The relationship of the traps to the surrounding vegetation, and the density of the vegetation, were determined from greatly enlarged vertical aerial photographs. The photographs were obtained during different years, but all in the same season, in order to assess the interaction between eolian transport and the vegetation (Gertner, 1989). These photographs were used to assist in interpreting the eolian sand trap data in the second year, since vegetation type and density are parameters determining the rate of sand movement.

2.1. Trap measurement

The traps were monitored every two weeks, or more often during storms (e.g. half-hourly). The sand level in each trap was measured; the trap liner was taken out and emptied; and a small sediment sample was saved. A 0.5 m iron rod was emplaced at each trap location. This rod was measured to determine the vertical changes in the dune during the study. Also, the wind velocity and

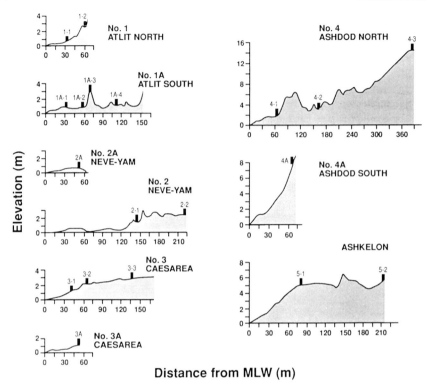

Distance from MLW (m)

Fig. 3   Dune profiles at sand trap location

direction were measured at each trap location. In addition to routine monitoring, one-day experiments were conducted at three locations and traps measured at half-hour intervals during periods of high winds. At the Neve Yam site, an additional eight traps were installed at regular intervals for this one-day experiment. More opportunities for such detailed monitoring during high wind conditions did not occur during the first year because of the unusually mild winter.

3. METEOROLOGICAL MEASUREMENTS

Wind velocity and direction were measured during each visit to the field using a Davis Instruments Electronic Wind Speed Indicator, at a height of 2 m at the front dune line. Wind velocity and direction were continuously measured at two locations—Ein Hacarmel (by a Velfle instrument at a height of 3.5 m) and at Ashdod port (by a Monro instrument at a height of 3.5 m)—by the Israel Meteorological Service. Later, the instrument at Ashdod port stopped working,

Trap No. 2.2.1

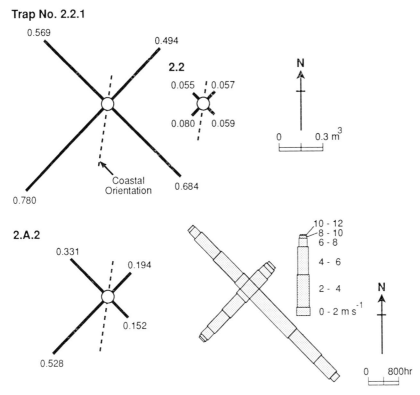

Fig. 4   Neve Yam sand transport and Ein Carmel winds during study

and the wind data were measured at Eshkol Power Station, Tel Aviv (at a height of about 40 m) by the Electric Company. These original data were digitized at one-hour intervals.

Relationships between these hourly data and the bi-weekly wind data measured at each trap location were established in order to apply the continuous wind data to the measured sand trap data (Table 1). The long-term wind data were divided into groups bounded by the dates and times that the eolian accumulations were measured in the traps at each location. These wind data were then used to calculate theoretical eolian transport using Bagnold's equation for comparison with the measured trap data. The hourly wind data were also divided into velocity groups, and by the four directions used in the trap arrangements, and wind roses were constructed for comparison with the trap data (Figs 4 and 5).

The extreme wind velocities were identified from the wind data. Since Q is proportional to U^3, these are the most significant sand-moving events.

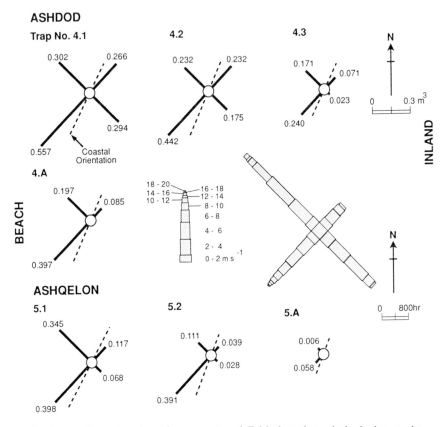

Fig. 5 Southern Israel sand transport and Eshkol station winds during study

4. ISRAELI CLIMATE

The Israeli climatic cycles are fairly simple, so transport events and net transfers can be reduced to predictable, well-defined weather events. In summer, the southern Mediterranean lies in the subtropical zone of the northern hemisphere. In winter, the entire system is displaced southwards, and this region becomes dominated by the middle latitudes belt of 'westerlies' or cyclonic winds. Of the 26 barometric lows that enter the Cyprus area, on a yearly average, 21 do so in winter and spring. Thus, this region is characterized by a 'Mediterranean climate' with a short rainy winter season and a rainless season from April to November. During the winter months in Israel, storm fronts from the west pass on an average of every 7 to 10 days, with rain lasting about 1 to 3 days (Goldsmith and Sofer, 1983).

During the summer, with the absence of cyclonic winds, the wind regime within the southeastern Mediterranean is restricted to breezes or regional winds such as the strong northwest winds over the southern Mediterranean. A sea breeze effect also is important in the summer. The daily onshore sea breeze is much stronger than the opposing land breeze at night. The easterly winds from the desert are common during the transition months (April to early June and October) and during the winter (Goldsmith and Sofer, 1983).

5. WIND DATA DURING STUDY

The study period was divided into three wind seasons: first winter (November 1985 to March 1986), summer (April 1986 to November 1986) and second winter (December 1986 to March 1987). The meteorological wind data were analyzed by dividing the data into directions which were the same as the trap directions. The calculations were made for Atlit and Neve Yam from wind data measured at Ein Hacarmel. The calculations for Ashdod and Ashqelon were made from wind data measured at Ashdod port and Eshkol Power Station (Fig. 1).

Low-velocity winds occur mostly from the southeast (Figs 4 and 5) and northeast, with velocities too low to cause eolian transport. Most of the high-velocity winds are from the southwest and northwest. Their durations are short, but they cause more eolian transport than the winds that blow from the southeast. The common direction of wind storms is from the southwest.

The winds in the south are stronger than in the north, and the duration of these higher velocity winds is longer. This may be attributed to the locations. In the north, the measurements were taken in Ein Hacarmel which is located farther inland than the southern locations. Also, northwest winds are more important in the south, which is due, in part, to the change in shoreline orientation.

6. DUNE MORPHOLOGY AT TRAP LOCATIONS

The trap locations were chosen according to the following criteria: (i) relatively natural areas representative of a variety of dune types and topography in both the north and south areas of the country (Fig. 3); and (ii) maximum inaccessibility to deter stealing and disturbance to the traps.

The coast at Neve Yam (Fig. 1) has a wide flat beach (100 to 150 m) backed by a narrow belt (Fig. 3) of low dunes (0.5 to 1.5 m in height). The dunes are covered with dense vegetation which causes the very gently undulating upper surface. The dunes are oriented to the northeast, one of the high wind directions during the winter storms.

The coasts of Ashdod and Ashqelon have narrow steep beaches (less than 50 m) backed by very wide areas of high hills (1 to 5 m in height). The dunes in the back beach are commonly vegetated, but less dense than in the north.

7. TRAP DATA

Eolian sand transport occurs from all four directions, but non-uniformly (Figs 4 and 5). Most of the sand moves to the northeast direction under the influence of the high winds from the southwest, and to a lesser extent to the southeast. The situation changes from site to site, according to the different local factors affecting the blown sand. These include vegetation, dune elevation, beach slope, beach width, coastline orientation and local topography. However, it is clear that net transport is onshore.

Although the second winter was more stormy than the first, the period of trap operation was shorter in the second winter. Thus, the sand transport measured by the traps in the second winter is not much larger. Sand transport is larger in the winter than the summer at Neve Yam, Ashdod and Ashqelon. Only at Atlit, which is a semi-enclosed bay with different shoreline orientations than the other areas, is the transport greater in summer.

7.1. Neve Yam

Eolian sand transport at Neve Yam is the largest of all the locations. In the first winter, sand transport was almost equal from all directions in both the front and back traps (Fig. 4). This sand transport was large because of the 250 m wide, gently-sloping beach (Fig. 3). The total amount of transported sand decreased landward (set no. 2.2 in Fig. 4), because of the dense vegetation cover at Neve Yam. The sand discharge recorded in set 2A is less than the discharge recorded in set 2.1 (both are at the front dune line), because of the local topography around set 2A which causes a decrease in the wind velocity.

In the second winter, transport from the southwest was the dominant direction because of the high winds. The storms in the second winter often included rain, which increased the moisture on the sand surface and caused trap flooding due to a rise of the groundwater table.

In the summer, despite the long measurement period, the blown sand is not very large because of the mild winds. The highest winds in summer were from the northwest, and also the southwest, and therefore the most sand is caught in these traps. Even though winds from the southeast are almost as significant, sand accumulation is insignificant from this direction because of the dense vegetation on this side.

7.2. Ashdod

Ashdod has a steep, narrow beach and therefore we should expect less sand transport into the dunes (Short and Hesp, 1982). In the first winter (Fig. 5), sand transport from the southwest was only slightly greater than from the other directions. Even though trap 4.2 was separated from the beach by a 6 m high foredune, and trap 4.3 was at an elevation of 16 m (Fig. 3), the decrease of sand transport inland was not significant. This is the opposite of Neve Yam, and is because of the less dense vegetation cover at Ashdod. In the second winter, the sand transport is low because of the short period of trap operation.

In the summer, the sand transport is less because of lower velocity winds, with the dominant directions being from the southwest and northwest. At the inland trap (set no. 4.3), the sand transport is almost zero. The calm winds and the slight vegetation cover prevent the sand transport.

7.3. Ashqelon

The beach at Ashqelon is even steeper (Fig. 3) than at Ashdod, so sand transport is also very low at this location. The dominant transport direction is from the southwest in the first winter, and in the second winter it is about equal from the southwest and northwest directions. The trap facing the southeast direction in set 4.1 accumulated less sand than the other southeast-facing traps because of a dense cover of vegetation in this direction.

In the summer, the dominant direction is from the northwest, as in the other locations. The rate of inland sand transport decrease is small between sets 5.1 and 5.2, and higher between sets 5.2 and 5.3, because of differences in the vegetation cover.

8. EXPERIMENTS DURING HIGH WINDS

During periods of high winds, detailed measurements of sand accumulation in the traps were made at half-hour intervals at Neve Yam, Ashdod and Ashqelon and compared with concomitant wind data. At Neve Yam, wind measurements were made during storms. During the storm of 24 January 1986 with winds from the east (offshore) with a velocity of $13 \, \mathrm{m\,s^{-1}}$, 22 cm of sand accumulated in trap no. 2.1.3 in 1 hr 25 min. ($0.012 \, \mathrm{m^3\,m^{-1}\,hr^{-1}}$) (Fig. 6). On 19 January 1986 at Neve Yam (Fig. 6), a total of 10 traps were temporarily deployed along the profile. Maximum transport occurred on the beach berm, and decreased landward, into the dunes, during onshore winds (from the west) of 11.1–$12.5 \, \mathrm{m\,s^{-1}}$.

On 19 December 1986 (Fig. 7), a SW storm of $5.3 \, \mathrm{m\,s^{-1}}$ was monitored at Neve Yam. While 92 cm of sand accumulated in 5 hours (which is $0.023 \, \mathrm{m^3\,m^{-1}\,hr^{-1}}$) in trap 2.1.1 (which faces SW in front of the dunes), only 5 cm accumulated in trap 2.2.1 (facing SW in the inner set). While 44 cm accumulated in 5 hours in trap 2.1.4 (trap facing SE in front of the dunes), no sand accumulated in trap 2.2.4 (trap facing SE in the inner set). These big differences occurred because of the vegetation.

A storm with westerly winds was recorded on 6 January 1987 with a wind velocity of $9.2 \, \mathrm{m\,s^{-1}}$. A total of 31 cm of sand accumulated in 5 hours

90

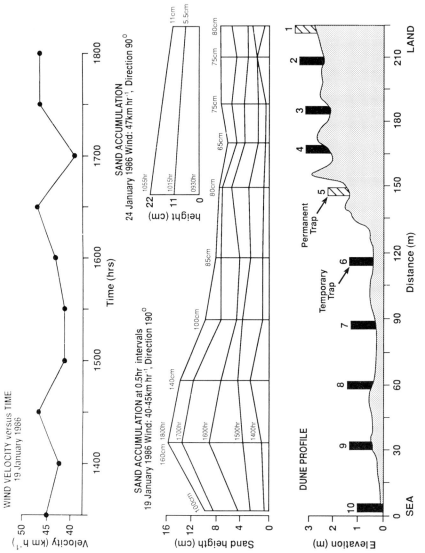

Fig. 6   Sand accumulation during 19 January storm at Neve Yam

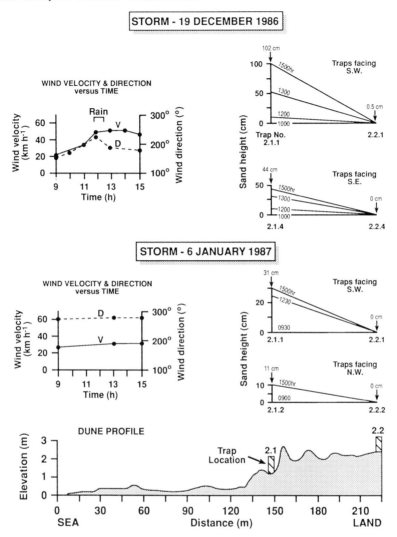

Fig. 7 Sand accumulation during 19 December and 6 January storms at Neve Yam

30 minutes $(0.004 \, \text{m}^3 \, \text{m}^{-1} \text{hr}^{-1})$ in trap 2.1.1 (facing SW) and 11 cm $(0.002 \, \text{m}^3 \, \text{m}^{-1} \text{hr}^{-1})$ in trap 2.1.2 (facing NW), in front of the dunes. No sand accumulated in the concomitant traps in the vegetated zone (Fig. 7).

The effect of moisture was observed in the storm of 19 December 1986. At 0700 hours it stopped raining and the sand immediately started to dry, and was transported normally by the winds. At 1200 (noon), heavy rain fell,for 20 minutes (Fig. 7). Wind velocity was the same both during and after the rain. The rate of eolian sand transport was also the same during the rain and after the rain stopped. Thus, it appears that the effect of moisture in decreasing sand transport is insignificant.

9. COMPUTATION OF EOLIAN TRANSPORT USING THE BAGNOLD EQUATION

9.1. Extreme winds from 1968 to 1986

Extreme wind events [$>10.2\,\mathrm{m\,s^{-1}}$ ($>20\,\mathrm{kt}$)] were studied from the available long-term data: Haifa port—1968 to 1984, Sde Dov, Tel Aviv—1977 to 1986, Ashdod port—1968 to 1984. These data were supplied by the Israel Meteorological Service (IMS). The number of high-velocity events for each year were summed and plotted (Fig. 8). No significant variations in extreme winds were discerned during the last 20 years. The winds at Haifa port seem to be the highest, especially the winds between 10.2 to $12.9\,\mathrm{m\,s^{-1}}$ (20 to 25 kt). These high winds (many of them from east to south directions) are caused by the effect of Mount Carmel (R. Ben Sera, IMS, personal com.). The high winds at Sde Dov, Tel Aviv, seem to be the least frequent of all locations, and Ashdod port high winds are in between. The frequency of high winds is greatest during December and January. The frequency of winds $<10.2\,\mathrm{m\,s^{-1}}$ is virtually zero in July, August and September.

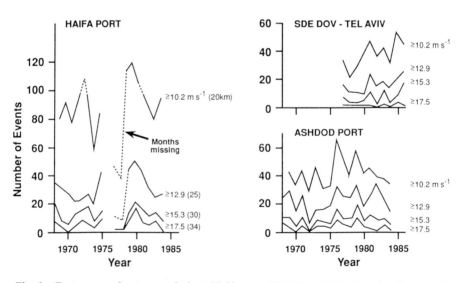

Fig. 8   Frequency of extreme winds at Haifa port, Sde Dov-Tel Aviv and Ashdod port in the years 1968 to 1986

9.2. Extreme winds during study

Extreme wind velocities at Ein Carmel and Ashdod were selected from the wind velocity data recorded by IMS during the study (September 1985 to March 1987). The first winter was extremely mild, with only a few events of extreme winds. The extreme winds are distributed uniformly during this winter. The second winter is much more stormy, with many events of high winds especially in December.

In the south, wind data were collected from two different locations, Ashdod port (at a 10 m elevation) and Eshkol power station (at about 40 m in elevation). Thus there are difficulties in comparing the two winters. In either case, the winds recorded in the south are much stronger. However, the northern and southern data are not completely comparable. The high elevation of the southern meteorological station results in higher wind velocities in the south, while the more inland meteorological station in the north acts to decrease the wind velocities here.

9.3. Comparison of Bagnold (theoretical) and measured transport

Wind data (velocity and direction) from September 1985 to March 1987 were supplied by IMS. A computer program was developed to calculate the sand discharge at the trap locations using the Bagnold equation. First, wind velocity and direction were measured at the trap locations during every visit. Then, a simple linear regression equation was used to 'correct' the hourly IMS wind data to the trap locations (Fig. 9). Quadratic equations were also attempted, but the results were less rewarding.

The r square values (i.e. correlations between the IMS winds and the wind measurements at the trap locations), which were calculated for all trap locations (the regression lines are shown in Fig. 9), are given in Table 1.

The corrected wind data were then used to calculate the sand transport by the Bagnold equation. The amount of sand transported during each season (first winter, summer and second winter) was calculated using the 'corrected' hourly wind data, and then compared to the amount of sand that was measured in the traps during the same time periods (Figs 10 and 11).

The second winter was chosen to examine the differences between theoretical and measured sand transport because of the abundance of storms and high wind events, as the first winter was a very mild one. Three sand roses are shown (Figs 10 and 11). One rose shows the sand transport measured by the front traps, a second rose shows the sand transport calculated by the Bagnold equation using the winds measured directly by IMS at the meteorological stations at Ein Carmel and Ashdod, and the third rose diagram shows the sand transport calculated by the Bagnold equation using IMS winds corrected to the Neve Yam and Ashdod trap locations.

WIND VELOCITY & DIRECTION-
SECOND WINTER CORRELATION
BETWEEN TRAP LOCATIONS & METEOROLOGICAL STATION

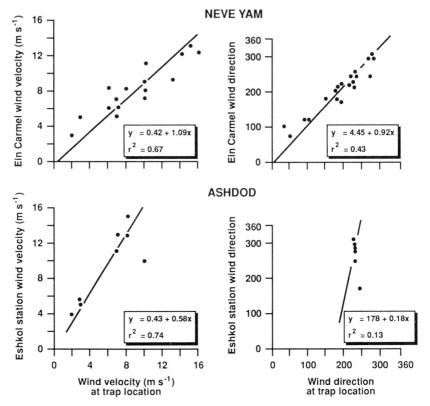

Fig. 9   Correlation between trap location and nearby met. station winds at Neve Yam
and Ashdod—second winter

9.4. Neve Yam—second winter (December 1986 to March 1987)

The measured sand transport at Neve Yam (Fig. 10), represented by trap set
2.1, indicates a dominant transport from the southwest. Hindcasted sand
transport using Ein Carmel winds shows a similar pattern but with proportionally
more transport from the northwest. This is based on the wind roses. This sand
transport appears to be about three times greater than the measured transport
since Bagnold's equation does not take into account the effects of beach slope,
local topography or vegetation. Also, the larger transport by Bagnold may be
accounted for by the traps filling up during storms. The differences in directions
might occur due to local topographic effects, such as the very wide sandy surface

Table 1   Linear regression for the second winter

Location	r square-vel.	r square-dir.
Atlit	0.57	0.68
Neve Yam	0.67	0.43
Ashdod	0.74	0.13
Ashqelon	0.41	0.19

facing the southwest, and the greater vegetation density in the northwest direction.

Calculated sand transport, using Ein Carmel winds corrected to the trap location, shows that the main sand transport is from the southwest and proportionally less transport occurs from the northwest than computed at Ein Carmel. Thus, the pattern of transport computed with the 'corrected' winds is more consistent with the measured transport. However, the transport computed at the trap location is greater than the transport computed at Ein Carmel. This is because the Ein Carmel winds are from an inland location, and thus the velocities are lower at Ein Carmel than at the trap location.

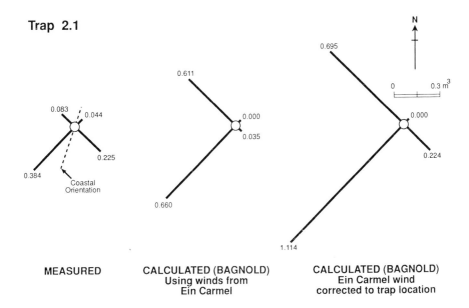

Fig. 10   Theoretical versus measured transport at Neve Yam—second winter

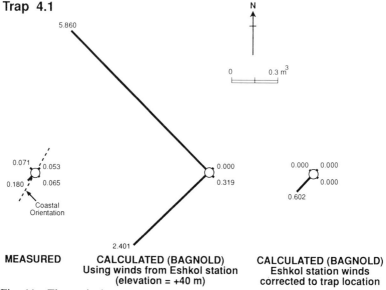

Fig. 11   Theoretical versus measured transport at Ashdod—second winter

9.5. Ashdod—second winter (January to March 1987)

Comparison between measured and theoretical sand transport at Ashdod is shown for the second winter in Figure 11. Only about half as much sand was measured at Ashdod than at Neve Yam. The transport calculated using the Eshkol station winds was found to be about five times larger than the measured transport, and is due both to the high elevation of the Eshkol station measurements ($+40$ m) and local factors such as vegetation, beach slope and local topography. The sand transport calculated using the Eshkol station winds corrected to the trap location is much reduced from that calculated at Eshkol station, but still is about three times greater than the measured transport.

   The differences in sand transport between the north and south locations using the meteorological station winds are very significant. This is mainly because the winds in the south were measured at a high elevation of 40 m and the northern Ein Carmel winds were at an inland location.

10. RELATION TO LONGSHORE SEDIMENT TRANSPORT

There is a net northerly longshore sediment transport along the eastern Mediterranean shoreline. The sediment is primarily input by the Nile Delta system to the south, which is reflected in generally wider beaches in southern

Israel, narrower beaches to the north, and rocky, sediment-starved beaches north of Akko (Goldsmith and Golik, 1980).

The relation between this longshore trend to eolian processes was determined by applying the net transport data to the adjoining lengths of coastline without cliffs and containing dunes. This was calculated by multiplying the length of Israeli coast with dunes (104 km) by the mean net eolian transport from the beach to the dunes (0.4 m^3 m^{-1} of beach) for the total period of measurements. The mean net transport was derived from the means of the net transport at the front traps at Neve Yam (0.35 m^3 m^{-1} of beach), Ashdod and Ashqelon (0.45 m^3 m^{-1} of beach).

The net eolian transport from the beach to the dunes, for the coast of Israel between Gaza and Akko, is 42,000 m^3 a^{-1} This annual loss of sand from the beach to the dunes, by the wind, along the coast of Israel, is approximately 20 to 30% of the longshore sediment transport estimate for the Israeli coast (Goldsmith and Golik, 1980). Thus, the loss from the beach to the dunes represents a significant loss from the longshore sediment transport system. Further, the longshore transport system may be providing a continuing input to the coastal dunes.

The longshore sediment transport system may also affect eolian transfers in that there is a fining of beach sediments to the north. The coarser mean grain size to the south requires a higher shear velocity for initiation and transport. This was examined by calculating the sand transport with the Bagnold equation for one month (December 1986—Eshkol station) using grain sizes and shear velocities for both the south and north; grain size = 0.024 mm in the north and 0.034 mm in the south, shear velocity = 0.24 m s^{-1} in the north and 0.28 m s^{-1} in the south. The amount of sand transport, based only on differences in grain size, was 10% higher for the north.

11. DISCUSSION AND CONCLUSIONS

The net total transport at Neve Yam (Fig. 4) is slightly onshore, with two exceptions: set 2.1 in the first winter, and set 2.2 in the summer, both of which have very small net transport offshore. The net transport at Ashdod and Ashqelon, in all sets and in all periods (Fig. 5), is onshore.

Most of the sand transport is onshore, with more toward the southeast direction in summer and to the northeast in winter. The highest transport of sand occurs at Neve Yam since the beach there is very wide, and has a gentle slope. The amount of sand transported inland from southern beaches was less than from northern beaches, perhaps because of the steeper beaches and the higher elevation of the traps.

In all locations, the differences between on-offshore transport in the seaward traps is smaller than the inner traps because of the effects of vegetation that decreases sand transport from the east to the west. The landward decrease of sand transport is more pronounced in the north because of the denser vegetation cover at these trap locations.

The amount of annual gross eolian transport measured along the Israeli coast (39 to 49 m^3 m^{-1} of beach) is similar to the net transport measured along the Oregon coast (34 m^3 m^{-1} a^{-1}) by Hunter, Richmond and Alpha, (1983), and that measured along the South African coast (15 to 30 m^3 m^{-1} a^{-1}) by Illenberger and Rust (1988). The difference being that along both the Oregon and the South African coasts, the southwest winds clearly dominate resulting in large net transports, whereas along the Israeli coast the net transport is small. However, both Israel and South African areas display significant decreases in eolian transport in the landward direction due to increased surface roughness caused by the presence of vegetation at these locations. Also, in Alexandria, South Africa, as along the Israeli coast, eolian sand deposition in the coastal dunes is deemed a significant sink in the coastal sand transport system. In Alexandria, the rate of deposition in the coastal dune fields is estimated to be 10 times as high as rates in continental sand seas.

Net eolian transport along the Israeli coast (0.1 to 0.2 m^3 m^{-1} a^{-1}) is comparable to eolian transport measurements made by Rosen (1979) on the Tabusintac barrier island system, New Brunswick, Canada and by Nordstrom and McCluskey (1985) at Fire Island on the south shore of Long Island, New York. Net onshore values of 0.3 m^3 m^{-1} at the dune base were reported for the May to September period. Higher values on the Canadian system were associated with areas of storm overwash, which did not exist in the Israeli setting.

Along the Israeli coast, the effects of vegetation and beach slope appear to reduce theoretical eolian transport by approximately 60%. In a comparison of transport between a vegetated and an unvegetated blowout at Island Beach, New Jersey, Gares (1987) measured eolian transport reductions of approximately 35% for the vegetated areas (his Table 6-15). Since the vegetation density in the New Jersey blowout is described as being only 'sparse to moderate', and since both blowouts had similar topography, these results are comparable to those on the Israeli coastal dunes. On Tabusintac Island, Rosen (1979), showed that landward sand transport was reduced an average of 88% upon crossing the vegetation line of an accretional dune.

The export of beach sand to dunes in Israel appears to play a significant role in the gradual diminution of beach sand in the northerly direction. This loss is estimated to be 20 to 30% of the total longshore transfer of sand. Application of the Bagnold equation to the regional fining of sediments to the north shows that 10% greater sand may be transported by wind in these downdrift, narrow beach areas.

Storms with wind velocities of 15.3 m s^{-1} which may be expected to occur 5 to 15 times per year at Ashdod, based on wind data for the years 1968 to 1986 (Fig. 8), will fill the 0.5 m high trap in about 100 minutes. This amounts to about 0.023 m^3 m^{-1} hr^{-1}. Thus, these storms appear to account for most of the eolian sand transport along the Israeli coast.

ACKNOWLEDGEMENTS

We would like to express our appreciation to Mr Danny Blumberg who wrote the computer programs, and assisted in many other ways. Appreciation is also expressed to Dr Haim Tsoar and Dr Abe Golik who graciously provided advice and other assistance from the inception to the conclusion of the project. The support from the US–Israel Binational Science Foundation is gratefully acknowledged.

REFERENCES

Bagnold, R. A. (1954) *The Physics of Blown Sand and Desert Dunes.* Will Morrow, New York, 265 pp.

Belly, P. Y. (1964) *Sand movement by wind.* US Army CERC, Technical Memo. No. 1.

Bowen, A. J. and Inman, D. L. (1966) *Budget of littoral sands in the vicinity of Point Arguello, California.* US Army CERC, Technical Memo. No. 19.

Bressolier, C. and Thomas, Y. (1977) Studies on wind and plant interaction on French Atlantic coastal dunes. *J. Sediment. Petrol.*, **47**, 331–338.

Chepil, W. S. (1945) Dynamics of wind erosion. The transport capacity of the wind. *Soil Sci.*, **60**, 475–480.

Danin, A. and Yaalon, D. (1982) Silt plus clay sedimentation and decalcification during plant succession in sands of the Mediterranean coastal plain of Israel. *Israel J. Earth Sci.*, **31**, 101–109.

Gares, P. A. (1987) Eolian sediment transport and dune formation on undeveloped and developed shorelines. PhD dissertation, Rutgers State University of New Jersey.

Gertner, Y. (1989) Vegetation effect on the rate of eolian sand movement and accumulation in the coastal dunes at Neve Yam. MA thesis submitted to Haifa University, 143 pp. (in Hebrew).

Goldsmith, V. (1973) Internal geometry and origin of vegetated coastal dunes. *J. Sediment. Petrol.*, **43**, 1128–1143.

Goldsmith, V. (1983) Dynamic geomorphology of the Israeli coast: a brief review. In Fabbri, P. and Bird, E. C. F. (eds), *Coastal Problems in the Mediterranean*, Commission on the Coastal Environment, Int. Geog. Un., pp. 109–124.

Goldsmith, V. (1985) Coastal dunes. In Davis, R. A. (ed.), *Coastal Sedimentary Environments*, 2nd edn. Springer Verlag, New York, pp. 171–236.

Goldsmith, V. and Golik, A. (1980) Sediment transport model of the southeastern Mediterranean coast. *Mar. Geol.*, **37**, 147–175.

Goldsmith, V. and Sofer, S. (1983) Wave climatology of the southeastern Mediterranean: an integrated approach. *Israel J. Earth Sci.*, **32**, 1–51.

Goldsmith, V., Rosen, P. and Gertner, Y. (1988) *Eolian sediment transport on the Israeli coast.* Final Report to US–Israel Binational Science Foundation, Jerusalem. 46 pp.

Gutman, A. L. (1977) Movement of a large sand hill: Currituck Spit, Virginia–North Carolina. In Goldsmith, V. (ed.), *Coastal Processes and Resulting Sediment Accumulation, Currituck Spit, Va., N.C.*, Virginia Institute of Marine Science, pp. 29-1–29-19.

Horikawa, K. and Shen, H. W. (1960) *Sand movement by wind action (on the characteristics of sand traps).* US Army CERC, Technical Memo. No. 119, 51 pp.

Coastal Dunes

Horikawa, K., Hotta, S., Kubota, S. and Katori, S. (1984) Field measurements of blown sand transport rate by trench trap. *Coast. Eng. in Japan*, **27**, 213–232.

Horikawa, K., Hotta, S. and Kraus, N. (1986) Literature review of sand transport by wind on a dry sand surface. *Coast. Eng.*, **9**, 503–526.

Hotta, S., Kubota, S., Katori, S. and Horikawa, K. (1984) Sand transport by wind on a wet sand surface. *Proc. 19th Coastal Eng. Conf.*, ASCE, New York, 17 pp.

Hsu, S. A. (1971) Wind stress criteria in sand transport. *J. Geophys. Res.*, **76**, 8684–8686.

Hsu, S. A. (1973) Computing eolian sand transport from shear velocity measurements. *J. Geol.*, **81**, 739–743.

Hsu, S. A. (1974) Computing eolian sand transport from routine weather data. *Proc. 14th Conf. Coast. Eng.*, pp. 1619–1626.

Hunter, R., Richmond, B. and Alpha, T. (1983) Storm-controlled oblique dunes of the Oregon coast. *Bull. Geol. Soc. Am.* **94**, 1450–1465.

Illenberger, W. and Rust, I. (1988) A sand budget for the Alexandria coastal dune field, South Africa. *Sedimentol.*, **35**, 513–521.

Johnson, J. W. and Kadib, A. A. (1965) Sand losses from a coast by wind action. *Proc. 9th Conf. Coast. Eng.*, pp. 366–377.

Jones, J. R. and Willetts, B. B. (1979) Errors in measuring uniform eolian sand flow by means of an adjustable trap. *Sedimentol.*, **26**, 463–468.

Kadib, A. A. (1964) Calculation procedure for sand movement by wind on natural beaches. US Army CERC, Misc. Paper no. 2–64.

Kutiel, P., Danin, A. and Orshan, G. (1979/80) Vegetation on the sandy soils near Caesarea, Israel. *Israel J. Bot.*, **28**, 20–35.

Leatherman, S. P. (1978) A new eolian sand trap design. *Sedimentol.*, **25**, 303–306.

McCluskey, J. M., Nordstrom, K. and Rosen, P. (1983) *An eolian sediment budget for the south shore of Long Island*, N. Y. Center for Coastal and Environmental Studies, Rutgers University, 78 pp.

Moreno-Casasola, P. (1986) Sand movement as a factor in the distribution of plant communities in a coastal dune system. *Vegetatio*, **65**, 67–76.

Nordstrom, K. F. and McCluskey, J. M. (1985) The effects of houses and sand fences on the eolian sediment budget at Fire Island, New York. *J. Coast. Res.*, **1**, 39–46.

Olson, J. S. (1958) Lake Michigan dune development. 1-2-3. *J. Geol.*, **56**, 254–263, 345–351, 413–483.

Pierce, J. W. (1969) Sediment budget along a barrier island chain. *Sed. Geol.*, **3**, 5–16.

Pye, K. (1983) Coastal dunes. *Prog. Phys. Geogr.*, **7**, 531–557.

Pye, K. and Tsoar, H. (in press) *Aeolian Sand and Sand Dunes*. Unwin Hyman, London.

Rosen, P. S. (1978) An efficient, low cost, eolian sampling system. *Current Research Part A. Geol. Surv. Canada*, **78-1A**, 531–532.

Rosen, P. S. (1979) Eolian dynamics of a barrier island system. In S. P. Leatherman (ed.), *Barrier Islands*. Academic Press, New York, pp. 81–98.

Short, A. D. and Hesp, P. A. (1982) Waves, dune and beach interactions in southeastern Australia. *Mar. Geol.*, **48**, 259–284.

Swart, H. (1987) Prediction of wind-driven transport rates. *Proc. 20th Coast. Eng. Conf.*, ASCE, New York, pp. 1595–1611.

Tsoar, H. (1974) Desert dune morphology and dynamics, El Arish (northern Sinai). *J. Geomorph.*, suppl. **20**, 41–61.

Tsoar, H. (1978) *The dynamics of longitudinal dunes*. Final Technical Report, Eur. Res. Off., US Army, London, 171 pp.

Tsoar, H. (1983) Dynamics processes acting on a longitudinal (seif) sand dune. *Sedimentol.*, **30**, 567–578.

Tsoar, H. and Moller, J. T. (1986) The role of vegetation in the formation of linear sand dunes. In Nickling, W. G. (ed.), *Aeolian Geomorphology*, Allen & Unwin, Boston.

Vallianos, F. (1970) Recent history of erosion at Carolina Beach, N.C. *Proc. 12th Conf. Coast. Eng., ASCE*, New York.

Wasson, R. J. and Nanninga, P. M. (1986) Estimating wind transport of sand on vegetated surfaces. *Earth Surf. Proc. Landf.*, **11**, 505–514.

Zhao, X. and Goldsmith, V. (1989) Surficial and internal geometry of carbonate eolianite in Fujian, China: Climatic and tectonic implications. *J. Coast. Res.* **5**, 765–776.

SECTION II

BEACH/DUNE
INTERACTION

There are more chapters in this section than in the other four, partially reflecting a greater interest among coastal scientists in process-response relationships at the meso-scale and a recognition of the need for studies that examine linkages between the beach and dune. The contributions by Pye and Gares in Section IV are also cast in this framework, but they are presented later because of their explicit treatment of human alterations.

Chapter Six, by Ritchie and Penland, presents the results of investigations in a rapidly transgressing coastal environment, where storm wave erosion and overwash restrict dunes to small-scale, transient features, with short-term cycles of accretion and erosion. Their results demonstrate both the vulnerability of dunes to wave modification and the rapid rate at which dunes can rebuild under natural conditions. The dunes in their study area are near one end point of a continuum based on the dominance of wave processes over wind processes. Their results provide an indication of changes that can occur in coastal areas where rapid sea level rise or a pronounced negative sediment budget allow for frequent storm wave attack. In Chapter Seven, Carter and Wilson identify dune characteristics in a location of sediment accumulation and dune construction that represents conditions closer to the other end of the wave-dominance continuum. Their study provides insight into the processes and constructional forms on accreting shorelines, such as spits and cuspate forelands. Chapter Eight, by Psuty, focusses on the spatial variation of the interaction between wave processes and the foredune in a barrier island environment where dune building and storm wave attack are more nearly balanced. There are a large number of studies in such environments, particularly on the barrier islands of the northeast coast of the United States (Hosier and Cleary, 1977; Schroeder, Hayden and Dolan, 1979; Godfrey, Leatherman and Zaremba, 1979). Psuty's study is conducted in an area where wave influence is confined largely to the seaward side

Coastal Dunes: Form and Process. Edited by K. F. Nordstrom, N. P. Psuty and R. W. G. Carter
©1990 John Wiley & Sons Ltd

of the foredune crest. Psuty goes beyond the traditional two-dimensional view of dune erosion and recovery to show how longshore wave periodicities may introduce periodicities in the dune crest line.

Much of the previous work on beach/dune interaction points to the significance of the beach as a buffer against attack of the dune by storm waves. In Chapter Nine, Davidson-Arnott and Law provide perspective on the significance of beach width as a variable affecting availability of dune sediment, while addressing some of the issues brought out in Section I concerning the relationship of wind characteristics to sediment transport. McLachlan places the exchange of material across the beach/dune interface in a different perspective by examining the transfer of salt spray. The role of groundwater and organic matter in nutrient exchanges is also identified. These exchanges alter the location and effectiveness of the biota that is part of the dune system. Chapter Ten thus reveals some of the complexity of the interactions between beach and dune and the interdependence of waves, winds, and biota in their impact on coastal dunes.

Some of the features created by wave erosion of coastal dunes are identified in the chapters by Ritchie and Penland, and Psuty. Their studies indicate that dunes may be considered a depositional element on an eroding shoreline, and many dunes may be losing volume in the modern environment. Erosional landforms are examined in greater detail in Chapter Eleven by Carter, Hesp and Nordstrom, who highlight the role that both wave and wind erosion play in the creation of depositional landforms. The chapter is a review of previous work, with added data from research now underway. The review format is used to provide the broadest perspective on erosional dune landforms.

REFERENCES

Godfrey P. J., Leatherman, S. P. and Zaremba, R. (1979) A geobotanical approach to classification of barrier beach systems. In Leatherman, S.P., (ed.), *Barrier Islands*. Academic Press, New York, pp. 99–126.

Hosier, P. E. and Cleary, W. J. (1977) Cyclic geomorphic patterns of washover on a barrier island in southeastern North Carolina. *Environ. Geol.*, **2**, 23–31.

Shroeder, P. M., Hayden, B. and Dolan, R. (1979) Vegetation changes along the United States east coast following the Great Storm of March 1962. *Environ. Mgt.*, **3**, 331–338.

Chapter Six

Aeolian sand bodies of the south Louisiana coast

W. RITCHIE
Department of Geography, University of Aberdeen

AND

S. PENLAND
Louisiana Geological Survey, Baton Rouge

1. INTRODUCTION

Various types of aeolian sand accumulations are found on the deltaic barrier coast of south Louisiana (Fig. 1). They exist as a continuum of landforms that ranges from small ephemeral mounds of sand on the upper beach to fully developed multiple dune ridges. As an integral part of the barrier sand bodies, dunes provide a significant part of the protection that these barrier islands and headlands provide for the Louisiana coastal wetlands. These aeolian accumulations are part of the total sediment store of the barriers and are part of the beach-dune-overwash cycle that is the crux of natural coastline change in this section of the Gulf of Mexico. These barrier dunes are normally lower (1.5 m maximum), and more vulnerable in their early stages than the typical, high-profile established dunes of the Atlantic coast (Godfrey and Godfrey, 1973; Leatherman, 1979). Nevertheless, in their essential relationship to upper beach and washover sedimentation processes, and in the importance of plant survival after sand burial, they are little different from the Atlantic coast barrier dunes which have been described in detail by Zaremba and Leatherman (1984). It is nevertheless suggested that the frequency of geomorphic events may be more rapid and the spatial density of overwash penetrations is greater than in the more northerly barriers.

The geomorphological processes that are involved in the development of these aeolian landforms have been observed and measured for more than 10 years along the Lafourche barrier coastline (Fig. 1), an area that is particularly

Coastal Dunes: Form and Process. Edited by K. F. Nordstrom, N. P. Psuty and R. W. G. Carter
© 1990 John Wiley & Sons Ltd

LOCATION MAP-SOUTH LOUISIANA COASTLINE

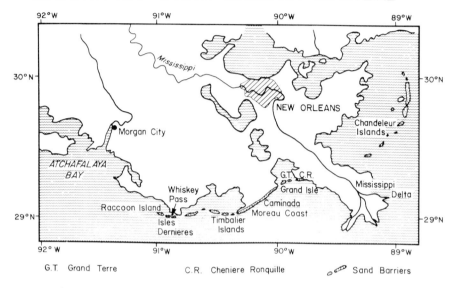

Fig. 1   Location map of deltaic barrier coastline of south Louisiana

suited to such study due to the rapid changes and the variety of coastal landforms in a relatively small area. Since these landforms exist within the rapidly retreating barrier coastline their lifespan is short. Few dunes can exist for more than 10 to 20 years before they are modified or destroyed by hurricane impact. The hurricane or severe tropical storm which has a probable return period, normally of 7 to 12 years, is the event which ends and begins the storm cycle model of barrier development along the Louisiana Gulf coastline. Effectively, the hurricane overwash planes coastal surfaces. Subsequently aeolian action and minor overwash events progressively build-up the barrier dunes (which also become progressively more resistant) until the next hurricane initiates a new cycle (Ritchie and Penland, 1988). In relation to these major and minor changes, elevation and position are the critical local factors; whereas sediment deficiency and sea-level rise due to differential subsidence are the over-riding regional factors.

The general purpose of this study is to evaluate the importance of dune development in the evolution of the barrier coastline of south Louisiana and to compare its development with the general model of barrier development which has been progressively refined from numerous studies on the Atlantic coast of the USA. Studies in Texas (Andrews, 1970; Clary, Stinson and Tanenbaum, 1974; Mathewson, Clary and Stinson, 1975) also provide valuable comparative models as they emphasise the importance of hurricanes and other storms under similar bioclimatic conditions to Louisiana. These Atlantic and Gulf Coast models apply to the south Louisiana coast although sediment deficiency, high subsidence rates and the peculiar geological setting within a major deltaic coastal plain complex (subject to the distinctive regional wave and tidal regimes of the Gulf Coast), suggest that the equivalent model for the Lousiana coast might be a regional sub-type of the standard forms that appear in most modern textbooks of coastal geomorphology. This regional sub-type of the dunes and other landforms of a hypothetical barrier island of south Louisiana is shown in Fig. 2 which provides a conceptual framework for the following detailed description and analysis of coastal aeolian landforms.

2. METHODOLOGY AND SOURCES

During the summer of 1987 the entire coastline (Fig. 1) was examined in detail. For most areas, background field studies and surveys had been completed, at different times, since 1979. Regional airborne video records of the entire coastline are available for 1984, 1985 (3 surveys), 1986, 1987 and 1988. Extensive stocks of aerial photographs dating from 1945 were also studied. Beach profiles for selected sites have been taken systematically since 1979, and for some areas additional ground information is available from reports and literature including dissertations, e.g. Gerdes (1982). However, a significant impediment for detailed morphological study is the absence of large-scale maps; the best scale available is 1 : 24,000. Moreover, the most recent USGS maps date from 1978 and most of the coastline has changed so quickly as to render these base maps of little value for field or office work.

In the field, profiles were surveyed by theodolite and reduced to mean sea level using tide gauges at Cocodrie for the islands to the west of the Caminada-Moreau Headland and at Grand Isle for the islands to the east (Fig. 1). Sediment samples were taken from beach, dune and landward washover deposits along the barrier shoreline and vegetation transects were also made.

3. SAND SUPPLY AND COASTAL PROCESSES

The source of sand for dune growth on the eroding coastline of south Louisiana is from reworking of old deltaic and beach ridge sand bodies. As the barrier islands retreat, distributary sand bodies from abandoned deltaic complexes become available for coastal processes as a result of shoreface erosion. At Caminada-Moreau, the beach ridges of Cheniere Caminada also provide a source of beach material (Fig. 3) with the landward retreating shoreface found at 6 to 9 m below sea level (Penland and Boyd, 1985). The dispersal of this sand along the foreshore depends on eastward longshore transport from a position in the centre of the Caminada-Moreau headland in the vicinity of Bay Champagne while to the west of Bay Champagne drift is westward but interrupted by the Belle Pass jetties. On Timbalier Island and Isles Dernieres

108

GENERALIZED LOCATION OF AEOLIAN ACCUMULATION FORMS

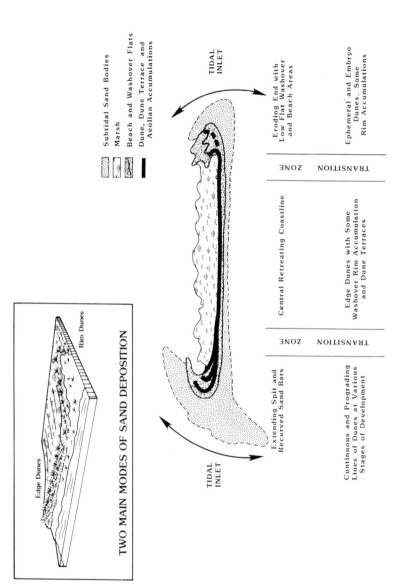

Fig. 2   Types of dunes and their typical locations

BEACH RIDGE PLAIN SAND SOURCE

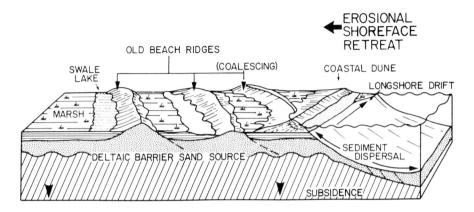

Fig. 3   Sources of sand supply for the coastline

the drift directions are to the west and the islands show elongation and tidal inlet migration in the same direction. Historically, Timbalier Island is migrating westward indicating a net westward sediment movement. The Isles Dernieres are being fragmented as a consequence of subsidence and marine erosion although the residual segments are accreting on either side of the dividing inlets, with sand derived from erosion of the island remnants.

Dune sand is reworked by the well-documented processes associated with overwash and coastal edge retreat. Neither process is unique to the coastline of Louisiana, but the low crest altitudes, variable resistance and local differences in the amount of vegetative cover create subtle changes in the response of the coastline to overwash. Like most dune systems there is a superimposition of old and new dune elements to form a mixture of remnant and active landforms. Residual ridges and new aeolian accumulations combine to provide the typical mosaic of coastal dune landforms. Characteristically, continuous lines of dunes and shore-parallel progradation are confined to downdrift spit formations (Fig. 4). Elsewhere, the coastline consists of alternating short sections of bare washover flats, oblique and shore-parallel ridges, amorphous hummocky dunes and residual areas of marsh and mangrove that have survived sand burial (Fig. 5). In some areas, the remnants of old deltaic landforms, such as levees and inter-distributary lakes emerge at the retreating coastal edge. It is this variety of small-scale coastal landforms, in such a relatively small area, that is the basis of the complexity of the evolution of this barrier coastline.

At the regional scale of south Louisiana, most coastlines are retreating but the rate of erosion varies spatially (e.g. for Caminada-Moreau the average

Fig. 4   The east end of Caminada Spit. This is an area of accretion at the distal end
of the barrier headland. Large continuous dune ridges develop from the extensive
downdrift beach and spit features

rate was 14 m and the maximum was 60 m over the period 1978 to 1985). This
retreat is partly due to a negative sediment budget where normal offshore/
onshore and longshore sand movements are not compensated by addition of
sand from old beach ridges and levees. There are also variable inputs of shell-
derived fragments. Other unknown sediment losses are through leaching,
offshore removal to deep water during storms and transport through the tidal
inlets into back bays and lagoons. Relative sea level is also rising at a rapid
rate (estimated at 6.3 mm a^{-1} for the Caminada-Moreau area). There is also
clear historical evidence for the reduction in size of barrier islands indicating
a substantial loss of sub-aerial sand. Thus, the positive input of sands both from
old beach ridges and the progressive exposure of old deltaic sand bodies
does not seem to be adequate compensation for the negative effects of
coastal subsidence and for sediment losses offshore and through tidal passes.
Accordingly, the sand dune depositional processes of the coast of Louisiana
are essentially based on reworking a limited and, probably, reducing quantity
of sediment. In the absence of sand-sized sediment input from the active
Mississippi, there is little prospect of change, so that beaches, dunes and
washovers will continue to develop on the basis of the recirculation and
reworking of a comparatively meagre sand volume. Thus, although at any one
time coastal dunes are intrinsically depositional landforms, the model of

dune-beach-overwash interaction for the coastline of south Louisiana is set essentially within an erosional and cyclic physiographic context.

4. VEGETATION

Typical distributions of the most important coastal plants are shown in Fig. 6 for Timbalier Island and Caminada spit. The most important single species is *Spartina patens*, a creeping, rhizomatous grass which grows up to 1.5 m tall, occurring in dunes, dune terraces and overwash flats. It seems to prefer moist sites and is associated with the typically low aeolian accumulations of the coast of south Louisiana especially to the west of the active delta. Its rate of colonisation and ability to survive sand burial make it the most common and versatile plant of the dune coastlines (Mendelssohn, 1985). In contrast, on Atlantic barrier dunes *Spartina patens* tends to occupy washover deposits inland from the main line of dunes, where the dominant plant is normally grass such as *Ammophila breviligulata* or *Uniola paniculata* (neither plant is common in south Louisiana). It is probable that the ability to recover after overwash burial is the most important reason for the dominance of *Spartina patens*; an ability recognised by Godfrey and Godfrey (1973) who described it as a sub-climax community, maintained by overwash.

On the foredunes, *Panicum amarum* may form dense, tall clumps, especially on foredunes. This plant is also rhizomatous, but unlike *Spartina patens*, does not appear to produce viable seed (Mendelssohn, 1985).

In fresh sand deposits and in brackish zones, primary colonisation is often by *Sporobolus virginicus, Paspalum vaginatum* and *Ipomoea stolonifera*. *Paspalum vaginatum* can withstand saltwater and seems to prefer wet dune areas. On the backshore and in some overwash areas on the Isle Dernieres, the low growing *Sesuvium portulacastrum* is an important pioneer species and may form isolated, rounded dune accumulations. Extensive low wash-out areas in the Isles Dernieres are covered by *Salicornia* spp. indicating salt water inundation. The runners of *Ipomoea stolonifera* cover ground with remarkable speed, especially on the backshore above the reach of storm tides although it is also found on washover terraces and in bare hollows between dune hummocks and low ridges. At Caminada spit, runners were observed to extend up to 12 m in a period of approximately three months on a backshore sand accumulation zone. Other locally dominant plants are the woody, shrub-like *Croton punctatus* and *Iva imbricata*, both of which tend to form hummocky topography especially near the shoreline. As shown on Fig. 6, numerous other plants fill ecological niches on the dune and dune terrace. From a geomorphological point of view the most important characteristic is rapid colonisation and growth. In most low dune and washover terraces overwash occurs regularly and rapid growth rates appear to be the most important characteristic of the typical dune plants of the Louisiana coast.

Fig. 5A   The north end of the Chandeleurs, showing a narrow part of the barrier with relatively high residual dune ridges. Farther north, the coast consists of wide flat washover deposits with dunes along the margins

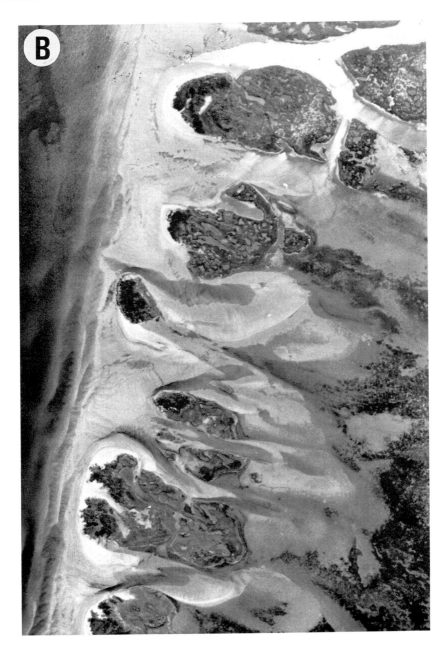

Fig. 5B   A post-hurricane, vertical aerial photograph of the north part of the Chandeleur barrier island chain. Note the extensive washover channels and associated sand deposits between residual areas of marsh and older dunes

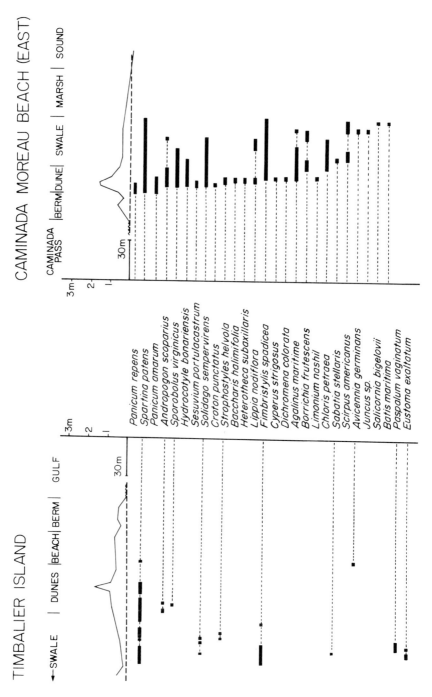

Fig. 6 Dune vegetation profile (from Mendelssohn, 1985) for Caminada Spit and Timbalier Island

5. CLIMATIC CONDITIONS AND SAND MOVEMENTS

Meteorological statistics for the coastline of south Louisiana are generally unavailable but data from New Orleans (80 km north) indicate that precipitation is on average more than 1600 mm a^{-1} with October being the driest month and July the wettest. Maximum daily temperatures range between 14°C in January to over 27°C in July and frost at the coast is unusual.

Unfortunately, reliable wind statistics for the barrier coastline are unavailable and useful calculations of aeolian processes cannot be made.

Sand samples from all the beach and dune areas have been collected and analysed, and median size values fall in the 2.10 to 2.60 ϕ range with more than 80% of all sand lying close to the median value. Thus even moderate winds can transport this fine to medium grade material if other factors such as surface wetness and local vegetation do not prevent entrainment. A very significant local variable is the presence of layers of relatively large shells on the beach and washover surfaces. In the Isles Dernieres, for example on Raccoon Island, these form extensive pavements which prevent sand transport. The most durable layers are formed of oyster shells which are derived from the reworking of reefs and beds exposed offshore by coastline retreat.

Previous research (Conaster, 1971; Murray, 1976; Ritchie and Penland, 1982) suggests that winds from the southeast are the most important in determining sand movements. However, limited data on regional wind velocities contain evidence that strong winds from the north may also be of significance, supported by evidence of sand accumulation on the backslopes of many dunes. Yet until meteorological equipment is installed in suitable locations along the coast, many basic research questions on aeolian transport dynamics will remain unresolved. Similarly, unlike the large areas of dunes such as those found on Atlantic barrier islands, there are no diagnostic morphological patterns of ridge crests, blowouts or sand waves to act as indicators of sand transport directions. The typical level terrace, low ridges and rounded forms of sand accumulation do not permit a systematic topological study of the aeolian relief from which vector information could be deduced.

6. DUNE TYPES

All dune ridges, hummocks and dune terraces grow by accretion in proportion to the availability of sand and the efficiency of vegetation in interrupting the flow of sediment. In Louisiana, there is an abundance of plants available to occupy the relatively narrow range of habitats in the coastal barriers. The warm wet climate, the proximity of the water table and the relative absence of salt spray, produce extremely fast growth rates. There is also an absence of grazing animals to inhibit growth and little evidence of burning and other adverse human impacts. The hot summer sun produces rapid surface drying, and numerous burrowing animals, especially land and beach crabs, loosen the surface.

116

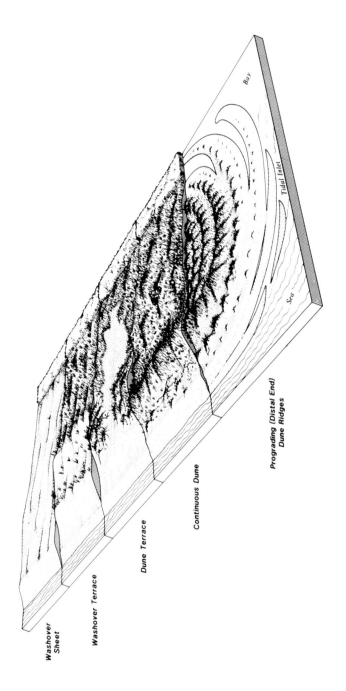

Fig. 7 Dune types of south Louisiana

Three factors control dune growth: (i) the frequency of overwash; (ii) the relative lack of sand in the total beach-dune system; and (iii) the proportion of the bare sand surface that is covered by a veneer of large shells. The dunes of the Louisiana coast are not high, rarely more than 2 m above beach level; they are discontinuous alongshore and rarely extend more than 50 m inland. Typically, dunes extend for short sections and are breached by washovers at various intervals. Mature dunes with shrubs and small trees are uncommon, except in a few areas of exceptional shore-parallel progradation or on old, inland ridges associated with past washover penetrations and stranded beach ridges.

A classification of barrier coastline types is given in Fig. 7 and provides the basic series of washover and dune forms for the south Louisiana coast. The six main types are as follows.

I. Washover sheet: this is a continuous sheet of sand produced by single or multiple washover events. It is low altitude and may be flat or gently undulating. The surface is frequently formed of shells. There are lines and patches of flotsam and dead vegetation (including seaweed). Dune forms are small, isolated or in localised groups. Due to the frequency of washover (often more than ten times per annum), dunes are essentially ephemeral. Nevertheless, being low, the surface is relatively close to the water table and vegetation colonisation can be rapid.

II. Washover terrace: the washover terrace is slightly higher than the washover sheet. The surface is normally covered by vegetation. Usually, the area is undulating or gently ridged. There are no distinctive shore-parallel or transverse dunes; the best description is 'amorphous'. The terrace suffers washover but the frequency is much less than type I, possibly once a year. Washover takes the form of a sheet of sand rather than narrow tongue-shaped penetrations and most of the sand is deposited near the coastal edge. If the washover cover is thin, most of the vegetation survives and continues to grow through the overburden.

III. Dune terrace: the dune terrace can be distinguished from the washover terrace by its greater height, relief amplitude and relative maturity of vegetation as indicated by density and species mixture. Whereas the washover terrace has amorphous but low aeolian forms, the dune terrace has distinctive ridges and low sand hillocks, usually concentrated near the shoreline. As a consequence of uneven elevations, most washover processes focus on the lower parts of the backshore. Accordingly, overwash is confined and develops higher flow regimes, leading to deep tongue-shaped or linear sand deposits which may extend across the terrace. Since these bare sand areas are often colonised by pioneer vegetation, there is a wide range of species on the dune terrace.

IV. Continuous dune: the essential characteristics of the continuous dune are that it is higher, fully vegetated, more or less unbroken, and only rarely breached, overtopped or overwashed. Sand accumulation is concentrated on the incipient foredune which, as it grows, prevents sand transport farther inland. There may be occasional low points in the foredune crest where high-level overwash penetrates but these tend to be rapidly sealed and revegetated. The foredune is shore parallel and exhibits normal seasonal profile changes in that winter storm surges scarp the dune face; in summer, sand from the upper beach transfers to the dune face and there is accretion and dune progradation. This progradational slope is colonised by pioneer vegetation and in some areas may form the basis of a more 'permanent' encroachment onto the backshore.

V. Prograding lines of dunes: a variation on the continuous dune type may be found in some parts of the south Louisiana coast where longshore sediment transfer produces a rapidly prograding beach platform, often with multiple nearshore bars and upper beach berms. As a consequence of this progradation, dunes develop as a series of shore-parallel ridges. These are normally curved and radiate outwards from a nodal hinge-point at the origin of spit curvature. The dune ridges may continue to curve around the end of the island as shore-parallel features which mirror the extension of the beach. The dunes are at different stages of development and the intervening depressions normally have finer, water-lain sediment and are usually bare of vegetation and occasionally filled with sea water. This progradational type V occurs exclusively at the distal end of islands and headland barrier beaches (Fig. 2).

VI. Artificial dune ridges: on Timbalier Island and Grand Isle there are examples of artificial dunes. These have been raised to provide protection to some low, vulnerable areas of the coastline. At Timbalier Island the dunes were created by sand-trapping fences (Mendelssohn, 1985) and at Grand Isle the ridges were shaped by earth-moving equipment. In both areas, vegetation cover was achieved by a combination of planting and natural processes.

In theory, the five types of natural sand accumulation form a sequence that could proceed from the washover sheet to the continuous dune stage. In practice, complete sequential development seldom takes place either because there is insufficient time to complete the cycle or the barrier headland or island dunes are eroding rapidly and being continuously overwashed. Several detailed examples of these dune types, and the sequence of change from one type to another, are given below. Repeated profile measurements are shown in Figs 9 to 13. These case studies demonstrate both the rates of change and the complexity of real, rather than model, dune and aeolian forms on this rapidly changing coast.

SAND DUNE DEVELOPMENT CYCLE   SOUTH LOUISIANA

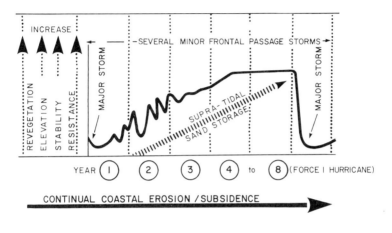

TYPICAL OVERWASH ELEVATIONS ARE:

MINOR COLD FRONT	10-15 PER YEAR	1·42 to 1·73 m
FORCE 1 HURRICANE	1 PER 8 YEARS	1·73 to 2·92 m
FORCE 5 HURRICANE	1 PER 32 YEARS	3·50 to 7·00 m

Fig. 8   Cycle of overwash as defined by washover and hurricanes

Any evolutionary sequence of dune growth has to be set within the general model of coastline development which, as shown on Fig. 8, is controlled by the return period of overwash embracing the 10 to 12 year cycle of major hurricane impact. In essence, the capacity for any dune area to grow and to consolidate is controlled by the frequency and intensity of overwash. For low, vulnerable areas such as washover sheets this may occur up to 15 times per year and substantial dune growth is impossible; for high continuous dunes only a major hurricane can interrupt the growth sequence but when such a storm does occur, the dunes are destroyed and the area is reduced to a semi-continuous sand plain.

Since the barrier shorelines of Louisiana are eroding, transgressing landwards and subsiding differentially, there are few areas of accretion, and it is logical to describe the cycle of dune development as being primarily destructive with elements of consolidation and temporary change to higher, more stable landforms. Therefore, the most rapid and dramatic change is from some form of vegetated dune to a bare washover sheet. In this destructive, major storm impact phase, the landforms do not pass progressively through intermediate stages but revert directly to the washover landforms. The major storm destructive phase is not a step-wise sequence, more a reversion to a common starting point—the washover sheet. Within this storm-punctuated model, there are subsidiary changes where minor storms may alter the dunes from one type to another and, in contrast, in the absence of major overwash, the sequence can produce relatively rapid sand accretion with associated vegetation cover, to create higher more continuous coastal landforms.

The modes of sand deposition in these terrace and dune forms are not unique to the Louisiana coastline. Plant survival after overwash burial, superimposition on remnant beach or dune ridges and primary pioneer colonisation on the upper beach or washover flat are common to most dune systems on all coastal barriers dominated by washover cycles. A frequent location for sand accretion is on the perimeter of a washover sheet. This is a variation of the survival type (III/IV) of incipient dune development. Frequently, the limit of overwash penetration is characterised by a slightly higher ridge of sand and beach debris, including root stocks and viable seeds, which are often stranded on this perimeter. Usually the ridge is less than 0.2 m high but this is sufficient to act as a zone of preferred deposition, especially as it marks the limit of vegetation beyond the overwash. Normally this ridge lies on the landward limit of the washover flat but it also occurs on the side (lateral form) and may be skewed according to the precise direction of the overwash surge. Field observations indicate that this 'rim dune' is a favoured area for relatively rapid dune growth because of the presence of vegetation to trap sand blown from the washover flat, the availability of seeds and rhizomes, the stimulation of sand burial and possibly enhanced nutrient availability.

Another favoured area for dune growth is the scarped edge of an old dune ridge or terrace. If run-up fails to surmount the barrier crest, it invariably cuts a low escarpment. When the water level falls, windbown sand is deposited along the scarp. Since only a small amount of sand moves beyond the scarp, there is relatively rapid deposition, although it rarely extends more than a few metres inland, giving rise to the distinctive concave transition between the scarp edge dune and the pre-existing terrace. Normally the vegetation continues across this transition, although the percentage of ground cover is greater on the older terrace surface. An important variant of this type is a ridge of shells, sometimes up to 0.4 m high and 4 m deep which may be left at the top of the wave-cut scarp and formed during storms. Observations at such times reveal that the relatively buoyant shells (and other flotsam) are trapped by the scarp topography especially where remnant vegetation provides an additional barrier. Plants take longer to colonise these shell ridges which are left as distinctive reminders of previous storms. Further inland, overwashed patches of shell are flatter, usually 5 to 20 m in length, and remain as zones of slow vegetation colonisation.

A distinctive feature of aeolian deposition on the Louisiana coast is the relatively short distance of primary and secondary sand transport. Vegetation growth is rapid and bare sand areas tend to seal relatively quickly, especially by rapidly growing creeping plants such as *Ipomoea stolonifera*. Ridges and mounds close to a sand source grow rapidly and only small quantities of sand

REDUCTION OF LOW DUNE RIDGE TO TRANSGRESSIVE WASHOVER FLAT PROFILE D (BAYOU MOREAU)

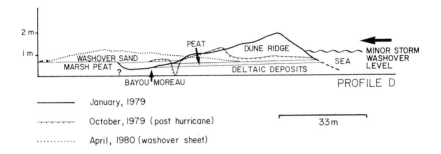

————— January, 1979

‧‧‧‧‧‧‧ October, 1979 (post hurricane)

············ April, 1980 (washover sheet)

RAPID CYCLE OF DESTRUCTION AND RECONSTRUCTION OF HIGH COASTAL DUNE RIDGE

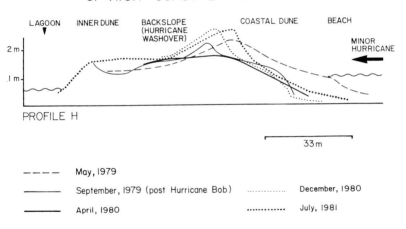

— — — — May, 1979

————— September, 1979 (post Hurricane Bob)   ············ December, 1980

————— April, 1980   ············ July, 1981

Fig. 9   Profile D and Profile H (Caminada-Moreau)

escape beyond the zone of first contact with vegetation. On days with very strong winds, repeated observations indicate a singular absence of long-distance sand movements. The landward encroachment of sand onto marshes, mangroves, backswamps and lakes is mainly by overwash and only rarely by wind.

7. EXAMPLES OF DUNE TYPES AND SEQUENCES OF CHANGE

The following examples are selected to show the various types of dunes and smaller aeolian accumulations of the barrier coastline of south Louisiana.

They also provide case studies of typical changes from one type of landform to another.

Profile D, Fig. 9, shows the conversion of a dune ridge to a washover sheet as a consequence of coastal erosion proceeding landwards to intersect the shore-parallel meander bend of Bayou Moreau (Caminada Coast). The major change occurred in 1979 to 1980. Since that period the area has remained as a washover terrace and coastal erosion has averaged $17\,\mathrm{m\,a^{-1}}$. The extension of the washover sheet has been greater, averaging $19\,\mathrm{m\,a^{-1}}$. Because of this high rate of change, the area has not been colonised by vegetation due to repeated overwashing, and any incipient dune developments are ephemeral.

Profile H, Fig. 9 (lower) shows an example of simple foredune accretion in an area of the Caminada Coast that is raised above the threshold of most storm surges. The rate of accretion can be gauged by comparing the April and December profiles for 1980. By July 1981 the dune was higher and the form had relocated a short distance to seaward. Nevertheless, the beach zone was relatively narrow and suffered severe damage when the three hurricanes crossed this area in 1985. The sea broke through to the lagoon to form an inlet which closed in 1986.

Profile A (at the west end of the Caminada Coast) (Fig. 10) was surveyed at frequent intervals to record the changes from dune terrace and ridges to hummocky dunes within a small area. The most important profile is July 1979, after Hurricane Bob. The overwash from this event destroyed the main dune ridge and transferred sand landwards as a 0.3 to 0.5 m thick unit. Part of this sand was derived from the upper beach. Within four weeks, dune ridges had reformed from vegetated remnants of the original dune field. The persistence of these clumps of vegetation was vital to rapid sand accumulation and dune rebuilding. In this area, patches of *Panicum amarum* and *Spartina patens* appeared to be stimulated where burial was of the order of 0.1 to 0.15 m. The spread of the surface creeping plant *Ipomoea stolonifera* was particularly important. Since *Panicum amarum* and *Croton punctatus* were vigorous on the frontal dune ridge, it was possible to confirm the view of Mendelssohn (1985) that these plants tend to produce higher, hummocky forms rather than continuous dune ridges. During a full year there was a net increase in supra-littoral sand storage at the expense of the beach zone, producing a net increase in height without extending the sand prism landwards.

Profile J (Fig. 10, lower) is located in an area of continuous dunes at the base of the Caminada spit and is reduced to two time periods to show simple loss and gain positions. After the hurricanes of 1985, the July 1986 profile was surveyed to show the substantial change that had occurred, along with the beginnings of revegetation and geomorphological recovery.

The second group of profiles illustrates local variations from other parts of the barrier coastline and which were surveyed in the Isles Dernieres, Timbalier and Grand Terre Islands.

SHORT TERM CHANGES IN A
CONTINUOUS DUNE AREA

PROFILE A

APRIL	MARSH FLAT	HUMMOCKY & COPPICE DUNES	DUNE	BEACH
JULY	MARSH FLAT	OLD 'DUNE' — UNDULATING WASHOVER SAND/REMNANT 'DUNE'		BEACH
AUG.	MARSH FLAT	OLD 'DUNE' — DEVELOPMENT OF LOW DUNES	NEW DUNE	BEACH
OCT.	MARSH FLAT	LOW DUNES	DUNE	EMB-RYO. BEACH

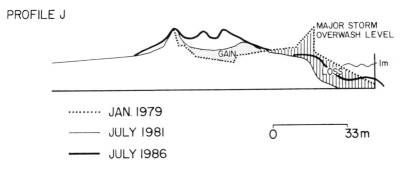

MINOR HURRICANE BOB

OVERWASH LIMIT OF 'BOB'

1m

```
........  APRIL 1979        ...........  AUGUST 1979
———  JULY (POST         -----  OCTOBER 1979
         BOB) 1979
```

0 ——— 33m

NET GAIN/LOSS PROFILES AND 1986 PROFILE
OF CONTINUOUS DUNE AREA (CAMINADA SPIT)

PROFILE J

MAJOR STORM OVERWASH LEVEL

GAIN

LOSS

1m

```
........  JAN. 1979
———  JULY 1981
———  JULY 1986
```

0 ——— 33m

Fig. 10   Profile A and Profile J (Caminada-Moreau)

In the Isles Dernieres, several profiles are available to show typical dune and other aeolian accumulation forms on a rapidly eroding series of low barrier Islands.

Profile 3 is located near the west end of the island to the east of Whiskey Pass (Isles Dernieres) and is an example of simple accumulation at the edge of a fully vegetated terrace. The vegetation is dominated by *Spartina patens* and the edge is only 0.4 m above the level of the upper beach so that in terms of height and width it is a typical frontal dune. It is also representative

WHISKEY PASS (EAST)-TYPICAL OVERWASH
EXTENSION WITH LATERAL RIM DUNE AND
SHORE-PARALLEL AND RESIDUAL DUNE SURFACES

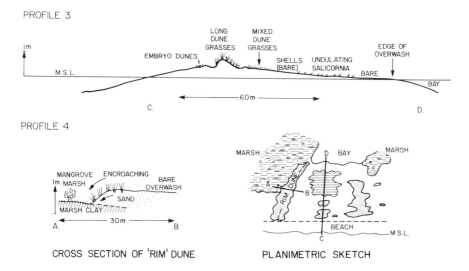

Fig. 11   Profiles 3 and 4 (Whiskey Pass, Isles Dernieres) and sketch of features near
Profile 3

of a rapidly changing area in that only 20% of the ground is vegetated, and
coastal erosion will ensure that full revegetation will not occur.

Further east on the same island, Profile 4 (Fig. 11) represents typical first-
stage colonisation of an extensive, partly shell-covered washover flat with rim
dune accumulation on the west margin (Profile 4). The sketch on Fig. 11 shows
the spatial relationships of these small-scale features. The vegetated hillocks
tend to be forming normal to the beach, but there are many individual clumps
of vegetation with rounded pioneer sand accumulations, rarely more than 0.4 m
high.

With the rapidity of change along the coastline of south Louisiana, the most
common dune type is some form of accumulation at the scarped edge of a dune
terrace. Since the coastline is retreating at rates that are never less than a few
metres per year, the frontal dune or dune terrace is being aperiodically undercut,
depending on its elevation and position on the barrier shoreline. The profile
from Timbalier Island (Fig. 12) is an actual example of the steep, narrow, sinuous
and continuously evolving edge dune shown in idealised form in Fig. 2.

At the ends of most islands, longshore drift causes sand to accumulate and
the amount of erosion is less. Within the barrier island model this location is
typically stable or accreting until a hurricane reduces the area to a washover

TIMBALIER ISLAND - RAPID EDGE - DUNE DEVELOPMENT WITH
RETREATING SHORE - FACE IN CENTRE OF ISLAND.

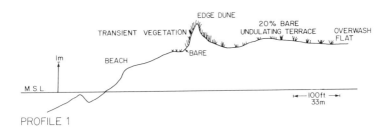

PROFILE 1

TIMBALIER ISLAND - PROGRADING LOW PARALLEL DUNE
RIDGES AT EXTENDING WEST END.

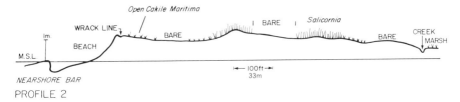

PROFILE 2

Fig. 12   Profiles 1 and 2 Timbalier Island (including progradational downdrift location)

surface, with sand coming from longshore transport and tidal inlets (Fig. 2).
Swash bars and recurved beach ridges form around inlet margins and dunes
develop on these higher sub-parallel ridges. Some islands, like Timbalier, are
migrating rapidly westward as a series of elongating beach ridges that provide
the foundation for prograding sand dunes (Fig. 12). In contrast, the west end
of Grand Terre is relatively stable due to the influence of the important,
navigational tidal inlet at Barataria Pass and the dunes here do not display the
characteristic series of low, young, immature ridges of the more dynamic islands.
This profile (Fig. 13) has a thick wedge of sand with distinct hummocky forms
leading to a well-developed sand ridge. The vegetation is different, being
dominated by shrubs (i.e. *Croton punctatus*) on the uneven frontal dune terrace
succeeding to a mature dune ridge with relatively large shrubs and small trees.
In contrast, the central part of Grand Terre has simple edge accumulation dunes,
with beach sand and reworked dune terrace sand encroaching onto the relatively
high, mature dune terrace that is characteristic of most of this flanking barrier
island (Fig. 13). This pattern is typical of the subsiding barrier islands east of
Grand Isle.

GRAND TERRE

MID-ISLAND DUNE TERRACE WITH EDGE ACCUMULATION

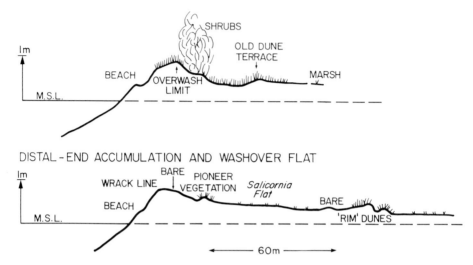

DISTAL-END ACCUMULATION AND WASHOVER FLAT

Fig. 13   West end and central part of Grand Terre

8. CONCLUSION

The aeolian coastal forms of the barrier coastline of south Louisiana are low, small, dynamic, short-lived, and characterised by rapid vegetational consolidation by a great number of indigenous species. They are little disturbed by man or animals. The sand is fine grained and easily transported. Aeolian processes are almost entirely constructive with deflation and blowout development almost unknown. Topographically, they are simple, in that there is an absence of older parallel or oblique or redepositional dune ridges. Complexity is introduced by sand accumulating on older non-aeolian landforms such as beach ridges or levees. Dune ridge or aeolian sand sheet migration is almost unknown. The life cycle of any dunescape is rapid with most dunes existing for 1 to 15 years before being overtaken by coast erosion or modified by overwash. Nevertheless, to describe these aeolian forms as being short-lived and easily destroyed is slightly misleading, in the sense that it would be better to describe them as being subjected to rapid changes. The south Louisiana dunes are the protection of the barrier coast. They are in a state of dynamic evolution both in terms of form and position, within the context of the exceptionally rapid sequences of erosion and migration that characterise all shorelines of this region of the Gulf of Mexico. These rapidly changing aeolian accumulations raise interesting conceptual issues especially when viewed across a range of geomorphological and geological scales. In one sense, they are fragile, easily destroyed and short-lived coastal landforms. In another sense, they are adaptive, persistent and effective coastal barriers. They exhibit an intrinsic strength common to most low, soft and unconsolidated coastlines with their adaptability and mobility epitomising a particular meaning of the term 'dynamic equilibrium'. Thus the dunes possess the freedom to move and reform continuously so that the landforms attain a net position of least disturbance in relation to the geomorphological environment.

ACKNOWLEDGEMENT

The author wishes to thank Elsevier Science Publishers for permission to reproduce, in a modified form, five figures which were originally published in Ritchie and Penland (1988), *Marine Geology*, **81**, 97–112.

REFERENCES

Andrews, P. B. (1970) Facies and genesis of a hurricane-washover fan, St. Joseph Island, central Texas coast. *Texas Bureau of Economic Geology*, **67**, 1–147.

Clary, J., Stinson, J. and Tanenbaum, R. J. (1974) Coastal aeolian fans—mechanism of landward sand transport. *Geol. Soc. Amer.* (abstract with program), 689–690.

Conaster, W. (1971) Grand Isle: a barrier island in the Gulf of Mexico. *Bull. Geol. Soc. Amer.*, **82**, 3049–3068.

Gerdes, R. G. (1982) Stratigraphy and history of development of the Caminada–Moreau beach ridge plain, southeast Louisiana. Unpublished MS thesis, Louisiana State University.

Godfrey, P. J. and Godfrey, M. M. (1973) Comparisons of ecological and geomorphic interactions between an altered and an unaltered barrier island system in North Carolina. In Coates, D. R. (ed.), *Coastal Geomorphology*. SUNY, Binghampton New York, pp. 239–258.

Leatherman, S. P. (1979) Barrier dune systems: a reassessment. *Sediment. Geol.*, **24**, 1–16.

Mathewson, C. C., Clary, J. and Stinson, J. E. (1975) Dynamic physical processes on a South Texas barrier island—impact of construction and maintenance. *IEEE OCEAN*, **75**, 327–330.

Mendelssohn, I. A. (1985) Sand dune vegetation and management. In Penland, S. and Boyd, R. (eds), *Transgressive Depositional Environments of the Mississippi Delta Plain*. Louisiana Geological Survey, Guidebook Ser. 3 Baton Rouge, pp. 203–233.

Murray, S. P. (1987) Currents and circulation in the coastal waters of Louisiana. *Coastal Studies Institute Tech. Rep. 210*. Sea Grant publication No. LSU-T-76-003 Louisiana State University, 33pp.

Penland, S. and Boyd, R. (eds) (1985) *Transgressive Depositional Environments of the Mississippi Delta Plain*. Louisiana Geological Survey, Guidebook. Ser. 3, 233pp.

Ritchie, W. and Penland, S. (1982) The interrelationship between overwash and aeolian processes along the barrier coastline of south Louisiana. *First Int. Conf. Meteorology and Air/Sea Interaction of the Coastal Zone*. American Meteorological Society, Boston, 358–362.

Ritchie, W. and Penland, S. (1988) Rapid dune changes associated with overwash processes on the deltaic coast of south Louisiana. *Mar. Geol. 81*, 97–112.

Zaremba, R. E. and Leatherman, S. P. (1984) Vegetative physiographic analysis of a U.S. northern barrier system. *Environ. Geol. Water Sci.*, **8** (4), 193–207.

Chapter Seven

The geomorphological, ecological and pedological development of coastal foredunes at Magilligan Point, Northern Ireland

R. W. G. CARTER AND PETER WILSON
Department of Environmental Studies, University of Ulster

The only dunes which have a special interest for the student of shore processes are those occurring in the form of parallel ridges on a beach plain. (Douglas W. Johnston)

1. INTRODUCTION

Foredunes are aeolian deposits fixed by vegetation landward of beaches. Hesp (1984, 1988) recognises two basic foredune types—incipient and established. Incipient dunes are those 'formed by sand deposition within pioneer vegetation species growing along the backshore of beaches' (Hesp, 1988, p. 18). As vegetation and soils develop and sand supply fails, incipient foredunes become 'established' foredunes. An essential characteristic of foredunes is that they form from windblown sediments, usually reflected in textural and structural indices. Thus the term 'foredune' is a generic, rather than a morphological one.

Short and Hesp (1982) introduced a morphodynamic foredune classification, derived from SE Australian examples and based on vegetation and morphology, which was linked to the reflective/dissipative continuum of beach-nearshore processes. They recognised five stages—in essence a spatial geobotanical classification—ranging from well-vegetated linear foredune ridges (stage 1), to sparsely vegetated, chaotic morphology (stage 5). While not definitive, this classification embodies a basic structure, which may be matched by the Irish dune systems (Carter, 1990b), particularly on a scale described by gross variations in sediment supply.

This paper concerns a rapidly prograding dune system at Magilligan Point in the north of Ireland, where recent reworking of Holocene beach ridges and

Coastal Dunes: Form and Process. Edited by K. F. Nordstrom, N. P. Psuty and R. W. G. Carter
©1990 John Wiley & Sons Ltd

dunes has allowed the creation of new foredunes. The paper aims to describe and discuss the processes of foredune formation, and stabilisation over a 35-year period (1950 to 1985), spanning the transition from incipient to established foredunes.

2. STUDY AREA

Magilligan is a triangular-shaped beach ridge plain or foreland at the mouth of Lough Foyle on the north coast of Ireland (Fig. 1). The beach ridge plain was formed largely during a mid- to late-Holocene sea level fall related to continuing isostatic delevelling in the area (Carter, 1982a). The ridges have been covered by up to three discontinuous aeolian sand deposits, and buried soils marking former stable land surfaces have been described by Wilson and Bateman (1986, 1987). These soils include inter-ridge peats, podzols, brown calcareous

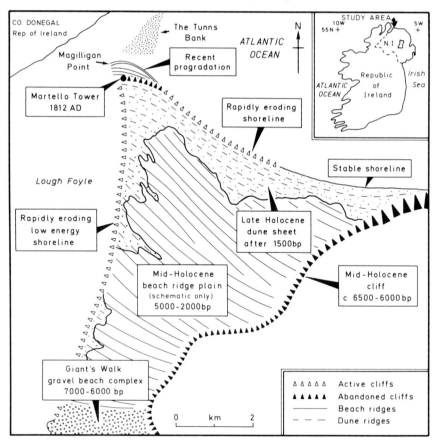

Fig. 1   Location map of the Magilligan area showing the essential geomorphological features

sands, and sand-pararendzinas. Carbon-14 dating of marine shells, peat and soil organic matter suggests that the ridges developed during the period 5000 to 2000 BP (Wilson and Farrington, 1989).

Magilligan Point, at the distal extremity of the Holocene foreland, has evolved between two intersecting wave fields. To the north and east, Atlantic swell-dominated waves move into the estuary of Lough Foyle, whereas to the west and south, locally-generated wind waves move seaward (Carter, 1979, 1980). The recurve of the Point is marked by strong process gradients of wave height, angle of breaker approach and wave energy, such that beach and nearshore sediment fluxes are high (probably in excess of $100\,000\,m^3\,a^{-1}$ at some locations). While some sediment moves along the beach, significant masses migrate in the nearshore zone as coherent longshore bars, periodically extending and welding onto the foreshore as Type II beach ridges (Carter, 1986). These bars form the main source of sand for dune building.

The beach is exposed to winds of all directions, although southwesterlies dominate. About 55% of all winds may be classed as 'offshore', and 45% onshore, although a significant proportion blow alongshore, and move material parallel to the coast. Carter (1975) and Carter and Stone (1989) have examined the wind records for Magilligan, in particular isolating the number of occasions when winds exceeded 20 knots $(10\,m\,s^{-1})$ for more than 5 hours. Over a 30-year period these wind records show a number of distinct peaks (late 1960s and early 1980s) which coincide with phases of oceanside dune cliff erosion (Carter and Bartlett, 1990) further reflected in the availability of sand for dune building. However, the incidence of winds capable of moving sand from the beach greatly exceeds the amount of sediment available for transport.

The influence of humans on the Magilligan foredunes is an important factor in both maintaining and promoting geomorphological instability. Much of the local diversity, in both plants and topography, stems from the activities of visitors, who walk (and ride) through the dunes. Deterioration of the path surfaces often leads to local redistribution of blown sand, so forcing ecological changes. The area is, however, ungrazed, so that the vegetation succession is not affected by animals.

3. METHODS

Since 1968 a number of investigations have examined the geomorphology, ecology, soils and palaeoecology of Magilligan. Field methods have been described in various papers, notably Carter (1975, 1979, 1980, 1982b, 1986), Wilson (1987) and Wilson and Bateman (1986, 1987). This study involved the repeated survey of a number of transects across the prograding dune system. First established in 1969, these survey lines were reoccupied at various intervals until 1983, when increased military activities resulted in access restrictions to much of the area. From 1969 to 1971, and 1978 to 1983, profiles were resurveyed at least every 14 days. Between 1971 and 1978 surveys were less frequent. At various times other projects were undertaken. These included an examination of cross-foredune windrun, using an array of Feuss cup anemometers in 1983, investigations of sedimentary structures and textures (mainly grain size and mineralogy) in the late 1960s and early 1980s, and soil surveys in the mid-1980s. Sampling of vegetation was undertaken on three occasions, in 1978, 1980 and 1983, and in 1987 a study of root biomass of *Ammophila* was completed. Almost all these studies have followed conventional techniques which need not be repeated here. Some further details of methods are included in the text where appropriate.

The investigations reported in this chapter centre on the foredune area to the west of profile line M6 (see Fig. 3), part of a designated Nature Reserve. The corresponding area to the east is part of a Ministry of Defence firing range, and not readily accessible.

4. SHORELINE CHANGES

4.1. Pre-1950

Ocean-side progradation of the main beach ridge plain probably ceased around 1500 years ago at a position somewhat seaward of the present shoreline, and was followed by a period of marine erosion which supplied sediment to form the high, transgressive dune sheet which covers the north and east of the foreland (see Fig. 1). In addition to forming high dunes, some sediment was dispersed alongshore to become trapped within the wave re-entrant at the mouth of Lough Foyle (Carter, 1980), accumulating as a wedge of tapering beach ridges about 15 ha in extent (Fig. 2). Carter (1975) showed that this area, Magilligan Point, has accreted and eroded several times since 1800. For a period of 30 to 40 years beach ridges and dunes prograde, but then undergo a phase of erosion before this growth/recovery cycle repeats itself. There is a strong inverse correlation over time between the volume of sediment at Magilligan Point, and the volume contained in the Tunns Bank, a submarine shoal immediately to the northeast (Fig. 1). It is probable that the area and shape of this shoal controls both the extent of shoreline erosion, via wave refraction, and the effectiveness of longshore sediment transport (Carter, Lowry and Stone, 1982). Thus the supply of sediment to Magilligan Point is limited by a major feedback between the offshore topography and the incident wave field.

4.2. Post-1950

The last major recession of the shoreline occurred in the early 1950s when a series of storms breached and then destroyed the foredunes that had existed

133

Fig. 2   An oblique air photograph of Magilligan Point (July 1987), showing the zone of recent progradation. The Martello tower in the centre of the photo was built in 1812, and in 1955 the coastline was just seaward of the tower

for at least 25 years. By 1953, the shoreline occupied a position just north of the Martello Tower, itself built near the water's edge in 1812 (Figs 2 and 3). Since 1953, the shoreline has prograded seaward, as a sequence of beach ridges welded to the Point. At first, blown sand appears to have accumulated against the early 1950s dune scarp, forming a high and wide dune rampart, but during the 1960s and 1970s, a series of shore-parallel dune ridges developed (Fig. 3), separated by three major interdune slacks opening to the West. These dunes accumulated at the landward margins of the beach ridges, usually along tidal litter lines. The upward growth of the dunes relied to a considerable extent on the occasional stranding of storm-fed bars (Carter, 1986). As each dune formed to the seaward, those behind became cut off from much of the sediment supply.

In the early 1980s the hitherto rapid progradation largely ceased. Storms in 1980 and 1981 truncated the easternmost ridges and sediment was transported alongshore to form a small divergent dune field between survey lines M6 and M7. Some eroded sediment contributed to raising the height of the seawardmost

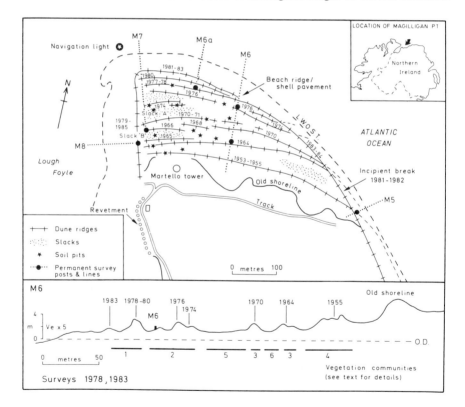

Fig. 3   The foredune ridges at Magilligan Point may be dated by reference to air photographs and, since 1968, ground surveys. The locations of sampling points and survey lines are given, and a cross-section along line M6 is shown

dune ridge (largely formed in 1964 near M5 and 1974 near M6), and this process acted to seal a major incipient breach immediately northwest of M5. In 1983 storms destroyed the outermost dune ridges west of M6. At present (1989), the outermost dune ridge has built up, and now forms a protective feature almost totally surrounding the prograded area. In places, especially near M8 on the west side of the Point, this dune is encroaching slowly across the 1960s and 1970s ridge ends. The area most vulnerable to inundation is near survey line M5.

5. ORIGIN AND GROWTH OF FOREDUNES

5.1. Beach morphology, sediment sources and supply

The essential control over foredune development at Magilligan is the availability of a sub-aerial source of medium to fine sand, most readily provided following the welding of a beach ridge to the upper shore (Carter, 1986). Sediment transport to the foredunes is controlled partly by the composition of the beach material, which is a mixture of relatively coarse, poorly-sorted carbonate particles and relatively fine, well-sorted quartz sand (Table 1). After dewatering, a stranded beach ridge will deflate. This process may be constrained to some extent by the formation of salt crusts (salcretes) or the presence of moist air, but at Magilligan deflation usually takes place within a few days of the ridge being emplaced. Studies by Carter (1976) and Carter and Rihan (1978) showed deflation of 0.2 to 0.3 m over a week across shore-normal distances of up to 80 m, representing a windblown sediment volume of 15 000 to 19 000 m^3.

Table 1   Mean sediment characteristics for beach and dune sands at Magilligan

Zone	No. of samples	Material[1]	Proportion %	Mean (phi units)	Sorting	Skewness	Kurtosis
Beach surface[2]	40	Quartz	90.0	2.54	0.34	0.20	0.60
		Carbonate	10.0	0.77	1.02	− 0.21	0.85
Deflation surface[2]	10	Quartz	82.5	2.50	0.37	− 0.03	0.48
		Carbonate	17.5	0.62	1.38	− 0.18	0.85
Foredunes	6	Quartz	87.9	2.57	0.33	− 0.01	0.47
		Carbonate	12.1	1.35	0.96	− 0.29	0.78
Inland dunes	30	Quartz	92.3	2.42	0.38	− 0.11	0.64
		Carbonate	7.7	1.72	1.15	− 0.32	0.78

[1]Excluding a small amount (< 1%) of heavy minerals.
[2]Excluding large shell fragments and whole valves.

The beach ridges contain a number of large shell valves (mainly *Bivalvia*) and large (>4 mm) fragments, concentrations sometimes exceeding 300 valves or fragments m^{-2} (Carter, 1976). Shell-rich units occur within the finer stratified ridge facies especially at the swash mark or along landward slip margins (Carter, 1986). This mix of immobile coarse and mobile fine elements is crucial in controlling the sediment supply (Fig. 4A). As deflation removes the finer materials so the coarser residue becomes concentrated at the surface as a shell pavement (Carter, 1976, 1978), shutting-off the supply and altering the boundary layer air flow (Fig. 4A and B), including the deepening of the viscous sub-layer tenfold from <0.0001 m to about 0.001 m, allowing remaining fine surface material to be unaffected by wind stress.

In addition to acting as a lower limit to deflation, the pavement also serves as a temporary reservoir for fine sand, which may accumulate as ripples or low relief sand waves that migrate downwind, very often parallel, or at a high angle, to the shoreline. If the wind veers or backs these fine bedforms may disperse rapidly into the foredunes. The pavement often provides the platform for the initial growth of new foredunes, especially along strandlines where tidal litter not only encourages deposition of sand but also includes viable root stock and

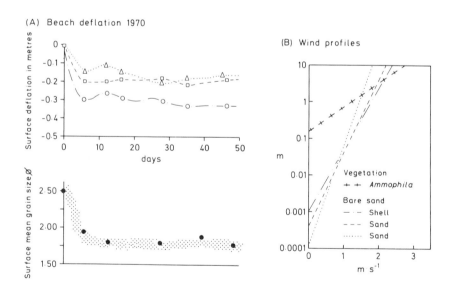

Fig. 4 A: Deflation curves (top) from three sites on the surface of a stranded beach ridge at Magilligan Point. Sand grain size variations as deflation proceeds (bottom). The finer material is removed leaving a coarser residue, as well as shell valves (Carter, 1976). B: Logarithmic wind profiles over *Ammophila* foredune vegetation and various bare sand surfaces

seeds, plus sufficient nutrients and moisture for germination and establishment of vegetation. At Magilligan, the litter lines are often marked by initial growths of *Atriplex* and other adventitious species, followed by *Elymus farctus* and *Ammophila arenaria*. It is these latter species that encourage deposition.

5.2. Initial foredune development

This section relies heavily on observations made between 1978 and 1983, in the area immediately west of profile line M6 (Fig. 3). A new survey line (M6a) was established in February 1980 to record the growth of two new foredune ridges. Foredune I was established in late 1978 (Fig. 5), and by the time of the first regular survey, in early 1980, stood about 1.40 m above the shell pavement and contained about 12.5 $m^3 m^{-1}$ shoreline of sand. Foredune II developed from a litter/vegetation line about 20 to 25 m seaward of Foredune I in early July 1980 during deflation of a beach ridge that welded on to the foreshore about 4 to 6 weeks earlier (see summary in Fig. 6).

Foredune II grew rapidly, at an average rate of over 0.05 $m^3 m^{-1} day^{-1}$ for the first 3 months of its existence, and by the autumn of 1980 was over 0.5 m high and about 10 m in width. The main locus of deposition was at the seaward margin of the *Elymus/Ammophila*, tapering downwind (see sketch in Fig. 7A). Continued upward growth was related to vegetation response; at first, as the vegetation began to grow, the dune experienced a very high sediment flux, and much of the accumulated sediment blew farther inland or returned to the beach. Fig. 6 shows the stages of foredune growth, in which crestal position (C in Fig. 7) migrated both landward and seaward, largely in response to wind variations and vegetation growth. The foredune comprised a series of landward and seaward dipping foresets, usually convex-upwards, with sub-parallel laminae often converging downwind. Bounding surfaces indicated phases of growth and migration. A typical Magilligan foredune cross-section, exposed by marine erosion in April 1981, is shown in Fig. 7B; here the foredune has developed on a deflated beach ridge (marked by a shell pavement). The overlying unit consists of gently convex strata, probably indicating initial *in situ* vegetation-related growth. In turn this unit is covered by overlapping aeolian foresets formed by vertical accretion as the dune crest shifts. This relationship between vertical growth and horizontal position is explored in Fig. 8, in which temporal variations in crest height and location provide distinctive signatures according to sediment availability, seasonality (especially associated with vegetation vigour) and excessively strong offshore winds causing dispersal. Particularly noticeable is the late-summer loop (top right), where the amount of sediment from deflating beach ridges engulfs the vegetation and leads to a pronounced landward shift of a rapidly rising dune crest. However, this material is unstable and tends to disperse. Other characteristic loops are those indicating winter dispersal—crest lowering, plus landward migration—and early spring vertical accretion as the

A

B

Fig. 5   Three views taken in 1979, 1981 and 1983 of incipient foredune development
at Magilligan

C

Fig. 5   (*cont.*)

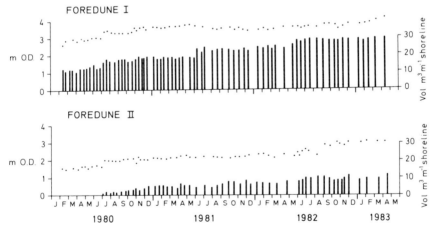

Fig. 6   Crest height (points) and sediment volume (bars) fluctuations on the incipient Foredunes I and II between 1980 and 1983. Foredune I began to form in 1978, while Foredune II started to develop around a tidal litter line and *Atriplex hastata* and *Elymus farctus* plants in early July 1980, fed by sediment from a newly welded beach ridge

vegetation renews growth and new shoots increase sward density. The formation and subsequent growth of Foredune II at site M6a had only a limited effect on the landward Foredune I. The vertical growth rate of Foredune II (0.0023 m day^{-1}) was greater than that at Foredune I (0.0017 m day^{-1}) over the two-year

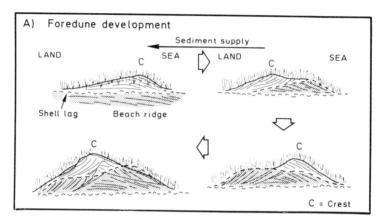

Fig. 7A   Schematic view of incipient foredune development at Magilligan

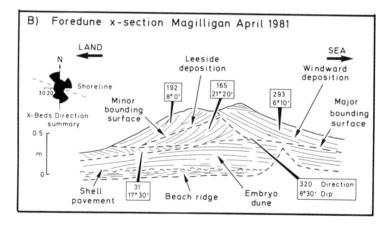

Fig. 7B   Foredune stratification exposed by marine erosion in April 1981

monitoring period, but the volume accumulation rate at Foredune II (0.018 m³ day⁻¹) was only 70% that at Foredune I (0.025 m³ day⁻¹). However, the volume rate at Foredune I fell from an average of 0.037 to 0.025 m³ day⁻¹ once Foredune II formed. The obvious conclusion is that Foredune II exercised some control over the rate of accumulation on Foredune I, but that it was not intercepting the entire beach supply. Over the three years of observation the two foredunes amassed between 7000 and 8000 m³ sediment. This is only about 55% of the total volume deflated from the adjacent beach. Of the remaining 45%, about 10% disperses into the rear dunes, leaving the rest to re-enter the beach/nearshore system during periods of offshore or alongshore winds.

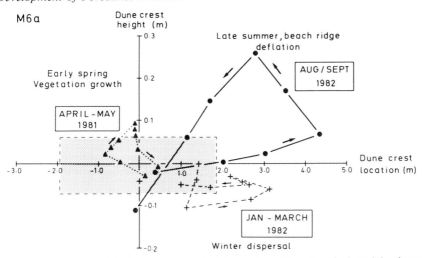

Fig. 8 The relationship between crest height and position (vertical and horizontal coordinates) for Foredune II over the period 1980 to 1982. The loops indicate a series of major seaward and landward shifts in the crestal locus, while the shaded area indicates the region of more prosaic changes (individual points not shown)

5.3. Consolidation of foredunes

Changes in a foredune formed in 1976 to 1977 were monitored on profile line M6. At the time of the first regular survey (February 1979), the crest of the dune stood at +4.25 m OD. Permanent survey markers were established at six points across this dune (although three were buried by blown sand within 12 months). The profile of the dune (Fig. 9A) exhibited a composite form, with a single crest, flanked by two small leeside crests and a convex seaward slope. The dune was well vegetated with *Ammophila*. During five years of study, the dune crest accreted little more than 0.3 m (Table 2; Fig. 9A), while at the rear of the

Table 2 Annual areas and standard deviations of front, crest and rear markers (heights in metres) for the established foredune at M6

Position		1979	1980	1981	1982	1983
Front	n	25	18	15	16	6
	x	1.98	2.29	2.43	2.57	2.63
	s	0.11	0.04	0.12	0.10	0.08
Crest	n	25	27	16	16	6
	x	4.40	4.52	4.61	4.62	4.81
	s	0.07	0.08	0.15	0.16	0.06
Rear	n	22	26	18	20	6
	x	1.92	1.95	2.01	2.00	1.99
	s	0.05	0.04	0.02	0.02	0.01

n = Number observations; x = Mean; s = Standard deviation

A

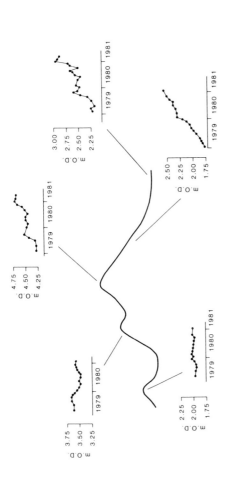

Fig. 9A   Surface heights over time for various parts of the foredune at M6

B

M5/M6   11 March 1983

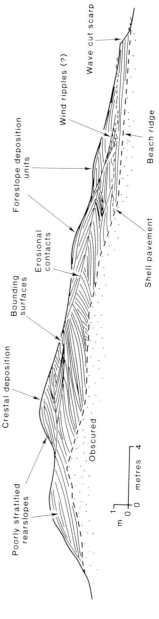

Fig. 9B   Cross-section of the same dune in 1983, showing various deposition units and structures

dune, vertical growth was minimal (<0.1 m). Most vertical accretion, and volumetric expansion, took place on the seaward flank of the dune, which built upwards and outwards throughout the period of observation, accumulating over 25 m³ m⁻¹ in four years. If this pattern of volumetric addition was maintained along the entire foredune, then about 8000 to 9000 m³ accumulated on the seaward flanks and a further 1000 to 1500 m³ at the crest or farther landward. On 11 March 1983 an easterly storm truncated the M6 ridge about 150 m southeast of the profile line. A sketch made at the time (Fig. 9B) showed a structure dominated by seaward-dipping, cross-beds, similar to those in Fig. 7B. The presence of a major erosional contact on the seaward flank preceded a phase of rapid deposition. The upper parts of the dune structure were often obscured by roots, but there appeared to be a small patch of coarse-grained wind ripples near the seaward edge. The lower structural limit of the dune was again marked by a shell pavement.

The data for crest height and crest location, both for the M6 foredunes and Foredune I on profile line M6a, were subjected to harmonic analysis. (Harmonic analysis is a mathematical technique for decomposing time series into sinusoidal components (Kendall, 1973). Each component comprises a given periodicity and amplitude reflecting the amount of variance explained by that harmonic.) In this example the calculation was based on an equally spaced series of 128, 9-day interval (34 months), points. Time series of both crest height OD and crest location (relative to survey markers) were analysed. Any significant linear trend ($p > 0.05$) was removed from the data before analysis. Harmonic analysis allowed the percentage variance, explained by various time periods, to be examined (Table 3).

In the case of dune height, most variability falls in the 6 to 12 month period, indicating an annual period of change, whereas in this location most variation is seasonal (3 to 6 months). However, substantial variation occurs throughout the spectrum.

As the foredunes grow, the near-ground airstream is progressively perturbed. Wind-run observations made in 1983 showed the marked compression of isovels at the foredune crests, with expanding, but rarely reversing, leeside flow. During offshore winds, extensive stagnant air 'pools' would form on the seaward sides of the dune ridges, enabling locally extensive avalanche and chute-fan structures to form. However, simple two-dimensional patterns of wind-run belie a more complex three-dimensional fluid flow, especially where the foredune is crossed by a gully or a low 'saddle'. These features, which often arise from gaps in the vegetation at the initial stages of foredune formation, serve to funnel the airstream through the dune, and in many circumstances create significant secondary aeolian mounds on the landward side (see Fig. 10). These gaps are often used by people and develop as wind-scoured routes through the dunes. Ultimately they may take the form of blowouts.

144

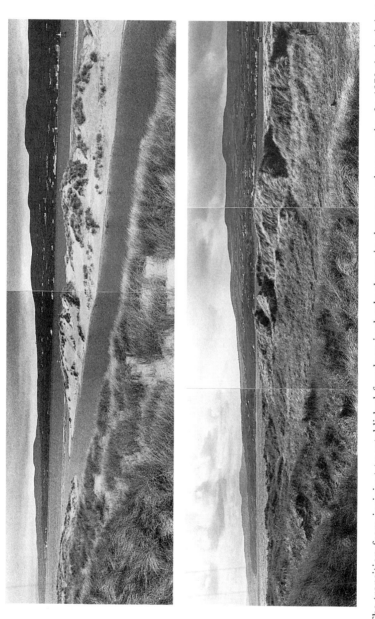

Fig. 10 The transition from incipient to established foredune is clearly shown in these two photographs. In 1970 the incipient dune is still receiving fresh sand, has relatively sparse vegetation, and is separated from the established dunes by a tidal slack. By 1987, the foredune is established. Despite the obvious increase in vegetation cover, the gross morphology of the dune has changed very little. Note the 'gap' towards the right of 1970 photography which has persisted throughout

Table 3   Foredune variability through time using Harmonic Analysis (data are % explained by each bandwidth)

Period months	M6 Crest height	location	M6a (foredune I) Crest height	location
>12	5	23	13	20
6–12	39	26	42	16
3–6	26	27	19	38
1–3	15	14	17	12
<1	15	10	9	14
Total	100	100	100	100

6. MATURE FOREDUNES

6.1. Inland dunes

6.1.1 Stabilisation

At Magilligan, the *Ammophila*-dominated foredunes capture about 50 to 70% of sand received from the beach. The remainder is dispersed by a relatively turbulent near-ground airstream, either into the inland dunes, where accretion is relatively slow (usually less than 0.01 m a^{-1}) or is returned to the beach or nearshore zones. Within 50 m of the beach it is not easy to detect any widespread annual sand accumulation. Inland sediment deposition usually occurs within vegetation, forming a structureless cover to the earlier dune bedding (Fig. 11).

The foredune stabilisation succession develops rapidly, within one or two years of a new incipient foredune forming. Thus the transition from an incipient to an established foredune takes about 5 to 8 years at Magilligan. Figure 10A shows the 1966 ridge, which remained 'active' until 1970, during its accumulation phase. The foredune comprises a single ridge, itself made up of hummocks (up to 4.5 m OD) separated by saddles. *Ammophila* tussocks flourish on both the seaward and landward slopes, but there are also large areas of bare, mobile sand. Between the 1966 ridge and the 1964 ridge (foreground) is a tidal slack, subject to spring tide inundation. The second photo (Fig. 10B) shows the same view in 1987. By now the beach ridge/foredune plain has extended 180 m seaward, and the 1966 dune has been 'inactive' for 17 years. The only significant topographic changes are in the secondary rear-slope dune which has built-up through the leeside deposition of sand transported through the gaps. Some of this material has spread onto, and raised slightly (by less than 0.2 m), the slack floor. (A small barrier dune at the entrance to the slack has prevented even occasional tidal inundations.) In the 1987 photograph, the 1966 foredune is cloaked in dense

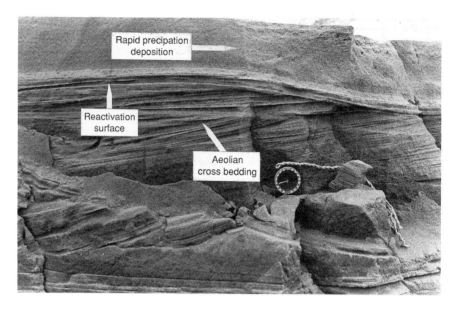

Fig. 11   The rapid deposition of a homogeneous sand unit on established foredunes
          usually arises from grain-fall deposition through vegetation

vegetation, including *Ammophila*, *Festuca*, *Agrostis*, *Lotus* and rhizocarpus mosses. There is virtually no bare sand, save on the steeper gully sides where slumping is occurring.

6.1.2 Vegetation communities and ecological succession

As progradation proceeds so a vegetation succession develops in response to sand deposition, shelter, pedogenesis and soil moisture. At Magilligan, the vegetation divides into four ridge and two slack communities, as follows (see Fig. 3 for locations):

I. Seaward accreting dunes dominated by vigorous *Ammophila arenaria*, with occasional patches of *Elymus farctus* and *Leymus arenarius*. Sand deposition on these dunes may exceed 0.3 or $0.4\,\text{m}\,\text{a}^{-1}$.

II. Landward ridges with deposition rates of less than $0.05\,\text{m}\,\text{a}^{-1}$. On those dunes where the supply has only recently declined, the vegetation comprises *Ammophila*, plus bryophytes like *Tortula ruraliformis*, *Barbula* sp. and *Bryum* sp. and low densities of herbs such as

Carex arenaria, Campanula rotundifolia, Lotus corniculatus and *Senecio jacobaea*. Within two or three years this community is often invaded by dune annuals like *Cerastium diffusum, Viola fasiculata* and *Vicia angustifolia* as well as bryophytes like *Byrum pedulum* which are less tolerant of engulfment by blown sand.

III. After 5 to 20 years, pleurocarpus bryophytes and herbs cover most of the 'fixed' landward foredunes. Apart from the dune crests, the ridges and inter-ridge hollows are thickly carpeted with the moss *Rhytidiadelphus squarrosus*, often in association with depauperate *Ammophila, Festuca rubra* and *Trifolium repens*.

IV. On older (>20 years) foredunes, especially to the south of the area, adjacent to Lough Foyle, a more open community develops, probably as a response to increasing exposure and internal redeposition of sand (the area is crossed by several unstable paths). Here *Ammophila* is associated with *Tortula ruraliformis, Camptothecium lutescens* (both bryophytes that are able to respond to burial by developing new shoots), as well as a variety of herbs, mainly *Lotus corniculatus, Galium verum, Festuca rubra* and *F. ovina*.

As well as the four ridge communities listed above, the dunes also contain, at times, well-developed tidal slack vegetation. In the late 1970s, two wet slacks were present, the first (shown in Fig. 10A) was inundated between 1966 (when it formed) and mid-1974, while the second (comprising two interlinked hollows) was often flooded from the time of its formation (1974) until 1980. During tidal inundation considerable quantities of flotant tidal debris accumulated in these slacks, forming a nutrient source exploited by many adventitious species. In 1978 and 1980 two slack communities were present:

V. In the seawardmost tidal slack (slack A on Fig. 3), the vegetation community was dominated by *Atriplex* sp. and *Tripleurospermum maritimum*, plus *Aster tripolium* and many isolated plants that presumably seeded from the tidal litter.

VI. In slack B (Figs 3 and 10), the open community was formed of many dwarf herbs, including *Plantago coronopus, P. lanceolata, P. maritima, Honkenya peploides* and *Agrostis stolonifera*, plus large areas of bare sand. This slack was still experiencing some groundwater flooding in the late 1970s.

Since 1980 both these slack communities have largely disappeared, often succumbing to invasions of the adjacent dune ridge species (Fig. 10). While such a change has been aided by the development of barrier dunes at the slack entrances, it is also partly due to accretion (both wind and water lain) across the slack floor, which has raised the surface to the point where groundwater seepage flooding no longer occurs.

6.1.3 Vegetation maturity indices

In 1978, 1980 and 1983, plant height, seed head (inflorescence) and root node lengths were measured in the *Ammophila* communities on successive ridges (Fig. 12). On each occasion, metre square quadrats were selected at random on both seaward and landward dune flanks, along three parallel transects across the dune field. (ANOVA testing showed no significant differences between flanks or transects, so the information has been pooled in this report.) All three indices show a variation with age across the dune ridges. The maximum height of *Ammophila* rises to a peak in 3 to 4 year-old dunes, falling thereafter to a mean of around 0.5 to 0.6 m. Seed heads are concentrated at the coast, numbers declining asymptotically landward, with very few (usually $< 5\,\mathrm{m}^{-2}$) found on the oldest dunes. Root node length—often taken as an indicator of sand burial (Gimingham, Gemmell and Greig-Smith, 1948, Ranwell 1972)—decreases slowly inland from around 0.06 m to 0.02 m.

In 1986 a series of above- and below-ground biomass *Ammophila* samples were collected for five 0.5×0.5 m quadrats on each dune ridge. The most recent 'active' ridge was 5 years old at the time of sampling, and it was not possible to acquire data from a newly formed foredune. Fig. 13 shows the results of this

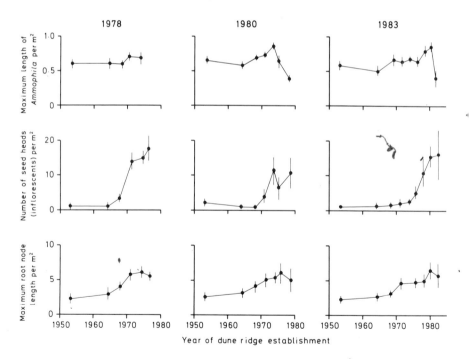

Fig. 12   *Ammophila* vegetation maturity indices for sequential ridges, taken on three occasions in the early autumns of 1978, 1980 and 1983

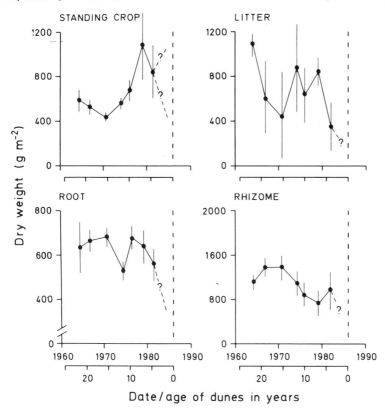

Fig. 13   Variations in *Ammophila* biomass on foredunes of different ages

exercise, expressed in terms of a square metre to provide comparability with earlier studies. The above-ground standing crop peaks at the 1979 dune (then 7 years old), falling by half on the 1971 (15 years old), 1966 (20 years old) and 1964 (22 years old) ridges. Vegetation litter shows a more complex pattern, which probably reflects not only production and decomposition differences, but also the collection of non-*Ammophila* debris among the samples, especially in vegetation communities 3 and 4. The below-ground biomass was divided into root (vertical) and·rhizome (horizontal) components (rhizome material included buds). Biomass of both increases over the first 10 to 15 years, but stabilises thereafter. The ratios of standing crop/litter, rhizome/root and above-ground biomass/below-ground biomass (Table 4) indicate clear trends with age, as more production is concentrated below ground, and litter increases, while the rhizome/ root ratio suggests a switch to greater rhizomatous activity as the dunes mature.

Table 4   Biomass ratios for dune vegetation across dated foredune ridges

Ratios	Dune date (Age in years)						
	1964 (22)	1966 (20)	1971 (15)	1974 (12)	1977 (9)	1979 (7)	1981 (5)
Standing crop/litter	0.54	0.86	0.91	0.63	1.01	1.25	2.38
Rhizome/Root	1.80	2.08	2.05	2.03	1.31	1.15	1.67
Above-ground/ below-ground biomass	0.34	0.25	0.20	0.35	0.42	0.76	0.53

6.2. Pedogenesis

Soil development on the Magilligan foredunes was assessed in 1986, using profile morphology and selected chemical properties as indicators of pedogenesis. Dune age and vegetation communities are the major factors influencing the degree of soil formation.

North of the 1970 foredune, substantial areas of unvegetated sand occur (vegetation communities 1 and 2). Profiles here display little, if any, evidence of pedogenic horizonation and show only slight colour variation from pale brown to light yellowish brown (10YR 6/3, 6/4) (Munsell Colours). The organic matter content is negligible ($<0.7\%$), pH values generally exceed 7.5, and $\%CaCO_3$ is high (10 to 20%) throughout.

Profiles examined on the 1970 and older foredunes (vegetation communities 3 and 4) possess thin (<0.08 m) surface accumulations of organic matter (3.5 to 9.5%) intermixed with dune sand. Weakly developed fine to medium granular aggregates are present in these Ah horizons, reflecting the ability of soil humus to bind mineral grains into structural units. Topsoil colours are greyish browns and browns (10YR 3/2, 4/2, 4/3, 5/3) in contrast to the pale browns and light yellowish browns (10YR 6/3, 6/4, 7/3, 7/4, 7/6) of the underlying Cu horizons. The pH of these soils is within the range 7.0 to 7.5, and although $\%CaCO_3$ is still high, values $<10\%$ occur in some of the Ah horizons, suggesting that decalcification is taking place.

More detailed analysis of temporal changes in organic matter and $CaCO_3$ contents was conducted using 10 samples representing depths of 0 to 0.02 m from each dated dune ridge.

Mean soil organic matter content increases from 0.3% after 9 years to $>7\%$ after 31 years, but the rate of organic build-up is non-uniform. After 12 years slightly more than 1% organic matter has accumulated; after 15 years the amount exceeds 2% and is $>4\%$ after 21 and 22 years. Sample ranges indicate wide variations in organic matter (e.g. 2 to 11% for the 1964 dune ridge).

Mean $\%CaCO_3$ shows a decline across the foredunes from *c.* 12% in the youngest dunes to 5.5% in the oldest. These changes are non-uniform.

Carbonate remains in excess of 11% during the first 20 years; thereafter a reduction to *c.* 8% at 21 years and *c.* 6% at 22 years occurs. Within-ridge variation in %CaCO$_3$ is evident, particularly in the 1978 to 1979 ridge (8 to 17%), although some ridges show much narrower ranges (e.g. 10 to 13% for the 1976 ridge).

7. DISCUSSION

The prograding foredunes at Magilligan are unusual, in as far as most sandy shorelines in the British Isles are eroding. However, it is the longshore redistribution of sandy sediments, within well-defined coastal compartments or cells, which supplied almost all the example of recent foredune development. The rate of recent foredune progradation at Magilligan (around 8 m a^{-1}), measured in terms of shoreline advance, is commensurate with similar sites, for example Tentsmuir in Scotland (Deskmukh, 1979), Gibraltar Point, eastern England (King, 1972) and Goulouven in northwest France (Hallégouet and Moign, 1976). At all these sites, incipient dunes are forming non-beach ridges at the downdrift end of littoral compartments. The supply of sediment is determined both by the incoming volume of drift, tidal exposure of the foreshore, wind effectiveness and sediment heterogeneity, so that foredune growth may range from near-continuous to episodic (as at Magilligan).

The general status of foredune development in Britain, Ireland and most of NW Europe between the Baltic and central-west France is strongly tied to Holocene sea-level histories and sediment budgets (Carter, 1990a). Pervasive beach ridge and foredune development probably ceased in the period 2000 to 5000 BP, since when erosion and sediment reworking has prevailed. The influence of man cannot be ignored, not only where littoral transport has been disrupted, but also where river mouth and estuary processes have influenced sediment recycling. On the Irish coast, long stretches of eroding dune systems are broken only by occasional foredune development (Carter, 1990b), usually adjacent to river mouth or estuary channels where land reclamation or drainage has reduced the tidal prism and allowed the shoreline to accrete (Carter, 1988).

The recent progradation at Magilligan Point has included the episodic development of incipient foredunes. Both the initial development and the growth of these foredunes are strongly controlled by (i) the stranding of tidal litter lines at, or even above HWOST and (ii) subsequent availability of fresh sand. The nucleus of a new dune ridge forms on the most recent-but-one beach ridge and is generally supplied from the most recent. The initial accumulations of sand may be pyramidal (Hesp, 1981) or elongate, depending on the expression and size of the litter line. The spread of sand-trapping vegetation, in this case mainly *Elymus* sp., promotes the accumulation of broad, shallow ridges, similar to those engendered by *Spinifex* in eastern Australia (Hesp, 1983), except for the fact that in the Magilligan examples the locus of deposition is at the *upwind* edge of the vegetation rather than down. This difference is ascribed to the individual sand-trapping abilities of the two plants. The efficiency of the sand-trapping and later sand-accumulating vegetation (mainly *Ammophila*) is manifest in the relative positional stability of the foredunes. However, the Magilligan foredunes are not the true 'Darss' type as described by Menez (1977) and Hallégouet and Moign (1976), as they experience some post-depositional mobility and are not forming on the most seaward ridge. The Magilligan foredunes grow continuously (Fig. 6), although this upward trend incorporates growth spurts which are largely associated with storm recovery beach ridge welding (Carter, 1986). The dissipative/reflective shore transition around Magilligan Point (described by Carter, 1986) has little noticeable effect on the dune type as adumbrated by Short and Hesp (1982) although in terms of morphology, the Magilligan dunes fall into Short and Hesp's class Fb foredunes—see Carter (1988, p. 314). Delivery of sand to the foredunes is controlled both by beach ridge textures (shell lags, salcretes, etc.) which serve to reduce asymptotically the volume of sand available for transport (Fig. 4), and also by the receptiveness of the vegetation to capture sand. Physiographic controls severely limit the amount of sand entrained by the wind, to the point of invalidating the use of empirical sediment transport volume equations of the type proposed by Kadib (1964), Hsu (1971), Horikawa, Hotta and Kraus (1986) and others. At attempt to use the Kadib formula (thought by Berg (1983) to provide the best results) led to a twenty-fold overestimation of annual sediment discharge at Magilligan. On several occasions between 1979 and 1983 sudden influxes of windblown sand led to the foredune vegetation being smothered to a depth of several hundred millimetres. Almost all this material was dispersed rather than trapped, much of it returning to the beach. Godfrey, Leatherman and Zaremba (1979) report that under favourable conditions *Ammophila* can grow at rates of $2 \, \text{mm day}^{-1}$ and recent research (Disraeli, 1984; Carter, 1988), would suggest that individual plants can survive a steady sediment input up to a maximum of around 0.5 to $0.6 \, \text{m a}^{-1}$. Above this the *Ammophila* plants are unable to respond, but in such circumstances it is likely that any loose material will be blown away, re-exposing the vegetation. At Magilligan, spatially impersistent, but relatively thick, deposits often occur on the leeside of dunes, (see Fig. 10A), and may account for the patchy distribution of *Ammophila* at such sites. The total volume and the eventual height in any foredune ridge depend entirely on how long it continues to remain 'active', at the seaward edge of the dune system.

Published records of sand accumulation on both natural and artificially encouraged dunes suggest that the maximum growth rates occur around sand fences with a porosity of 0.4 to 0.5 (Phillips and Willetts, 1978; Carter, 1988, p. 494), but that these rates can be almost matched by vegetation in favourable conditions (Alsop, 1973; Dahl, Fall and Otteni, 1975; Dahl *et al.*, 1983). For example, accumulations around sand fences on the Outer Banks of North Carolina averaged between 8 and $9 \, \text{m}^3 \, \text{m shoreline}^{-1} \, \text{a}^{-1}$ (Savage and

Woodhouse, 1968), while deposition on experimental planted dunes on Padre Island, Texas (Dahl, Fall and Otteni, 1975) ranged from 9 to $12\,m^3\,m$ shoreline$^{-1}\,a^{-1}$, all values very similar to the natural foredunes at Magilligan.

High sediment flux between beach and dune during times of net foredune accumulation accounts for the structures described in Figs 6 and 8. The variability both of the wind and of the local near-ground airstream leads to non-uniform deposition and the development of distinct aeolian sand flow avalanches in the lee of the foredune crest, and further inland, grainfall deposits. Thus the dune builds-up as a series of overlapping cross-stratified units, separated by distinct bounding surfaces (non-deposition) and erosion planes.

Morphology is often determined in the very early stages of foredune growth. The variations in dune crest height along newly formed foredune ridges tend to persist as the dunes mature (see Fig. 10). Under some circumstances an inherited gap may provide the nucleus for a blowout, especially if it is used as a path or track. Gares and Nordstrom (1988) also point to gaps as being important precursory foci for blowouts. Earlier results by the same workers (Gares and Nordstrom, 1987) also point to the importance of such gaps in allowing sediment to move inland. At Magilligan, sediment transported through these gaps tends to be deposited in secondary ridges or aeolian fans behind the primary ridge. What is noticeable, though, is that the gaps tend to serve to transport sediment inland as it rarely returns through them to the shore.

The seawardmost foredune at Magilligan acts as an effective barrier to landward transport of sediment. Budgetary calculations suggest that only about 10 to 15% of the net flux passes beyond the incipient foredune, and much of this may move through gaps. The low sediment volume plus the dispersal mechanism within a turbulent boundary layer appear to lead to slow, widespread accumulation of structureless grainfall deposits. The relatively abrupt fall in sediment supply is mirrored both in the ecology as a number of ground cover species colonise this zone, and to a lesser extent in the soils where organic litter accumulates. Stabilisation of the Magilligan foredunes occurs rapidly, within a few years of them becoming inactive. As in many other systems, the trend towards stabilisation is marked by an increase in herbaceous species, and a marked decline in the vigour of *Ammophila*. Olson (1958) noted in the Lake Michigan dunes that the number of both *Ammophila* shoots and infloresences declined inland; at Magilligan, there is a similar pattern as well as marked reduction (>50%) in standing crop of *Ammophila* like that described by Deshmukh (1979) at Tentsmuir in Scotland and by Wallen (1980) at Sand Hammeran in southern Sweden. Furthermore, *Ammophila* in the Magilligan Point foredunes shows an increasing tendency (Fig. 12) to sub-surface productivity, noted by both Wallen (1980) and Disraeli (1984). The rate of these changes in the Magilligan dunes is slightly slower (5 to 20 years) than the 3 to 12 years reported by Wallen (1980).

The trends displayed by soil morphology and chemical properties are similar to those identified in other British and Irish coastal dunes (Salisbury, 1922, 1925; Ranwell, 1959; Wilson, 1960; Ball and Williams, 1974; Wilson, 1985). However, the speed of soil development has been considerably faster at Magilligan than other locations. Between-site variations in soil-forming factors (climate, relief, vegetation and parent material), and the use of different field sampling procedures and/or the accuracy attached to estimates of dune age by earlier workers are probably responsible for the slower rates of dune soil formation recorded elsewhere. Data compiled by Wilson (1987) highlight some of these between-site variations.

Ball and Williams (1974) suggested that a major influence on the speed of organic matter incorporation and horizon differentiation in dune sands was the initial $CaCO_3$ content of the parent material. Low-carbonate dunes were thought to encourage more rapid pedogenesis, and while there is some additional evidence to support this claim (Wilson, 1960), the Magilligan data indicate clearly that rapid leaching of minor quantities of $CaCO_3$ and the development of an acid reaction are not essential for substantial humus accumulation.

The effects of between-site variations in vegetation on dune pedogenesis are difficult to evaluate because the available information relates only to species diversity rather than to percentage vegetation cover and the contribution made to surface organic horizons by different species. The expansion of pleurocarpus mosses at Magilligan at the expense of *Ammophila* is evident, especially on the older dunes. The ability of these mosses to retain moisture and add organic matter to the soil partly accounts for the distinctive profile morphology and properties. As yet the full contribution made by mosses to dune soils has not been considered.

Climatically, Magilligan is a wetter area than most other studied sites, with moisture surplus 21 to 560% more than the other dune soils studied (cited above), and is probably the dominant factor in encouraging moss growth and expansion.

Age estimates for these other dunes have been based on documentary evidence and their positions relative to coastal defence structures. Such dates may refer to earlier phases of composite dune construction, making the start of modern soil formation a more recent event and thus underestimating the speed of pedogenesis.

The future of the Magilligan Point foredunes is hard to predict. Most of the observations reported in this paper occurred during a phase of rapid progradation. Since 1983, the shoreline has retreated, eliminating Foredune II at M6a. The seawardmost ridge, now an amalgam of the 1977, 1979 and 1981 ridges, has accumulated a large volume of sand and now stands at over 5.5 m OD, presenting an even more impenetrable boundary to the dune field for windblown sediment. It seems likely that if shoreline retreat continues, this barrier dune will slowly migrate inland, further increasing in height. However, if the volume of sand available from the beach should increase significantly, then it may engulf the active foredune vegetation, and lead to a major destabilisation of the entire Magilligan Point dune system. We await these possible changes with interest.

8. CONCLUSIONS

The prograding foredunes at Magilligan Point comprise a relatively unusual shoreline landform in northwest Europe. In this example, the foredunes are fed from sediment eroded from adjacent sand dune cliffs. This material is transported alongshore in nearshore bars which eventually weld onto the beach. Although deflation of these welded bars proceeds rapidly it is halted within a few days by the formulation of an immobile shell lag. While the episodic nature of the sand supply is reflected in both the vertical and volumetric growth of the foredune, constant redistribution of material in response to vegetation growth allows sustained accretion over periods of months to years. When a new incipient foredune begins to form, sediment supply to the older incipient foredune begins to fall, and the dune begins to be stabilised. Vegetation cover increases and diversifies and the soil profile develops. The Magilligan foredunes show a relatively rapid sequence of stabilisation spanning 5 to 20 years.

ACKNOWLEDGEMENTS

We would like to thank Mary McCamphill for typing the manuscript, Kilian McDaid for cartography and Nigel McDowell for photography. Reviews of the original manuscript by Karl Nordstrom and Norb Psuty were greatly appreciated.

REFERENCES

Alsop, P. F. B. (1973) The stabilization of the coastal sand zone. *Proc. Conf. Eng. Dynamics of the Coastal Zone*, Austral. Inst. Eng., 140–147.

Ball, D. F. and Williams, W. M. (1974) Soil development on coastal dunes at Holkham, Norfolk, England. *Proc. 10th Int. Congr. Soil Sci. (Moscow)*, **6**, 380–386.

Berg, N. (1983) Field evaluation of some sand transport models. *Earth Surf. Proc. Landf.*, **8**, 101–114.

Bird, E. C. F. (1985) *Coastline Changes*. Wiley, Chichester, 219pp.

Carter, R. W. G. (1975) Recent changes in the coastal geomorphology of the Magilligan Foreland, Co. Londonderry. *Proc. R. Ir. Acad.*, **75B**, 469–497.

Carter, R. W. G. (1976) Formation, maintenance and geomorphological significance of an eolian shell pavement. *J. sediment. Petrol.*, **46**, 418–429.

Carter, R. W. G. (1978) Ephemeral sedimentary structures formed during aeolian deflation of beaches. *Geol. Mag.*, **115**, 379–382.

Carter, R. W. G. (1979) Recent progradation of the Magilligan Foreland, Co. Londonderry, Northern Ireland. *Pub. CNEXO, Act. Coll.*, **9**, 17–27.

Carter, R. W. G. (1980) Longshore variations in nearshore wave processes at Magilligan Point, Northern Ireland. *Earth Surf. Proc.*, **5**, 81–89.

Carter, R. W. G. (1982a) Sea-level changes in Northern Ireland. *Proc. Geol. Assoc. Lond.*, **93**, 7–23.

Carter, R. W. G. (1982b) Some problems associated with the analysis and interpretation of mixed carbonate and quartz beach sands, illustrated by examples from north-west Ireland. *Sed. Geol.*, **33**, 35–56.

Carter, R. W. G. (1986) The morphodynamics of beach ridge formation: Magilligan, Northern Ireland. *Mar. Geol.*, **73**, 191–214.

Carter, R. W. G. (1988) *Coastal Environments*. Academic Press, London, 617pp.

Carter, R. W. G. (1990a) Coastal processes in relation to geographic setting, with special reference to Europe. *Senckenberg. Marit.*, **21**.

Carter, R. W. G. (1990b) The geomorphology of coastal dunes in Ireland. *Catena Suppl.*, **18**, 31–39.

Carter, R. W. G. and Bartlett, D. (1990) Coast erosion in northeast Ireland: Part I: Beaches, dunes and river mouths. *Ir. Geogr.*, **23**, 1–16.

Carter, R. W. G. and Rihan, C. L. (1978) Shell and pebble pavements on beaches: examples from the north coast of Ireland. *Catena*, **5**, 365–374.

Carter, R. W. G. and Stone, G. W. (1989) The mechanics of dune erosion at Magilligan, Northern Ireland. *Earth Surf. Proc. Landf.*, **14**, 1–10.

Carter, R. W. G., Lowry, P. and Stone, G. W. (1982) Ebb shoal control of shoreline erosion via wave refraction, Magilligan Point, Northern Ireland. *Mar. Geol.*, **48**, M17–M25.

Dahl, B. E., Fall, B. A. and Otteni, L. C. (1975) Vegetation for creation and stabilization of foredunes, Texas coast. In Cronin, L. E. (ed.), *Estuarine Research: Volume II. Geology and Engineering*. Academic Press, New York, pp. 457–470.

Dahl, B. E., Cotter, P. C., Wester, D. B. and Drbal, D. D. (1983) *Post-hurricane survey of experimental dunes on Padre Island, Texas*. US Army, Coast. Eng. Res. Center, MR 83–8.

Deshmukh, I. K. (1979) Fixation, accumulation and release of energy by *Ammophila arenaria* in a dune succession. In Jefferies, R. L. and Davy, A. J. (eds), *Ecological Processes in Coastal Environments*. Blackwells, Oxford, pp. 353–362.

Disraeli, D. J. (1984) The effects of sand deposits on the growth and morphology of *Ammophila breviligulata*. *J. Ecol.*, **72**, 145–154.

Gares, P. A. and Nordstrom, K. F. (1987) Dynamics of a coastal foredune blowout at Island Beach State Park, New Jersey. *Proc. Coast. Sediments '87*, *ASCE*, 213–221.

Gares, P. A. and Nordstrom, K. F. (1988) Creation of dune swale blowouts by foredune accretion. *Geogr. Rev.*, **78**, 194–204.

Gimingham, C. H., Gemmell, A. R. and Greig-Smith, P. (1948) The vegetation of a sand dune system in the Outer Hebrides. *Trans. Bot. Soc., Edinb.*, **35**, 82–96.

Godfrey, P. J., Leatherman, S. P. and Zaremba, R. (1979) A geobotanical approach to classification of barrier beach systems. In Leatherman, S. P. (ed.), *Barrier Islands*. Academic Press, New York, pp. 99–126.

Hallégouet, B. and Moign, A. (1976) Histoire d'évolution de littorale dunaire: la baie de Goulven (Finistère). *Penn ar Bed.*, **10**, 263–276.

Hesp, P. A. (1981) The formation of shadow dunes. *J. sediment. Petrol.*, **51**, 101–111.

Hesp, P. A. (1983) Morphodynamics of incipient foredunes in New South Wales, Australia. In Brookfield, M. E. and Ahlbrandt, T. S. (eds), *Eolian Sediments and Processes*. Elsevier, Amsterdam, pp. 325–342.

Hesp, P. A. (1984) Foredune formation in southeast Australia. In Thom B. G. (ed.), *Coastal Geomorphology in Australia*. Academic Press, Sydney, pp. 69–97.

Hesp, P. A. (1988) Morphology, dynamics and internal stratification of some established foredunes in southeast Australia. *Sediment. Geol.*, **55**, 17–41.

Horikawa, K., Hotta, A. and Kraus, N. C. (1986) Literature review of sand transport by wind on a dry sand surface. *Coast. Eng.*, **9**, 503–526.

Hsu, S. A. (1971) Wind stress criteria in eolian sand transport. *J. geophys. Res.*, **76**, 8684–8686.

Kadib, A. A. (1964) *Sand movement by wind.* US Army, Coast. Eng. Res. Cent. Technical Memo, 64–62.

Kendall, M. (1973) *Time Series* (2nd edn). Charles Griffith, London.

King, C. A. M. (1972) *Beaches and Coasts* (2nd edn). Arnold, London.

Menez, S. (1977) Les crêtes successives dunifiées de type Darss de la Côte sud-ouest de la baie de Goulven (Finistère). *Norois*, **96**, 593–599.

Olson, J. S. (1958) Lake Michigan dune development. 2. Plants as agents and tools in geomorphology. *J. Geol.*, **66**, 345–351.

Phillips, C. J. and Willetts, B. B. (1978) Selective literature review on sand stabilization. *Coast. Eng.*, **2**, 133–147.

Ranwell, D. S. (1959) Newborough Warren, Anglesey, 1. The dune system and dune slack habitat. *J. Ecol.*, **47**, 571–601.

Ranwell, D. S. (1972) *Ecology of Salt Marshes and Sand Dunes.* Chapman & Hall, London, 258 pp.

Salisbury, E. J. (1922) The soils of Blakeney Point: a study of soil reaction and succession in relation to the plant covering. *Ann. Bot.*, **36**, 391–431.

Salisbury, E. J. (1925) Note on the edaphic succession in some dune soils with special reference to the time factor. *J. Ecol.*, **13**, 322–328.

Savage, R. P. and Woodhouse, W. W. (1968) Creation and stabilisation of coastal barrier dunes. *Proc. 11th Conf. Coastal Eng. ASCE*, 671–700.

Short, A. D. and Hesp, P. A. (1982) Wave-beach-dune interaction in southeast Australia. *Mar. Geol.*, **48**, 259–284.

Wallen, B. (1980) Changes in the structure and function of *Ammophila* during primary succession. *Oikos*, **34**, 227–238.

Wilson, K. (1960) The time factor in the development of dune soils at South Haven Peninsula, Dorset. *J. Ecol.*, **48**, 341–359.

Wilson, P. (1985) Dune soils of North Bull Island. In Carter, R. W. G. and Mulrennan, M. (eds), *The Coastal Environments of Co. Dublin.* Field Guide No. 1, Geographical Society of Ireland, Dublin, pp. 28–32.

Wilson, P. (1987) Soil formation on coastal beach and dune sands at Magilligan Point Nature Reserve, Co. Londonderry. *Ir. Geogr.*, **20**, 43–49.

Wilson, P. and Bateman, R. M. (1986) Nature and palaeoenvironmental significance of a buried soil sequence from Magilligan Foreland, Northern Ireland. *Boreas*, **15**, 137–153.

Wilson, P. and Bateman, R. M. (1987) Pedogenic and geomorphic evolution of a buried dune palaeocatena at Magilligan Foreland, Northern Ireland. *Catena*, **14**, 501–517.

Wilson, P. and Farrington, O. (1989) Radiocarbon dating of the Holocene evolution of Magilligan Foreland, Co. Londonderry. *Proc. R. Ir. Acad.*, **89B**, 1–23.

Chapter Eight

Foredune mobility and stability, Fire Island, New York

NORBERT P. PSUTY

Institute of Marine and Coastal Sciences, Rutgers University U.S.A.

1. INTRODUCTION

1.1. Coastal foredunes

The general model for the establishment of coastal foredunes ascribes the accumulation of the shore-parallel sand ridge to the stabilizing effects of the pioneer vegetation that occupies the stressful environment niche at the upper reaches of storm waves (Carter, 1988). Some authors relate aspects of the dune's dimensions to the particular assemblages of plant species and to the unique nature of propagation practiced by the plants (Godfrey, Leatherman and Zaremba, 1979). Wind regime has also been put forward as a controlling mechanism for dune development, assuming that a greater persistence of wind moving across the beach to the foredune will correlate with greater dune dimensions or rate of sand accumulation (Pye, 1983). Whereas there is little argument about the importance of wind and vegetation, given an adequate sediment supply, there is some difference of opinion as to the temporal development of coastal dunes related to the general sediment budget. On a broad temporal scale, some suggest that coastal dunes are a product of a regressive phase of coastal development and therefore related to net positive sediment budgets (Bird and Jones, 1988). Others relate dune development to a transgressive phase during which the dune form is a locus of sediment accumulation or has a positive budget at a time of general shoreline displacement inland caused by a negative beach budget (Psuty, 1986; Orme, 1988).

In many shoreline studies, dune development is poorly linked with, or completely separated from, the processes operating on the lower portion of the dune/beach profile. Too often, the foredune is viewed as a static phenomenon

Coastal Dunes: Form and Process. Edited by K. F. Nordstrom, N. P. Psuty and R. W. G. Carter

that gains or loses volume, that enlarges or attenuates, but that is not spatially dynamic. The dune form is presented as a feature that is virtually isolated in its spatial setting; it happens to co-exist in a dynamic beach environment but it is not considered to participate in the cyclic or seasonal variations in form, volume, and position. More recent inquiry into the temporal and spatial dynamics of coastal dunes reports on the mobile and transgressive nature of the coastal foredune and its interaction with beach processes during its developmental history (see Psuty, 1988).

1.2. Foredune dynamics

The sequential development and location of the coastal foredune during periods of shoreline displacement is not often presented in the literature. The foredune may be incorporated in the erosional or depositional variations by inference and assumed to exist in some static form or it may be completely ignored in the assessment of the shoreline change record of the past several centuries to millennia.

Several authors have proposed that dune erosion occurs primarily during storm events when storm surge and wave action work on the dune face and remove sediment from this portion of the dune to the offshore zone (Hughes and Chui, 1981; Vellinga, 1982). The dune face is scarped in this process and the dune form is essentially shifted inland by the removal of the seaward margin of the dune form. Whereas the general thrust of this line of inquiry has been directed toward establishing wave action as a mechanism for sediment loss from the upper portion of the profile to be relocated at the position of the offshore bar, it has been approached as a two-dimensional component of change. That is, the sediment exchange is viewed as an onshore-offshore translocation of sediment within the closed-system beach-dune profile without a spatial shift of the total profile.

Research by Fisher (1984) on the various characteristics of the Rhode Island shoreline does identify the temporal onshore-offshore shift in the dune form, both its crest and toe position, for a 50-year period. He relates the dune changes to those of the beach and suggests that they are occurring at more or less the same rate of change but that the dune form has a finite capacity for change because it is losing volume and eventually will cease to exist in the profile. His data show that the dune form is migrating inland as the beach is shifting inland. Although he does not report on foredune volumes, he suggests they are diminishing through time. In his portrayals of the alongshore component of change, it is apparent that there is some periodicity in the amount of displacement as well as the direction of displacement. Thus, even though the shoreline and dune forms are generally shifting inland, there are sites and episodes when the dune crest is displaced seaward. Fisher does not comment on this situation in the article but, characteristically, privately brought the spatial variation to this author's attention (Fisher, *pers. comm.*). A similar alongshore pattern of foredune height and width change is noted in the Okracoke dunes (Stephenson, 1978). Once again there is some recognition of the scale of dune dimensional change relative to the beach change, but there is little notice of the alongshore repetition of crestline displacement. Yet the data sets provide an indication of a pattern of increasing and decreasing amounts of crestline displacement. And, similar to the Rhode Island foredune displacement, there are instances of seaward crestal movement.

Thus, there is some recognition of the offshore-onshore displacement of the coastal foredune and the non-linear alongshore distribution of the shifts. There is also a portrayal of some periodic repetition of spatial displacement in the published data sets. In a few instances there has been some attempt to track changes in both dune and beach form and to establish an interrelationship between these two components of the profile (Stephenson, 1978; Short and Hesp, 1982; Fisher, 1984). Recently, there have been attempts to show a dimensional relationship between nearshore circulation cells and the alongshore undulations of the foredune crestline (Psuty and Allen, 1987; Allen and Psuty, 1987). They point to alongshore periodicities that appear to be present in both the beach and the foredune planimetric views. Further, Psuty, Allen and Starcher (1988) have pointed to the application of harmonic analysis in dune crestline study and its potential for looking at spatial variation through time. This chapter builds on that earlier study by providing a more detailed analysis of the crestline sinuosity and by incorporating a third crestline data set. The sequence of displacement of the coastal foredune is seen to incorporate the traditional components of sediment loss and gain but also to involve a spatial association that includes a series of harmonic periodicities that describe the alongshore foredune crestline.

2. STUDY AREA

The area of investigation is the central portion of Fire Island, a barrier island located along the southern margin of Long Island, New York State (Fig. 1). Of the 26 km length, the primary focus is on the 19 km that extend from Kismet to Ocean Ridge, within the developed portion of the island. Fire Island extends in a NE-SW direction and is exposed to long-period ocean swells from the east and south. Mean tidal range is 1.3 m, reaching 1.6 m at spring tide.

The island has a multitude of jurisdictions placed on it. There are 17 communities (falling within two county townships), the Fire Island National Seashore, a county park, as well as New York State exercising some sort of natural resource management within the study area. Dune management and shoreline erosion are key issues of concern to these manifold jurisdictional levels. Over the years, the Fire Island foredune has been subjected to a variety of practices applied by the several management levels, all dedicated to dune protection. This has been interpreted to mean maintaining the spatial position of the dune form, virtually independent of the changes in the beach portion of the dune/beach profile. Of course, this goal has not been achieved

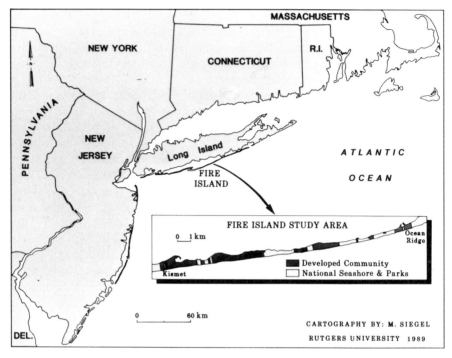

Fig. 1   Study area, Fire Island, New York. Distribution of community and Park holdings
on the island

(Fig. 2). Further, questions related to the processes of dune development, destabilization, and mobility as well as the spatial variation of dune characteristics present on the island continue to thwart attempts at dune management. There are no structured methods of monitoring the effects of the existing dune management practices and as a consequence there are no bases to evaluate proposed dune management strategies.

3. DATA COLLECTION

3.1. Background

In an effort to determine the spatial character of the coastal foredune, large-scale aerial photos, 1:2400 and 1:1200, were flown initially in 1976, and repeated in 1981 and 1986. Dune features were interpreted from each of these three stereo photo sets and registered to a common map base. Among the variables plotted and measured were foredune crestline, seaward toe of the foredune, landward toe of the foredune, and the inland extent of the dune forms. The coastal foredune crestline was considered to be the seawardmost dune ridge crest

Fig. 2   Numerous failed attempts at stabilizing the foredune crestline. Several approaches are positioned in the beach; others are in the face of the foredune and toward its crest

which was high enough to be above spring tide elevations and which was colonized by vegetation, at least on the landward slope of the ridge. Essentially, the mapped foredune was more than an ephemeral ridge created by sand accumulation adjacent to a sand fence. The foredune ridge had aspects of permanence because of elevation, location, and colonization by vegetation.

There are many variables at work along Fire Island which affect the temporal and spatial location of the crestline. Numerous cultural practices are employed to protect and stabilize the dunes in the communities as well as in the Park holdings. There are few areas that are without some attempts to control the location of the foredune. The Park places sand fences at breaches in the foredune and along dune faces which are severely scarped. The communities and the private landowners embark on a variety of schemes such as fence lines, Christmas trees, bulkheads, sandbags, planting and irrigation of dune grass, emplacement of brush heaps, and bulldozing of sand from the beach to the dune face as a means of creating sand accumulation in desired locations (Fig. 3). The construction of the dune form is treated as the prescription to remedy a manifestation of the problem rather than to address a symptom of the problem, which is a negative sediment budget within the developed portion of the island. The effect of these cultural actions is to mask the forms produced by the natural processes and to artificially reposition the dune crestline. The resulting modified dune

Fig. 3   Fencelines installed to accumulate sand on the profile line in a developed area (A) and in an undeveloped area (B). Fences are successful in trapping sand in the short term. Colonization by vegetation is necessary if the accumulation is to survive the next storm event

crestline has a more linear plan view and is apt to be sited farther seaward than the natural crestline. Thus, the mapped dune crestline will probably be skewed toward the seaward margin of the foredune and its planform will tend to be a straight line extending the length of the island. Comparisons of crestline positions in subsequent years would incorporate the natural variations produced by the processes affecting the dunes balanced against the persistence of the cultural practices attempting to hold the crestline in place. Presumably, coastal foredunes undergoing landward displacement will receive more management effort than areas of seaward displacement and thus there will be some bias toward reducing the dimensions of the landward shifts.

3.2. Dune crestlines

The dune crestline has been interpreted and recorded for three different times, at five-year intervals. Each crestline was digitized and recorded as a separate data set for subsequent comparison and analysis. These crestlines incorporate all of the natural and cultural modifications that have shaped the dune form during this period. There was no attempt to bias the crestline portrayal to emphasize either the cultural or the natural processes. Over the ten-year period, it may be expected that nearly all of the developed area and much of the Park area will have experienced some attempts at stabilization. Specifically, the efforts at stabilization should create a more uniform, linear sand-ridge in the alongshore direction. Thus, blowouts spanned by fences along their seaward margin might be mapped either as the crest of the parabolic feature or as the straight line crestal form. The distinction is whether the artificial form appears to be more than an ephemeral feature; that is, would it survive winter storms and function as the dune in the dune/beach sand-sharing system.

3.3. Profiles

Since 1981, 26 dune/beach profiles have been surveyed annually in order to record the changing crestline positions as well as to provide information of cross-sectional area variations. The 26 sites were chosen to represent particular conditions, such as dune accretion or erosion in developed or undeveloped areas. In some instances, profiles were run in stable areas as well. The result is an uneven spacing of profiles that emphasize locations of change. A comparison of the profiles reveals the influences of the cultural practices on the natural form and especially shows the contrasts among the several combinations of displacement and development. Areal changes as well as crestline position changes are portrayed through this data set.

3.4. Five-year and ten-year displacements

A comparison of the three interpreted dune crestlines yields information on the temporal displacements of these lines for the entire length of the study area in two time durations. Variations are recorded for each of the five-year intervals and for the ten-year period. Through replication tests of the accuracy of the interpretation and mapping technique, it was determined that differences less than 1.5 m were within the range of error. Thus, only variations greater than 1.5 m were recorded as displacements. Linear differences along the crestline are categorized as seaward displacement, landward displacement, or stable (less than 1.5 m change). The absolute distance of the displacement was measured for the length of the study area at 76 m intervals to create a data set of total displacement in each category as well as net displacement in a linear mode. In addition, each polygon created by the intersecting crestlines was measured to determine areal displacement by category as well as net areal displacement.

4. ANALYSIS OF FIRE ISLAND FOREDUNE CRESTLINE

4.1. Crestline shifts

The simplest measure is that of the relative positioning of the dune crestlines between two mapping exercises. This measurement reveals that despite the manifold attempts at stabilization, more than 44% of the shoreline was displaced inland more than 1.5 m during each of the five-year intervals (Fig. 4). Further, over the ten-year period, over 62% of the dune crestline shifted inland more than 1.5 m. These data suggest that both the developed as well as the undeveloped portions of the island share a similar proportion of the landward displacement.

Comparing the linear and areal dimensional changes (Fig. 5), it is observed that the five-year values are not in complete agreement on a relative scale. Specifically, for the 1976 to 1981 period, the linear and areal values are not in the same direction. However, the ten-year values are similar in both relative magnitude and direction. A review of the aerial photos shows that the large positive values of the first time period were created by the construction of fence lines across the openings of several large parabolic dune forms. The fence line was successful in creating a linear dune ridge positioned seaward of the crenulated parabolic dune and thus caused a net positive areal shift in the crestline. This situation occurred several times in the Park holdings and resulted in a large positive shift in the areal budget. The second time period did not have a similar magnitude of dune form modification and thus there were no major additions to the areal balance. Yet all time periods do record some examples of areal growth. Thus, although the dune crestline is being displaced inland along a large proportion of the study area, some portion of the dune line is being displaced seaward. This indicates that the foredune is not shifting unidirectionally but that there is some spatial variation in the displacements. Further, the accretional displacements are occurring in developed as well as in undeveloped areas. This suggests that there are natural accretional processes operative in addition to the cultural practices.

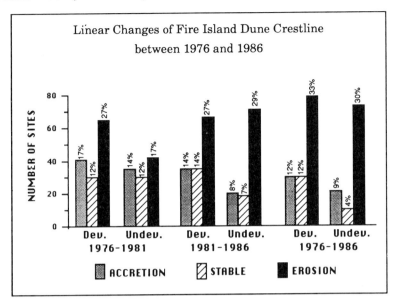

Fig. 4   Foredune crestline shifts, expressed as a percentage of the total crest length.
Changes are greater than 1.5 m per time period

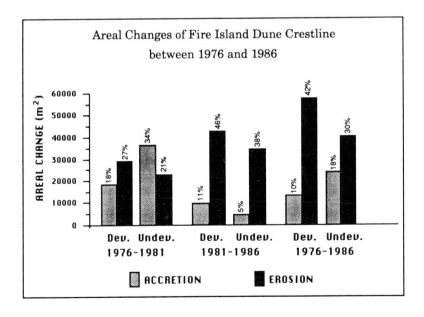

Fig. 5   Areal dimensions of foredune crestline shifts

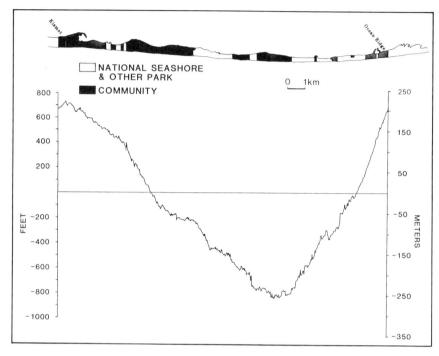

Fig. 6 Digitized 1976 foredune crestline configuration, interpreted from aerial
photography

4.2. Spatial periodicity

Representation of the digitized 1976 dune crestline includes the general
curvilinear trend of the island as well as the many other variations in the dune
crestline along the island (Fig. 6). It is apparent that the trend line is composed
of a fundamental harmonic plus a number of higher frequency harmonics that
make up the total curve. The overall curve can be decomposed by using harmonic
analysis to determine the amplitudes and periods of the sequential series of
curves. Once identified, the effect of the island's general curve can be removed
so as to isolate and emphasize the shorter period waveforms. Subtraction of
the island trend highlights the presence of a second-order waveform with a period
of about 2800 m and an amplitude of about 15 to 20 m (Fig. 7). There are many
other smaller waveforms present in the smoothed trend line and they can be
isolated using the same procedure. However, using spectral analysis, a grouping
of harmonic numbers can be determined that will identify the periodic wave
lengths in the foredune crestline. A further simplification of this approach
identifies groupings of wave lengths in terms of a weighting of the component
harmonic numbers and a smoothing of the resulting curves (Fig. 8). In this figure,

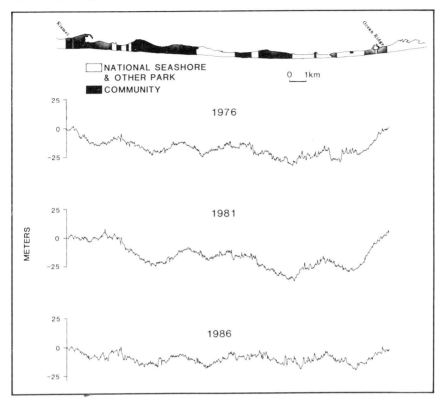

Fig. 7   Trends of the foredune crestline, minus the effect of the island's curvature, 1976,
1981, and 1986

the dominant effect of the island trend has been removed by deleting the first
group of harmonic numbers, the long wave lengths that describe the island,
and then smoothing the curve by using a seven-unit running average. There is
considerable subjectivity in the smoothing technique, because it is an attempt
to strike a balance between reducing the variation in the data set while retaining
the general sequence of peaks and valleys. The waveform heights are especially
distorted by this technique. Whereas it may prove of value to focus on the
specific dimensions of the repetitive waveforms, it is here noted that each of
the three crestlines has a series of peaks that appear to have similar wavelength
dimensions.

4.3. Alongshore shifts

The alongshore migration of the wave form in the foredune is derived by
comparing the three crestlines, essentially providing only two periods of

GROUPED   SPECTRA

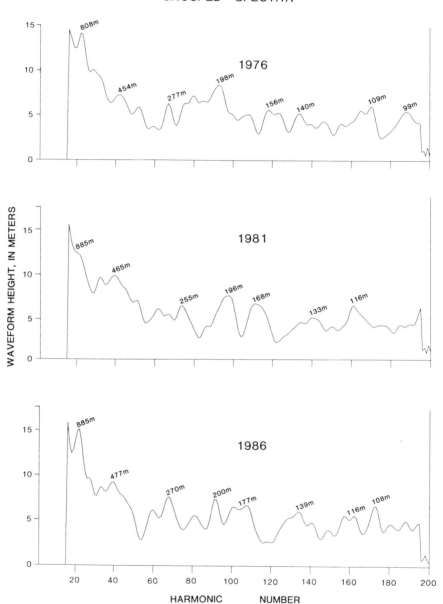

Fig. 8   Weighted groupings of harmonic numbers employing a seven-unit running average
to represent the relationship of waveform lengths and waveform heights, 1976, 1981,
and 1986. The numbers above the individual peaks are the wavelengths in meters
corresponding to the harmonic scale

comparison. However, the crestline is a fairly conservative landform, not subject to short-term oscillations. It is likely that the foredune face is scarped only under the more severe storm conditions and that, once scarped to the extent that the foredune crest is displaced, the crest will continue to occupy that position as shorter term episodes of accretion and erosion occur at the dune toe. Thus, the foredune crest responds to high-magnitude, low-frequency events that scarp the dune face and cause the foredune crestline to be displayed landward in stepwise fashion. On the other hand, a seaward displacement of the foredune crestline will occur gradually as the face of the dune is built up and out by the slow accumulation of sediment low on the dune profile. Thus, the five-year periods may be a reasonable sampling interval to record the crestline shifts due to infrequent high-magnitude events.

If the foredune crestline were being eroded persistently on the seaward margin, a temporal comparison of its position should show an inland displacement along the linear feature. If the foredune were retreating landward at differential rates, a temporal comparison would reveal a curvilinear form that might or might not have periodicity. However, each of the five-year comparisons shows an alongshore variation that incorporates both landward and seaward displacements (Fig. 9). Several observations can be made from these crestal positions. First, despite the many attempts to control the position of the foredune, there are natural mechanisms operating that are causing retreat and advance of the crestline. Certainly, human efforts could be responsible for the seaward shift because of sand fences and other manipulations that cause sand to accumulate on the seaward face of the foredune. Likewise, the eroding areas could be locations outside the developed areas. However, the displacements are not related to the distribution of the communities. It is evident that the temporal comparisons have an alongshore periodicity that transcends the Park/community dichotomy.

What is apparent from the five-year and ten-year comparisons is that there is an alongshore displacement of erosional and depositional nodes superimposed on the general landward shift of the dune crestline. This indicates that the dune crestline is sharing in the erosional and depositional episodes occurring on the beach and it is also sharing in the alongshore variation of exposure to storm surge. As noted by Dean (1976), the presence of offshore topography is important in determining exposure to storm surge and dune scarping. It is likely that the primary sinusoid that describes the curvature of Fire Island is related to the submarine topography of the barrier. The first-order seaward bulge is the product of a large offshore outwash deposit whereas the smaller features may be determined by migrating sand waves and longshore bar circulation cells as suggested by Allen and Psuty (1987).

The trend shown by the comparisons indicates that there is a displacement wave traveling alongshore which can be described as varying from low inland displacement values, to high inland displacement values, to low inland

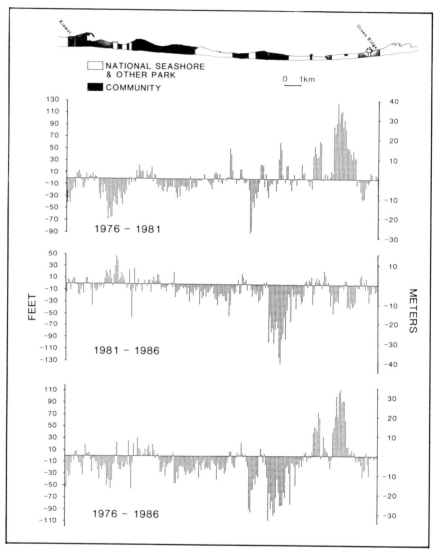

Fig. 9   Displacement of the foredune crestline, 1976 to 1981, 1981 to 1986, and 1976
to 1986, showing an alongshore periodicity of crestline shift

displacement values, and even to low seaward displacement values. Thus, each
foredune area will pass through the varying amounts of displacement as the
cycles travel alongshore. The mechanism appears to be scarping during storms
associated with breaks in the sand waves and nearshore bars. As the nearshore
systems shift, the foredune form is scarped in some proportion to the alignment

Fig. 10   Inland displacement of foredune crestline is associated with offshore topography and higher runup on the beach face (photo by J. R. Allen)

of deeper and shallow water topography. An example is shown in Fig. 10 which portrays the scarping of the dune crestline in association with a break in the bar and differential wave set up. As the break migrates downdrift the scarping of the dune face will follow. But as the degree of exposure waxes and wanes due to the changing nature of the nearshore bar, the degree of dune scarping will likewise vary. The passage of one of the scarping episodes will cause the dune crestline to be displaced in a rhythmic sequence, creating an arcuate dune crestline that will be retained until the next episode.

4.4. Dune profiles

Annual profiles document the magnitude of the cross-sectional change as the dune crest is displaced inland or landward (Fig. 11). They show the continuing inland displacement of the dune form as it loses volume by scarping along its seaward margin but gains volume on its leeside by eolian transport over the seaward crest. Scarped dune faces are invariably accompanied by a layer of freshly deposited sand at the crest and to the lee of the crest (Fig. 12). In part, this scarping process resulted in the inland displacement of the crestline. But more importantly, the free dune face provides a transport surface for eolian processes to move sediment up the unobstructed seaward face to the dune crest and beyond. This action indicates that the dune form can maintain its sand

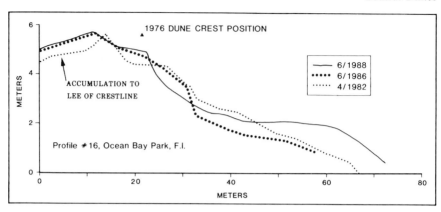

Fig. 11   Beach/dune profile data portraying: (i) location of 1976 interpreted dune crestline in horizontal relationship to survey datum; and (ii) 1982, 1986, and 1988 surveys showing positive and negative variation low on the profile whereas the dune crest has been shifting landward and accumulating sediment. Profile Line No. 16, Ocean Bay Park, Fire Island

Fig. 12   Accumulations of sediment at and to the lee of the foredune crestline associated with episode of scarping of the foredune face

volume as the beach is being eroded. Or, that the dune form has a sediment budget which is a distinct subset of the total dune/beach budget. Under this condition, the dune could conceivably increase, decrease, or remain constant in mass as the shoreline shifts inland. Presumably the difference between these outcomes is the balance of sediment supplied to the dune by wave and eolian processes versus the amount removed by scarping.

5. CONCLUSION

During the ten years of investigation, the dune crestline in the central portion of Fire Island has been undergoing a variety of natural changes intermixed with cultural manipulations of the morphology. Whereas the crestline has generally shifted inland in an erosional mode, there is spatial variation in the magnitude of the shift. Despite the many attempts at stabilization, the dune crestline is passing through an alongshore migrating wave of greater and lesser in-land displacement, sometimes accompanied by landward displacement. The responsible processes are related to the effects of offshore topography affecting set up and storm surge which scarps the dune face in waxing and waning sequences. The dune form is capable of retaining its dimensions as it is displaced inland by transfers of sediment from the seaward to the landward face. The foredune sediment budget can be positive, negative, or in equilibrium even while the total dune/beach profile is negative and the shoreline is eroding. The presence of coastal dunes may indeed be part of a negative shoreline budget scenario. Certainly, many eroding shorelines continue to retain foredunes in their profiles even after centuries of documented landward displacement.

REFERENCES

Allen, J. R. and Psuty, N. P. (1987) Morphodynamics of a single-barred beach with a rip channel, Fire Island, NY. *Coastal Sediments '87. ASCE* 1964–1975.

Bird, E. C. F. and Jones, D. J. B. (1988) The origin of foredunes on the coast of Victoria, Australia. *J. Coast. Res.*, 6, 181–192.

Carter, R. W. G. (1988) *Coastal Environments.* Academic Press, London, 617 pp.

Dean, R. G. (1976) Beach erosion: cause, processes and remedial measures. *CRC Critical Reviews in Environmental Control*, 6, 259–296.

Fisher, J. J. (1984) Regional long-term and localized short-term coastal environmental geomorphology inventories. In Costa, J. E. and Fleisher, P. J. (eds), *Development and Applications of Geomorphology.* Springer-Verlag, Berlin, pp. 68–96.

Godfrey, P. J., Leatherman, S. P. and Zaremba, R. (1979) A geobotanical approach to classification of barrier island systems. In Leatherman, S. P. (ed.), *Barrier Islands.* Academic Press, New York, pp. 99–126.

Hughes, S. and Chui, T.-Y. (1981) *Beach and Dune Erosion During Severe Storms.* Univ. Florida, Dept. Coastal and Oceanographic Engineering Report UFL/COEL-TR/043, 290 pp.

Orme, A. R. (1988) Coastal dunes, changing sea level, and sediment budgets. In Psuty, N. P. (ed.), *Dune/Beach Interaction. J. Coast. Res., Special Issue No. 3*, 127–129.

Psuty, N. P. (1986) Principles of dune-beach interaction related to coastal management. *Thalassas*, **4**, 11–15.

Psuty, N. P. (ed.) (1988) *Dune/Beach Interaction, J. Coast. Res., Special Issue No. 3*, Charlottesville, Virginia, 136 pp.

Psuty, N. P. and Allen, J. R. (1987) Analysis of dune crestline changes, Fire Island, New York, U.S.A. In Gardiner, V. (ed.), *International Geomorphology 1986*. Part 1. Wiley, Chichester, pp. 1169–1183.

Psuty, N. P., Allen, J. R. and Starcher, R. (1988) Spatial analysis of dune crest mobility, Fire Island National Seashore, New York. In Psuty, N. P. (ed.), *Dune/Beach Interaction. J. Coast. Res., Special Issue No. 3*, pp. 115–120.

Pye, K. (1983) Coastal dunes. *Prog. Phys. Geogr.*, **7**, 531–557.

Short, A. D. and Hesp, P. A. (1982) Wave, beach, and dune interactions in southeast Australia. *Mar. Geol.* **48**, 259–284.

Stephenson, R. A. (1978) Coastal changes on Okracoke Island, North Carolina. *Proc, Special Session on Marine Geography*, Annual Meeting, Association of American Geographers, Psuty, N. P. (ed.), 39–43.

Vellinga, P. (1982) Beach and dune erosion during storm surges. *Coast. Eng.*, **6**, 361–387.

Chapter Nine

Seasonal patterns and controls on sediment supply to coastal foredunes, Long Point, Lake Erie

ROBIN G. D. DAVIDSON-ARNOTT AND MARK N. LAW
Department of Geography, University of Guelph

1. INTRODUCTION

Coastal sand dunes characteristically develop landward of most sandy beaches as a result of aeolian sediment transport by onshore winds. Generally most of the sediment transported landward from the backshore is trapped initially by vegetation colonizing the area just beyond the limit of wave action, leading to the development of an incipient foredune parallel to the shoreline; exceptions to this occur where vegetation is sparse because of limited moisture, or where sediment supply is so large as to prevent the establishment of vegetation. With continued sediment supply, the foredune grows in height and width and, on a progradational shoreline, a sequence of transverse dunes may form. Sediment supply to the dune represents a loss to the beach so that over the longer term (years or tens of years) foredune growth necessitates a positive beach sediment budget; otherwise a negative feedback cycle is initiated through narrowing of the beach and erosion of the dune by wave action (Psuty, 1988). The controls on dune growth on this time scale are thus complex, and require a knowledge of sediment supply to the coastal reach and sea-level trends, in addition to those factors related directly to aeolian sediment transport.

Over the short term (weeks or months) the rate of foredune growth is dependent primarily on the volume of sand transport from the beach. This in turn is controlled by the wind climate and by factors such as sediment size and mineralogy, precipitation and moisture content, salt crusting and beach width which influence the threshold of sediment motion and sand transport rate. Following the work of Bagnold (1941), laboratory and field studies have greatly improved our understanding of, and ability to predict the instantaneous

Coastal Dunes: Form and Process. Edited by K. F. Nordstrom, N. P. Psuty and R. W. G. Carter
©1990 John Wiley & Sons Ltd

threshold of sand motion and, to a lesser extent, the volume rate of sand transport by wind action (e.g. Horikawa *et al.*, 1984; Sarre, 1987). As a result of this work, a number of empirical and semi-theoretical deterministic equations have been formulated to predict instantaneous sediment transport rates. However, because of the spatial and temporal variability of factors such as beach form, sediment size and sorting, and moisture content on natural beaches, it is not possible to utilize these directly to predict sediment supply to the foredunes over periods of weeks or months.

   Several studies have attempted to develop schemes for predicting rates of sand supply based on standard climatic data such as hourly wind speed and direction, and precipitation (e.g. Fryberger and Dean, 1979; Pye, 1985) but these have generally been regional in scale and limited by the relatively poor quality of the measurements of sand transport rates or volumes of sediment deposition. The purpose of this study was to obtain accurate measurements of sediment supply from the beach to the foredune over a period of several months, and to relate these to relatively simple measurements of several key controls, including wind speed and direction, water level and beach width, for which data are readily available on the temporal and spatial scale required. The ultimate goal is to develop a model that can be used widely to predict sediment supply to coastal dunes. The results presented here are more limited in scope and are designed to demonstrate that this approach is feasible. They are based on a field experiment carried out from May to early December 1987 at two sites on a large barrier spit.

2. APPROACH

The underlying premise of the study was that a model to predict sediment supply for foredune growth on a scale of months or years should be based on a few key controlling variables and utilize data that are either readily available or can be measured simply and quickly. The key variables selected were wind speed and direction, sediment size, and mean beach width. Hourly measurements of wind speed and direction are widely available from meteorological stations; sediment size and mean beach width can be measured relatively simply in the field.

   The accurate measurement of sediment transport from the beach was essential to the study, and this poses much greater difficulties. There are a number of problems with utilizing sediment traps, including the difficulty of establishing their efficiency, and their tendency to clog when the sand is wet. Traps require frequent maintenance to clean and empty them and this poses logistical problems in a study extending over a period of months. Finally, because traps sample only a small cross-sectional area, elimination of the effects of spatial variations in beach width, slope and sediment characteristics over a finite length of beach would necessitate the deployment of a large number of traps.

   Therefore, it was decided to take an alternative approach and to measure sediment transport from the beach indirectly, by determining the volume of deposition in the incipient foredune area. The basic assumption here is that, provided the vegetation is sufficiently high and dense, all sediment transported from the beach will be deposited within a few tens of metres of the beginning of the vegetation line. Accurate measurements of changes in the bed elevation in this area over a sufficient number of points will then account for the natural variability associated with the pattern of vegetation height and density. Recently Sarre (1989) has described a similar approach in a study of sediment transport to foredunes on the southwest coast of England.

A second objective of the study was to determine the effect of beach width on the volume of sediment supply to the dune area. While the effects of sediment size, shape and mineralogy, moisture content and salts on the volume of sediment transport by aeolian action have received considerable attention in the last several decades, limited attention has been paid to determining the minimum length of surface over which the wind must blow in order to achieve full sediment transport for a given wind speed. In many coastal dune fields the width of the beach supplying sand may be only a few metres, particularly because periods of strong onshore winds are usually accompanied by storm surge and high wave action. If the minimum length of surface required to achieve full sediment transport is appreciably greater than the beach width available, this will be an important conrol on the actual amount of sediment supplied. This problem was addressed in two ways: (i) in the field experiment two sites were monitored, one opposite a narrow beach and the other opposite a wide beach; (ii) a series of field measurements using traps were carried out to measure the relative change in sediment transport rates with distance onshore from the swash limit.

3. STUDY AREA

The study was carried out on Long Point, a 40 km long barrier spit on the north shore of Lake Erie (Fig. 1). The spit represents the sink for a littoral drift cell that extends some 85 km to the west, with sediment being supplied by erosion of bluffs developed in glacial and glacio-fluvial sediments. The total volume of sediment supply is large, and the sand and gravel component has been calculated to be 1.0×10^6 m^3 a^{-1} (Rukavina and Zeman, 1987). The sediment budget over the 10 to 12 km long distal end of the spit is positive and a progradational sequence of dune ridges and intervening dune slacks over 3 km wide has developed.

Two study sites about 1 km apart were established at the distal end of the spit (Fig. 1). This area is characterized by a main foredune ridge aligned parallel to the shoreline, which ranges in height from 2 to 8 m, and by development of a well-vegetated incipient foredune zone between it and the top of the beach. The area is managed as a nature reserve by the Canadian Wildlife Service, and public access is severely restricted, thus eliminating problems due to human interference. However, access to the sites necessitates a 20 km trip by boat.

180

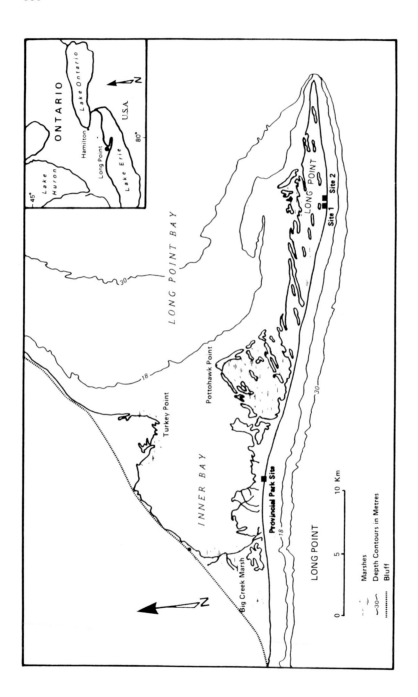

Fig. 1   Location map of Long Point and study sites

The dominant colonizing plants on the incipient foredune and the windward side of the main foredune at the two study sites are American beach grass (*Ammophila breviligulata*) and switch grass (*Panicum virgatum*). The growth form of *Ammophila* on Great Lakes beaches has been described in detail by Maun (1985). American cottonwood (*Populus deltoides*) which is also an important colonizer in this area, was rare within the two study sites.

Beach width varies both temporarily and spatially with changes in lake level and sediment abundance (Davidson-Arnott, 1988). In addition to storm surges, which may exceed 2 m, there are annual changes in mean monthly lake level of 0.3 to 0.6 m resulting from seasonal variations in basin runoff and evaporation, and long-term changes of 1.0 to 1.5 m related to climate cycles typically of 10 to 20 year duration. The annual change is characterized by a peak in June or early July, with low levels in the fall and winter months; although in some years a secondary peak occurs in November and early December. Beach width also varies seasonally because of onshore-offshore sediment transport associated with landward migration and welding to the beach of inner bars during the summer months, and erosion and flattening of the beach by the more frequent and intense storms in the fall and spring (Stewart and Davidson-Arnott, 1988). In addition, beach width varies periodically alongshore on two scales: (i) variations in width of 10 to 20 m, on a length scale of 100 to 150 m, associated with giant cusps, rhythmic inner bar topography and rip cell development; (ii) variations in width of 50 to 90 m, on a length scale of 600 to 3000 m, associated with longshore sandwaves (Stewart and Davidson-Arnott, 1988). The sandwaves migrate eastward in the direction of net sediment transport at rates of 150 to 300 m a^{-1}, producing a cycle of erosion associated with the embayment, and protection and foredune accretion opposite the sandwave itself (Davidson-Arnott and Stewart, 1987).

3.1. Site characteristics

Two study sites were established about 1 km apart on a sandwave near the distal end of the spit (Fig. 2A), with the western site (Site 1) located at the narrow updrift end of the sandwave and the eastern site (Site 2) located at the wide downdrift end. At Site 1 the beach was initially about 15 m wide with an inner nearshore bar located about 25 m offshore (Fig. 2B, 3A). During the summer this bar migrated landward and welded to the beach (Fig. 3A) and this, combined with the seasonal drop in lake level, resulted in an increase in the beach width in the late summer and fall to nearly 40 m. At Site 2 the beach itself was over 40 m wide with a prominent berm, but the backshore was characterized by a shallow depression resulting from the incomplete infilling of a trough during downdrift migration of the sandwave (Fig. 2C, 3B). This trough remained damp over most of the summer, trapping sediment blowing inland from the berm and reducing the effective width of beach supplying sand to the dune to less than 20 m. However, by mid-August a combination of infilling by blowing sand and the drop in lake levels increased the effective beach width to nearly 50 m (Fig. 3B).

Fig. 2   A: Oblique aerial photograph of study sites. B: Photograph of Site 1 looking west.

C

Fig. 2 (*cont.*)   C: Site 2 looking east; the steel rods are corner posts of the bedframe stations

3.2. Winds

Both the prevailing and the dominant winds are from the southwest and west at an oblique angle to the beach azimuth of 200°. The strongest winds are associated with the passage of depressions, and these are more frequent and intense in the spring and fall than in the summer. However, moderate onshore winds in the summer occur frequently due to diurnal heating and cooling. Wave action is curtailed for three to four months each year, beginning in mid-December, by the growth of an ice foot and the development of a nearly continuous ice cover over Lake Erie. Aeolian transport on the beach and dunes is also restricted at this time by snow cover and by the freezing of the sand.

Winds for the period of study were weaker than the long-term average, but the pattern was similar. Winds exceeding $19 \, km \, hr^{-1}$ (taken as the threshold for sand movement) are summarized for the summer period June to August, and the fall period September to November (Fig. 4). Winds with an onshore component include SE through to W, but the SE direction is insignificant in both periods. In the summer months there were 245 hours of winds from the S, SW and W exceeding $19 \, km \, hr^{-1}$, but fewer than 10 hours with speeds above $39 \, km \, hr^{-1}$. In the fall (Fig. 4B) there was an increase in both the frequency and the magnitude of winds from these three directions, with 720 hours of winds exceeding $19 \, km \, hr^{-1}$ and 148 hours of winds exceeding $39 \, km \, hr^{-1}$. The fall months also show a much greater increase in the frequency of winds from the south and west.

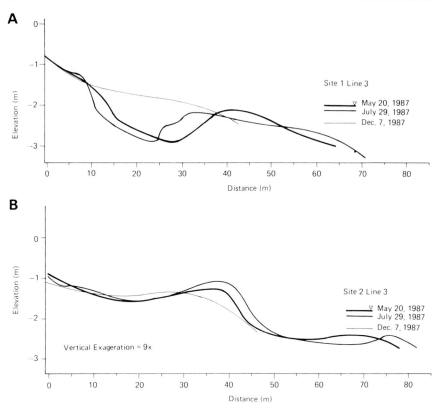

Fig. 3   Profiles across the beach and inner nearshore. A: Site 1. B: Site 2

4. METHODOLOGY

Hourly mean wind speed and direction at a height of 10 m were obtained from a standard meteorological station located at the lighthouse at the tip of the spit some 10 km east of the study sites. Wind speed was measured in kilometres per hour and wind direction in degrees from true north. Sediment size was obtained from settling velocity analysis of samples taken in the field. Beach width was measured directly by profiling in the field, and by extrapolation from water level gauges maintained by the Canadian Hydrographic Service at two harbours nearby.

Measurement of sediment deposition was accomplished using a bedframe device (Fig. 5), consisting of a 2 m by 2 m square angle iron frame. Thirteen sections of pipe, 20 cm long and with an internal diameter of 1 cm, are mounted on the frame in a regular pattern so as to provide uniform coverage of the area within it. Each pipe forms a vertical sleeve for 1 m long aluminium rods, which have a 4 cm disk on the bottom to prevent penetration into the sand. At each

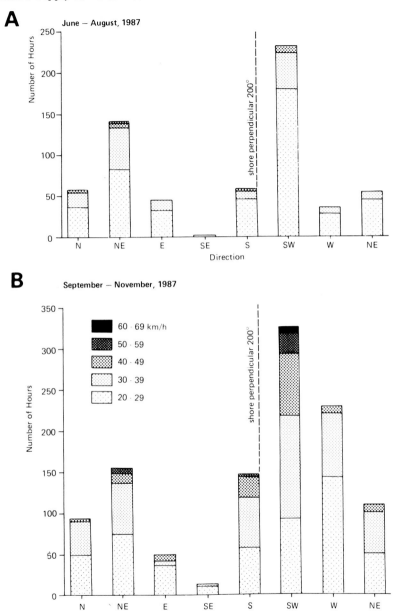

Fig. 4 Summary of wind speed and direction by octants for winds exceeding 19 km hr^{-1} (the threshold of motion for fine sand). Data are summarized from hourly wind measurements made at a standard meteorological station maintained by the Atmospheric Environment Service of Transport Canada at the lighthouse 10 km east. A: Summary for June, July and August. B: Summary for September, October and November

Fig. 5   Photograph of bedframe

measurement station four steel rods (0.8 cm in diameter and 1.6 m long), marking the corners of the station, are driven into the sand for approximately half their length and their tops levelled. In order to make measurements, the frame is placed over the measurement station so that the four corner posts engage in corresponding sleeves on the frame. The aluminium rods are allowed to drop to the bed and the distance from the top of the disk to the frame measured for each of the thirteen rods. Changes in the bed elevation through time are determined by subtracting the distance to the bed at each rod from the distance measured previously. The frame is moved from one station to the next to obtain the measurements, and is stored away from the site, so that the stations themselves are marked only by the four corner posts.

   The establishment of a large number of stations within the incipient foredune and foredune area on a section of beach ensured sufficient accuracy in the estimation of net sediment deposition in this zone. At each study site a total of 36 bedframe stations were established along five lines perpendicular to the beach and spaced 10 m apart (Fig. 6). The first station of each line was established just lakeward of the vegetation line in early May, with five more stations spaced at 6 m intervals within the incipient foredune zone. On two of the lines, three additional stations were set up; one on the windward slope of the main foredune ridge, the second on the leeward slope, and the third in the dune slack. The purpose of these stations was to verify the assumption that

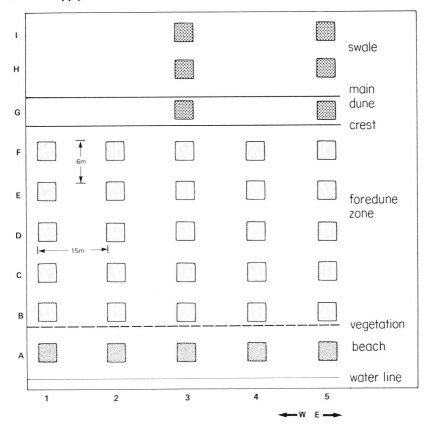

Fig. 6   Layout of bedframes at study sites

all sediment blown inland from the beach was trapped within the incipient foredune zone. During the study all except one of these stations recorded less than 3.0 cm of accretion, and the accretion at that station could be traced to sand derived from a small blowout near the ridge crest landward of the primary measurement area.

Measurements of sediment accumulation were made at roughly two-week intervals during the summer, but difficulties of access restricted these to monthly during the fall. During each measurement, vegetation height and density within each station were estimated and at least two profiles at each site were surveyed from the water line across the main foredune ridge.

As part of the effort to determine the effects of beach width on sediment supply, field experiments to measure instantaneous sediment transport were carried out at a site within the Provincial Park (Fig. 1). Sediment transport was measured with traps modified after a design by Leatherman (1978) and wind speed by an array of three anemometers mounted at heights of 0.5 m, 1.0 m,

Table 1   Net change in sand volume (m³ per metre beach width) for each of five lines at each study site for the 11 measurements made between 7 June and 7 December. (Mean changes for each interval and for each line are also shown. The data in Fig. 8 were obtained by grouping the bi-weekly measurements to produce seven roughly equal monthly periods.)

Site 1 Line	1	2	3	4	5	Mean
Measurement period						
05/05–20/05	0.114	0.246	0.384	0.414	0.366	0.305
20/05–06/04	0.042	0.036	0.060	0.018	0.084	0.048
06/04–06/16	0.096	0.120	0.090	0.138	0.150	0.119
06/16–07/01	0.084	0.204	0.210	0.102	0.216	0.163
07/01–07/15	0.144	0.048	− 0.006	0.078	0.162	0.085
07/15–07/28	− 0.198	− 0.174	− 0.186	− 0.054	− 0.120	− 0.146
07/28–08/25	0.156	0.228	0.258	0.216	0.324	0.236
08/25–09/17	0.066	0.114	0.060	0.180	0.048	0.094
09/17–10/01	0.282	0.252	0.234	0.126	0.114	0.202
10/01–11/02	0.792	− 0.018	0.312	0.342	0.336	0.353
11/02–12/07	0.558	1.050	0.924	1.056	1.152	0.948
Total	2.136	2.106	2.340	2.616	2.832	2.406
Site 2 Line	1	2	3	4	5	Mean
Measurement period						
05/05–20/05	0.366	0.330	0.294	0.330	0.210	0.306
20/05–06/04	− 0.066	− 0.078	− 0.090	0.060	0.108	− 0.013
06/04–06/16	0.306	0.228	0.102	− 0.018	0.024	0.128
06/16–07/01	0.102	0.234	0.210	0.276	0.168	0.198
07/01–07/15	− 0.108	− 0.042	− 0.030	0.024	− 0.096	− 0.050
07/15–07/28	0.114	0.006	0.006	0.156	− 0.036	0.049
07/28–08/25	0.240	0.228	0.270	0.336	0.222	0.259
08/25–09/17	0.288	0.162	0.324	0.324	0.216	0.263
09/17–10/01	0.720	0.474	0.408	0.270	0.288	0.432
10/01–11/02	1.710	2.022	2.400	2.082	2.238	2.090
11/02–12/07	2.928	2.154	3.318	2.628	1.344	2.474
Total	6.600	5.718	7.212	6.468	4.686	6.137

and 2.0 m. Experiments were carried out on three days when wind speeds exceeded the threshold for sand movement, with a number of runs being carried out on each day. In each case, several days of dry weather prior to the experiment had produced a thick (> 0.1 m) layer of dry sand on the exposed beach. During the experiments, nine traps were emplaced in the beach at 5.0 m intervals along a line parallel to the wind direction, with the first trap located at the boundary between wet and dry sand marking the swash limit. At the start of each run, covers over the traps were removed and the traps allowed to collect sediment for 3 to 10 minutes, depending on the wind speed. The covers were then replaced in the same order that they were removed, and the sediment collected in each trap transferred to sample bags for weighing later.

5. RESULTS

5.1. Sediment supply to the foredune

Data for all the measurement periods for the five lines at each site are summarized in Table 1. Each line consists of five bedframe stations within the foredune zone (see Fig. 4), except for lines 1 and 2 at Site 1 where station B, closest to the beach within the foredune zone, appeared to have been influenced by wave action during several storms. The data thus reflect measurements at 299 points for Site 1 and 325 points for Site 2. The average change in elevation within each station between measurement dates is converted to a volume, and the change over the whole line expressed as a volume change per metre width of beach. In general over the whole study period there was very little scatter between the lines at each site. At Site 1 the mean for the whole study period was 2.406 m^3 with a range from 2.106 to 2.832 m^3, while that for Site 2 was 6.137 m^3 with a range of 4.686 to 7.212 m^3. At each site the sign and magnitude of change from one period to the next was also very consistent. Thus, measurements at Site 1 on 20 May 1987 showed moderate deposition on each line, measurements on 28 July 1987 showed small losses along each line, and measurements on 7 December 1987 showed large positive values for each line.

Changes along Line 3 at each site over three time periods are shown in Fig. 7. Station A is located on the beach just lakeward of the vegetation line and at both sites there was erosion at this location during the study. At Site 1 most of this erosion occurred during the fall and resulted from a combination of storm wave action and deflation. At Site 2 there was nearly continuous deflation over the summer and fall months, but these losses were compensated by accretion in the late fall. It can be seen that at both sites there were only small changes in the foredune zone over the summer period, with most deposition occurring close to the beach. At Site 1, by the end of the primary study period on 7 December, there was considerable deposition at stations B, C and D, but little change at stations E and F. Accretion at G, located just landward of the crest of the foredune, could be traced to a small blowout on the lakeward side of the crest. After 7 December, most deposition occurred at stations E and F where the vegetation had not yet been buried, and some sand transport occurred right over the foredune. The pattern at Site 2 was similar, but there was greater overall deposition in the incipient foredune zone and the main zone of deposition extended 10 m farther inland to station E. The volumes of sand transported and depth of deposition in the incipient foredune and foredune zones are similar to those measured along single lines at these two sites the previous year (Davidson-Arnott, 1988). These data show that the assumption that all the sediment transported from the beach was trapped within the incipient foredune zone was valid until the end of the main study period on 7 December, but that beyond this date it probably became increasingly invalid as the vegetation died back or was buried.

In order to compare changes over roughly similar time periods the data have

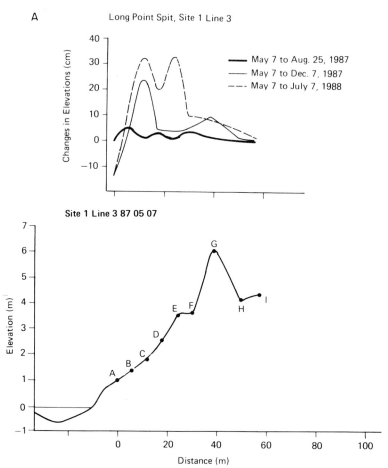

Fig. 7   Changes in mean elevation at bedframe stations along one line over three time
periods in 1987–88. A: Line 3, Site 1. B: Line 3, Site 2

been grouped into monthly intervals as indicated on Table 1, and the data
presented in Fig. 8 are based on these exact intervals. Volume changes at each
site for these monthly periods are summarized in Fig. 8A. It can be seen that
at Site 1 there was moderate accretion of between 0.25 and 0.35 m^3 each month
from May to the end of October, except during July when there was slight
erosion. Only in November does deposition increase substantially to 0.95 m^3.
At Site 2 the magnitude and pattern of deposition is similar through the end
of August. In September transport at this site increased to nearly 0.7 m^3, and
in October and November it exceeded 2.0 $m^3\,m^{-1}$ beach width.

   The pattern of much higher transport in the fall than in the summer reflects
the much greater occurrence of strong onshore winds in the fall (Fig. 4). In

B

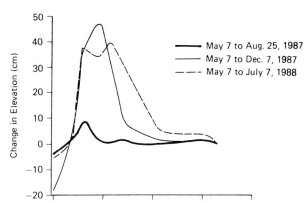

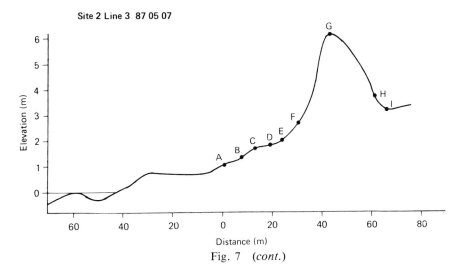

Fig. 7 *(cont.)*

order to bring out the relationship more clearly, the onshore component of wind velocity was calculated for each of the periods corresponding to the measurement of sediment deposition using the equation:

$$U_{tot} = \Sigma \cos\alpha_i (U_i - U_t)^3 \tag{1}$$

where U_{tot} = onshore component of wind velocity for the period
$\quad\alpha_i\quad$ = angle of wind to shore normal
$\quad U_i\quad$ = hourly average wind speed at time i
$\quad U_t\quad$ = threshold wind speed (20 km hr^{-1})
$\quad n\quad$ = total number of hours in period of measurement.

Most research on sand transport by wind (e.g. Bagnold, 1941; Kawamura, 1951;

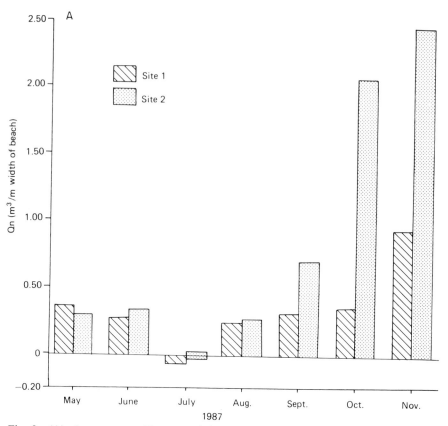

Fig. 8   (A): Average monthly rates of deposition at the two study sites. The data are
   derived from Table 1 and represent the actual grouping of dates shown there

Lettau and Lettau, 1977) suggests that the transport rate is some function of
the cube of a velocity term. A threshold velocity is subtracted in many equations
so that predicted transport is zero when the wind speed drops below this. The
cos α term is included because as the wind becomes more oblique, the sand being
transported through a unit width perpendicular to the wind direction is spread
over an increasing width of dune. Ultimately, when the wind is blowing parallel
to the shoreline none of the sand in transport along the beach is deposited in
the dune. The calculated values for the onshore component of wind velocity
are plotted for the monthly periods in Fig. 8B. The values are low for the months
of May, June and July, increase steadily through August and September, and are
relatively high in October and November. These trends thus reflect the increase
in both frequency and magnitude of onshore winds during the fall period.
   The other factor thought to influence sediment transport into the dune was
beach width. The average beach width at each site for the periods corresponding

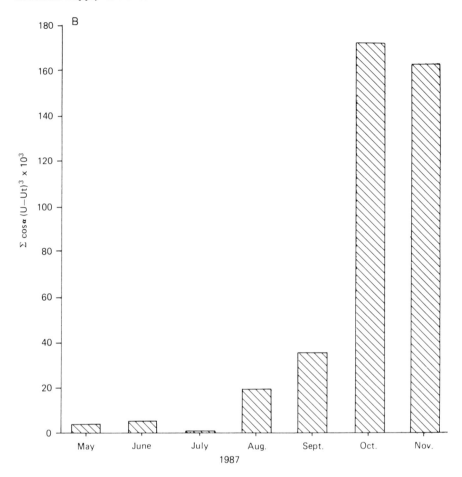

Fig. 8 (*cont.*)   B: Summary of the onshore component of wind velocity for the monthly periods used in Fig. 8A, derived from hourly wind data using equation (1)

to those for measurements of deposition and onshore winds was determined from surveys at the two sites and from daily water level records (Fig. 8C). The steady increase in beach width at Site 1 reflects both beach progradation and the seasonal drop in lake levels from mid-summer (Fig. 3A). The changes in beach width at Site 2 from May to August reflect the gradual infilling and drying up of the remnant runnel in front of the site, which initially limited the effective beach width supplying sand to the incipient foredune area. Once the runnel was infilled, sediment could be supplied from the full beach width; hence the large increase in sediment deposition in September.

A comparison of Fig. 8A with 8B and 8C suggests that the measured volumes of sediment deposition are correlated with the onshore component of wind

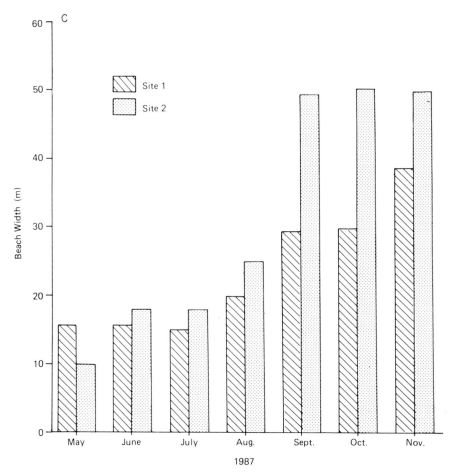

Fig. 8 (*cont.*)   C: Variations in mean monthly beach width at each site

velocity determined by equation (1), and with changes in beach width. This was tested for each variable separately using simple linear regression as well as with a mutiple regression combining the two variables (Table 2). The correlation coefficients with the onshore component of wind velocity as the independent variable were significant at the 0.01% confidence level for both sites separately, and for the two sites combined. However, the correlation is lower for Site 1 than for Site 2, suggesting that some other variables, such as beach width, may be important. An examination of the regression plots (Fig. 9) shows that much of the error in the regression for Site 1 can be attributed to one data point— that for the month of October when values for deposition were much lower than predicted. It can also be seen that the slope of the regression line is lower for

Table 2 Results of correlation analysis of sediment accumulation versus the onshore component of wind (determined from equation 1) and mean beach width

Dependent variable	Independent variable	No. of observations	R^2	d.f.	F ratio (0.01)	F critical
Site 1						
Qn	U onshore	11	0.80	1.9	16.12[1]	10.6
Qn	Beach width	11	0.85	1.9	22.5[1]	10.6
U onshore	Beach width	11	0.912	1.9	44.5[1]	10.6
Qn	U onshore, Beach width	11	0.72	2.8	10.3[1]	8.65
Site 2						
Qn	U onshore	11	0.99	1.9	419.0[1]	10.6
Qn	Beach width	11	0.68	1.9	7.77	10.6
U onshore	Beach width	11	0.69	1.9	8.05	10.6
Qn	U onshore, Beach	11	0.98	2.8	186.0[1]	8.65
Sites 1 and 2						
Qn	U onshore	22	0.90	1.20	36.6[1]	8.1

[1]Significant at 0.01 confidence level.

Site 1 than for Site 2, again suggesting that some other variable is acting to restrict the rate of sediment transport from the beach with increasing wind speed.

Sediment deposition in the embryo dune was also tested against beach width. Correlation of net deposition with beach width was significant at the 0.01% level for Site 1 but not for Site 2. This suggests that beach width was a limiting factor at Site 1. However, the pattern of beach width there reflected closely the lake level fluctuations and onshore-offshore sediment movement, which have a seasonal pattern that is very similar to that of onshore wind speeds. Indeed, as can be seen in Table 2, there is a significant correlation between the two. Thus, it is difficult to determine from these data whether the correlation with beach width has a true physical meaning, or whether it merely reflects the correlation of beach width with the seasonal wind pattern. Some insight into this can be obtained from the results of experiments designed to measure the effects of beach width on the instantaneous sediment transport rates.

5.2. Effect of beach width on instantaneous sediment transport rate

Field experiments to determine the effects of beach width on instantaneous rates of sediment transport were carried out on three days. In order to facilitate comparison with data in the preceding section, the wind speed at 10 m height has been derived from the measured velocity profile near the bed. Sediment transport has been standardized for the width of the trap opening and is expressed as a flux per metre width of beach. No attempt was made to determine the efficiency of the traps, so the absolute values should be treated with caution. It was assumed that the traps had the same relative efficiency, thus differences in the amounts collected in each trap represent real differences in the amount of sediment being transported.

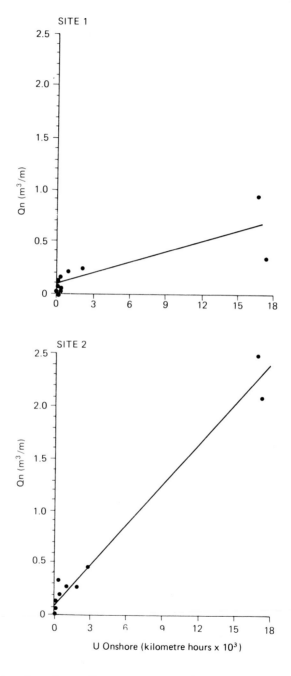

Fig. 9   Regression plots of net sediment supply versus onshore component of wind speed and beach width

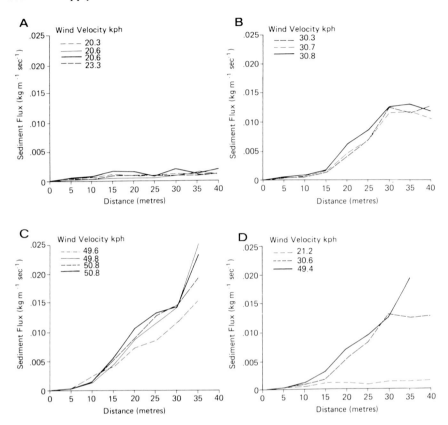

Fig. 10 Variations in sediment transport rate with distance from the top of the swash: A: wind speed 20 km hr^{-1}. B: wind speed 30 km hr^{-1}. C: wind speed 50 km hr^{-1}. D: average values for three wind speeds

Examples of the variation in sediment collected with distance from the top of the swash zone for each of the three experiments are shown in Figs 10A, B, and C. During the first experiment wind speeds were just above the threshold for motion; during the second experiment they were just over 30 km hr^{-1}; and during the third experiment they were about 50 km hr^{-1}. A total of 4 runs was conducted during the first experiment, 6 during the second and 8 during the third. Average values for each of the days are presented in Fig. 10D.

With winds just above the threshold of motion sediment transport increases slowly away from the top of the swash and appears to reach a maximum at a distance of 15 m. With wind speeds of 30 km hr^{-1} sediment transport increases slowly to 15 m then increases rapidly, but appears to reach a maximum at 30 m. The sediment transport rate at 30 m is roughly 8 times that at 15 m. In the third experiment, with winds of about 50 km hr^{-1}, the maximum width of dry beach available was only 35 m (due to storm surge) and it appears that maximum sediment transport had not yet been achieved. However, the sediment transport rate at 35 m was approximately 12 times that at 15 m.

The curves for each wind speed appear to be similar up to the point at which maximum sediment transport is achieved (Fig. 10D). This suggests that prediction of the sediment transport for a given beach width and wind speed may be relatively simple if the maximum value for that wind speed is known, or can be predicted. The data presented here are too sparse to permit that, and further data collection will be needed to establish the maximum rate of transport for typical beach situations. The data indicate that the distance required to initiate full sediment transport on a beach is important, particularly for higher wind speeds. Thus, beach width is an important control on sediment supply to the foredune.

6. DISCUSSION

The technique used to measure sediment deposition in the incipient foredune zone appears to give a good measure of sediment transport from the beach over the time period considered. However, by the beginning of December the effectiveness of the vegetation in trapping sediment within this zone diminished rapidly because of the death of the plants, loss of foliage through breakage, and because of burial by sand. Unless the ground is frozen or snow-covered, substantial quantities of sand can be transported through the incipient foredune zone onto and over the main foredune ridge. In addition, the marked reduction in vegetation cover during the winter and early spring months permits transport of sand back onto the beach and ice foot by offshore winds. Thus, the final measurements made in June 1988 (Table 1) show net deposition in the incipient foredune area over the period of a year, but they are no longer a good measure of net transport from the beach.

The good agreement between the magnitudes of deposition recorded at the two sites in the spring and summer and high correlation with the onshore component of wind velocity may in part be attributed to the similarity of beach widths during this period and the generally low wind speeds. Likewise, the difference in magnitude of deposition at the two sites in the fall may be attributed in part to the greater beach width at Site 2. However, the differences in mean beach width appear to be too small to account for the magnitude of the difference in sediment deposition. One explanation for this is that the mean beach width is based on calm conditions. Storm surge produced by strong onshore winds and the effects of higher waves will reduce the actual beach width, and the extent of this reduction will vary with the local beach slope and elevation. At Site 2 in the fall the beach was characterized by the development of a wide, high berm with a very steep slope at the lakeward edge. The decrease in beach width would have been very small here, except during a very high-magnitude storm, compared to the decrease in beach width at Site 1 which had a relatively smooth, gentle slope. The effective width of beach available for sediment transport at Site 2 during this time was much greater than is indicated by the mean beach width value.

While the regression models show a significant correlation between sediment transport and the two dependent variables, it is difficult to determine whether they are applicable to other means. The magnitude of the sediment transport to the dunes was also affected by several other variables, most notably rainfall and the development of a surficial lag deposit in the late fall which led to armouring of the surface. The effects of these may not have been of sufficient magnitude to change the linearity of the relationships; or their effects may have varied over a limited range.

The concept of an equilibrium distance or minimum fetch needed to achieve a fully saturated sand flow has been stated by a number of authors, but there appears to be very little data available from field measurements on natural beaches. Svasek and Terwindt (1974) suggest a minimum fetch of 10 to 20 m based on a few observations. Horikawa *et al.* (1984) note the significance of determining the equilibrium distance but present no data from their field experiments. Gares (1988) specifically included beach width as an independent variable in a correlation study of factors controlling sediment transport on coastal dunes but the correlation was not significant, perhaps because beach widths exceeded the critical length for most wind speeds. A recent study by Sarre (1989) found that the water table level in the adjacent dune system was the most important factor explaining differences between predicted and observed sand supply to the foredune system. It seems possible that in this case the water table was an important control on the width of dry sand at the top of the beach.

Finally, this study has demonstrated that beach width is a significant control on the potential volume of sediment transport for a given wind velocity. An important corollary to this is that it may reduce the importance of the higher velocities shown by the cubic relationship between sediment transport and wind velocities and increase the significance of moderate wind speeds. This is because of the correlation of storm surge height and the limit of swash uprush with onshore wind velocity, so that the highest onshore winds will always be associated with much narrower beaches. In the study area the most effective winds for sediment transport from the beach to the dunes appear to be of moderate magnitude approaching at some angle to shore perpendicular, because storm surge and wave height are much lower than for the same magnitude winds approaching shore perpendicular.

7. CONCLUSIONS

The conclusions of the study to date may be summarized as follows:

(i) Sediment accumulation in the incipient foredune zone can provide a good measure of sediment transport rates from the beach over a period of weeks or months provided the vegetation cover is sufficiently dense to trap all sediment leaving the beach.

(ii) The onshore component of wind velocity (defined by equation 1) is a good predictor of sediment supply to the foredune and is the most important variable controlling the rate of sediment transport.

(iii) Beach width is a limiting factor for sediment transport on a beach. The distance needed to achieve maximum sediment transport increases from 10 to 15 m for winds just above the threshold to over 40 m for winds exceeding 50 km hr^{-1}.

(iv) On narrow beaches oblique winds may transport more sediment to the dunes than winds of the same magnitude blowing directly onshore, because of the greater storm surge and wave activity associated with the latter condition—i.e. the effect of the reduced beach width may be greater than the effect of spreading sediment over a greater length of dune front associated with the oblique winds.

REFERENCES

Bagnold, R. A. (1941) *The Physics of Blown Sand and Desert Dunes*. Methuen, London.

Davidson-Arnott, R. G. D. (1988) Temporal and spatial controls on beach/dune interaction, Long Point, Lake Erie. *J. Coast. Res.*, *Spec. Issue No. 3*, 131–136.

Davidson-Arnott, R. G. D. and Stewart, C. J. (1987) The effects of longshore sand waves on dune erosion and accretion, Long Point, Ontario. *Proc. Can. Coast. Conf.*, National Research Council of Canada, Ottawa, 131–144.

Fryberger, S. G. and Dean, G. (1979) Dune forms and wind regime. In McKee, E. D. (ed.), *A Study of Global Sand Seas*. US Geol. Soc. Prof. Pap. 1052, 137–169.

Gares, P. A. (1988) Factors affecting eolian sediment transport in beach and dune environments. *J. Coast. Res.*, *Spec. Issue No. 3*, 121–126.

Horikawa, K., Hotta, S., Kubota, S. and Katori, S. (1984) Field measurement of blown sand transport rate by trench trap. *Coast. Eng. Japan*, **27**, 213–232.

Kawamura, R. (1951) Study of sand movement by wind. University of Tokyo Rept. of the Inst. of Sci. and Tech. 5, no. 314.

Leatherman, S. P. (1978) A new aeolian sand trap design. *Sedimentol.*, **25**, 305–306.

Lettau, K. and Lettau, H. (1977) Experimental and micro-meteorological field studies of dune migration. In Lettau, K. and Lettau, H. (eds), *Exploring the World's Driest Climate*. University of Wisconsin Press, Madison.

Maun, M. A. (1985) Population biology of *Ammophila breviligulata* and *Calamovilfa longifolia* on Lake Huron sand dunes 1: Habitat, growth form, reproduction, and establishment. *Can. J. Botany*, **63**, 113–124.

Psuty, N. P. (1988) Sediment budget and beach/dune interaction. *J. Coast. Res.*, *Spec. Issue No. 3*, 1–4.

Pye, K. (1985) Controls on fluid threshold velocity, rates of aeolian sand transport and dune grain size parameters along the Queensland coast. Proc. Int. Workshop on the Physics of Blown Sand, *Univ. Aarhus Mem. 8*, 3, 483–509.

Rukavina, N. A. and Zeman, A. J. (1987) Erosion and sedimentation along a cohesive shoreline—the north-central shore of Lake Erie. *J. Great Lakes Res.*, **13**, 202–217.

Sarre, R. D. (1987) Aeolian sand transport. *Prog. Phys. Geogr.*, **11**, 157–182.

Sarre, R. D. (1989) Aeolian sand drift from the intertidal zone on a temperate beach: potential and actual rates. *Earth Surf. Proc. Landf.*, **14**, 247–258.

Stewart, C. J. and Davidson-Arnott, R. G. D. (1988) Morphology, formation and migration of longshore sand waves; Long Point, Lake Erie, Canada. *Mar. Geol.*, **81**, 63–77.

Svasek, J. N. and Terwindt, J. H. J. (1974) Measurement of sand transport by wind on a natural beach. *Sedimentol.*, **21**, 311–322.

Chapter Ten

The exchange of materials between dune and beach systems

ANTON McLACHLAN

Institute for Coastal Research and Zoology Department, University of Port Elizabeth

1. INTRODUCTION

The coast of southern Africa is exposed to strong wind and wave action and characterised by extensive sand transport, mostly in the northerly direction. Approximately 70% of the coastline is sandy, taking the form of high-energy sand beaches backed by a variety of dune types. The dunes have recently been reviewed by Tinley (1985) and range from the vast shifting sands of the Namib coast in the west to large forested dunes along the subtropical east coast of Natal. Beaches are mainly of the intermediate morphodynamic type, i.e. neither reflective nor dissipative (Short and Wright, 1983). Dunes cover a wide range of morphologies, including barchanoid, transverse, reversing, linear, star and vegetated dunes such as hummock dunes, beach ridges, precipitation ridges and

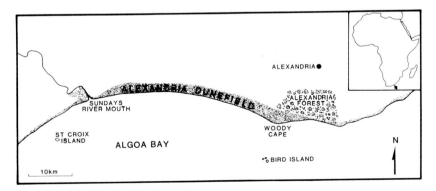

Fig. 1   Map of the study area

Coastal Dunes: Form and Process. Edited by K. F. Nordstrom, N. P. Psuty and R. W. G. Carter
© 1990 John Wiley & Sons Ltd

202

Fig. 2   Oblique view of the Alexandria dunefield looking landward. The dunefield is about 3 km wide (photograph by Werner K. Illenberger)

various blowout and parabolic types. Dunes related to topographic features, such as climbing and falling dunes and headland-bypass dunes, are also commonly encountered.

This shoreline is subject to a semi-diurnal, mesotidal (2.1 m range at maximum springs) regime and covers both east and west coast swell environments according to the classification of Davies (1972). Along most of the coast predominant winds come from the southwest, although there can be strong seasonal variations, often resulting in reversing dunes.

On the southeast coast, the shoreline takes the form of numerous half-heart (zeta) bays opening from southwest to northeast. Around each bay is a gradation of beach and dune types, from low-energy beaches with small foredunes in the southwest, through high-energy beaches backed by large transgressive dune sheets in the northeast, as exposure increases. This represents a similar dune/beach coupling to that described in Australia by Short and Hesp (1982).

The area covered by this chapter is the Alexandria dunefield on the northern shores of Algoa Bay (Figs 1 and 2). This dunefield backs a 40 km long intermediate to dissipative sand beach and is the largest active coastal dunefield in the area. There is an important ecological distinction between these two parallel ecosystems: the marine ecosystem comprising the beach and surf zone and the terrestrial ecosystem comprising the active dunefield. Beaches have often been viewed as interfaces and only recently has their ecosystem status been recognised (McLachlan and Erasmus, 1983). Similarly, dunes have been seen as part of a coastal sediment transport system (Davis, 1978, McKee, 1979) rather than as ecosystems in their own right.

This chapter summarises previous and ongoing ecological studies on the Alexandria dunefield and its adjacent beach/surf zone. The aim is to evaluate the current status and dynamics of the beach and dune systems in terms of the exchanges of materials between them. The materials are sand, salt spray, groundwater and organic matter. Geomorphologists tend to see these two systems only in the light of sediment exchanges, thereby regarding them as a single entity. This chapter will attempt to give the ecologists' perspective.

2. THE BEACH/SURFZONE ECOSYSTEM

The Alexandria coast supports a high-energy intermediate beach characterised by continuous heavy wave action. The modal state is the longshore bar-trough configuration (Short and Wright, 1983) where waves first break on the outer bar about 250 m from the beach. The inner surf zone, or terrace, is characterised by longshore and rip currents. The larger rip currents discharge through the outer bar to form a rip head zone which extends 200 to 300 m outside the breakers (Talbot and Bate, 1987a). This water is recirculated, being advected back into the surf by wave action, and the surf zone circulation consists of a series of interconnecting cells or gyres with rapid exchange of water between

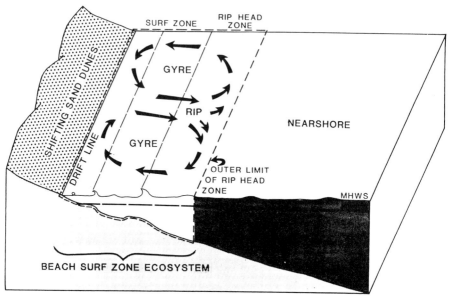

Fig. 3   Water circulation in the beach/surf zone ecosystem

them. The seaward limit of the surf circulation system is marked by the outer limit of the rip head zone which averages about 500 m offshore (Talbot, 1986).

The turnover time of water in the surf zone is in the order of a few hours while the turnover time of water in the surf and rip head zones as a whole is several days. The beach and surf zone is a discrete marine ecosystem with a landward boundary at the driftline, the landward limit of wave uprush, and a seaward boundary at the outer limit of surf circulation cells (McLachlan, 1987). Between these boundaries, the system comprises (i) the sand body of the beach and surf zone and (ii) the overlying water envelope of the surf and rip head zones (Fig. 3).

Despite the seemingly barren nature of this system, it supports rich plant and animal life and high productivity. Absence of stable substrate precludes macroscopic plants and the main source of primary production is surf zone phytoplankton, in particular a single species of diatom, *Anaulus australis*. This diatom has a daily cycle of migration between the sediments at night and the water surface during the day. It forms a semi-stable foam on the water surface during late morning when cell densities can reach 10^{-6} ml^{-1} (Campbell, 1987). These small diatoms, together with other diatom species and flagellates, result in high primary production (Campbell and Bate, 1988) and the generation of much food material in the surf zone. The development of surface accumulations of these diatoms is controlled by wave energy, the richest foam formation occurring during and after storms when the sediments are turned over to great depth and the bulk of diatom population stirred into the water column.

Surf zone phytoplankton is a characteristic feature not only of this beach but of high-energy beaches on all five continents (Lewin and Schaefer, 1983). Its production fuels three distinctive food chains: macroscopic, interstitial and microbial.

The macroscopic food chain includes all the larger animals living on and in the sediment (the benthos), also the zooplankton (shrimps and prawns in the surf zone) as well as birds and fishes, the top predators. Huge populations of filter-feeding clams develop in the intertidal zone and rich swarms of shrimps and prawns occur beyond and in the surf zone, where they feed directly on the surf phytoplankton. Because the phytoplankton is concentrated over rip currents by the physical circulation of the surf zone during the day (Talbot and Bate, 1987b), many of the macroscopic animals concentrate in these areas during daylight hours. At night, when the phytoplankton sinks and disperses, these grazing swarms may also break up.

The second food chain, the interstitial system, includes the micro-organisms living in the sand: bacteria, protozoans and small metazoan animals up to 2 mm in length. Bacteria average $10^8 \, g^{-1}$, protozoans $10^3 \, g^{-1}$ and meiofauna $1 \, g^{-1}$ of sand. This food chain is fuelled by dissolved and particulate organic matter (mostly originating from the phytoplankton) pumped through the sand by wave and tide action. In the intertidal zone, swash flushing by wave action pumps about $10 \, m^3$ of water through each lengthwise metre of beach each day (McLachlan, 1979), while in the surf zone, pressure changes caused by waves passing overhead pump about $50 \, m^3 \, m^{-1}$ through the bed each day. During this process, the organic materials are mineralised by the microfauna, and nutrients are recycled to the sea.

The third food chain, the microbial loop, comprises micro-organisms in the water column: bacteria ($10^6 \, ml^{-1}$), small flagellates ($10^3 \, ml^{-1}$) in the 3–5 μm size range and large ciliates and microzooplankton ($0.1 \, ml^{-1}$) forms that cover a range of sizes (Romer, 1990; McGwynne, 1990). Dissolved compounds released by phytoplankton (Campbell, Fock and Bate, 1985) are rapidly taken up by the bacteria. Fine particulate material in the surf zone waters is also utilised by these organisms. The rapid turnover of bacteria and the speed with which they utilise this material make the microbial loop an important food chain.

There is little trophic interaction between these food chains and they can be seen as quite distinct entities culminating in separate predators. Each food chain constitutes a carbon sink, mineralising the organic materials it receives, but recycling nutrients such as nitrogen and phosphorous. In terms of relative importance, the microbial loop is the most important of the three food chains, followed closely by the interstitial systems and the macroscopic food chain (McLachlan and Bate, 1985) (Fig. 4).

The surplus of primary production, after driving these three food chains, is exported across the seaward boundary of the system to supply nearshore waters. The three food chains recycle nutrients which, supplemented by a nutrient supply from groundwater (McLachlan and Illenberger, 1986), help to maintain high nutrient levels in the surf zone waters at all times. Nutrients are therefore not limiting to primary production, rather primary production is limited by wave energy.

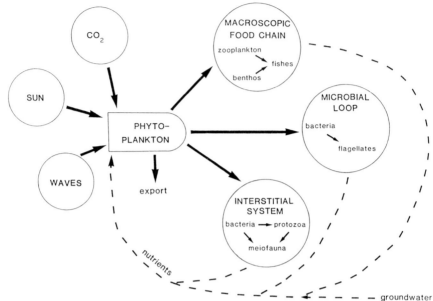

Fig. 4   Trophic structure of the beach/surf zone ecosystem. Broken lines indicate fluxes
of nutrients (after McLachlan, 1987)

3. THE DUNE ECOSYSTEM

The Alexandria coastal dunefield covers 120 km² and embraces a variety of
dune types. It is 80% unvegetated and consists of reversing transverse dunes
acted on by the predominant southwesterly winds. Because strong easterly winds
develop during the summer months, the dunes reverse temporarily at this time.
The dune ridges migrate eastwards at an average sand transport rate of
$20 \, \text{m}^3 \, \text{m}^{-1} \, \text{a}^{-1}$ sand. Depending on the height of the dune ridges, the rate of
advance ranges 1 to $10 \, \text{m} \, \text{a}^{-1}$. Other types of morphology found in the system
include barchanoid dunes, star dunes, blowouts and a major precipitation ridge
where the dunefield is encroaching slowly inland. The maximum height reached
by the dunefield is 150 m above sea level near its landward edge. As both the
easterly and westerly winds have an onshore component, the dunefield is
transgressive and has an average landward advance of $0.25 \, \text{m} \, \text{a}^{-1}$.

About 20% of the dunefield is vegetated, embracing five major habitat types
(McLachlan, Sieben and Ascaray, 1982). From the sea moving landwards these
are: (i) foredune hummocks backing the beach, colonised by *Scaevola thunbergii*
and other pioneer plants; (ii) wet slacks immediately landward of the beach where
a sparse vegetation develops, limited by sand movement and salt spray; (iii)
pockets of bush that develop in the landward part of the dunefield and average
about one hectare in extent; (iv) the major slipface or precipitation ridge (averaging
30 to 40 m in height) mostly well-vegetated by a variety of species; and (v) a

stabilised area around the mouth of the Sundays River where introduced vegetation has halted sand movement over an area of approximately 1 km².

The dunefield vegetation is a mix of east coast subtropical thicket, macchia and karroid forms with foredune pioneers fronting the sea. Species diversity increases landwards from three species on beach hummocks to more than forty species on the main slipface 2 to 3 km inland. Accompanying this increase in diversity inland is an increase in cover (from 10 to 50%), height (from 0.2 to 5 m at the base of the precipitation ridge) and woodiness. Faunal impact on the vegetation is limited and the latter develops primarily in response to physical gradients across the dunefield. Birds are, however, significant as seed dispersal mechanisms.

The dunefield has an arid fauna, limited mainly to the vegetated habitats. Insects, arachnids, air-breathing crustaceans, molluscs, amphibians, reptiles, birds and mammals all occur.

Crustaceans dominate the foredunes (80% of biomass) but give way to insects inland (90% of biomass), these two taxa being equally important in the slacks (60% and 56% of biomass respectively (McLachlan, Ascaray and Du Toit, 1987)). Unique species in this area include sand frogs, Damara terns and gerbils.

Macroscopic vegetation provides the primary food source. Primary production has not been quantified but is known to supply three food chains, of generally similar structure to those in the surf zone system. These are a food chain comprising macroscopic grazers, one consisting of macroscopic detritivores and the third the interstitial fauna in the sand (Fig. 5).

A comparatively small amount of primary production is grazed directly, usually by insects and various herbivorous mammals or parasitic nematodes feeding on the roots. Although no figures are yet available, probably less than 10% of the total primary production is cropped in this fashion.

The second food chain, macroscopic detritivores, consists primarily of insects that feed on vegetation litter. Most of the dune vegetation decomposes to litter before becoming available to animals. This is blown around the dunefield, collecting on slipfaces and in the lee of vegetation hummocks, where it is consumed by coleopterans, dermapterans and other insects. Preliminary information suggests that only a small proportion of total primary production is utilised in this way.

The third food chain, the interstitial system, is the most important, both because of the high proportion of living vegetation biomass (69%) that occurs below ground in the form of roots, and because of the rapid sand-movement rates which bury much of the vegetation and litter—97% of detritus is below ground in the slacks (McLachlan, Ascaray and Du Toit, 1987). The fate of most primary production is to be buried by sand, whence it is utilised by interstitial organisms—bacteria, protozoans, meiofauna and fungi, forming a distinct and important food chain.

Primary production by vegetation in the dunefield is controlled by a number of factors; the availability of water, salt spray and sand movement rates. Although vegetation communities generally become more diverse and vegetation

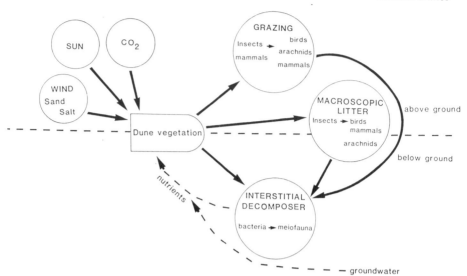

Fig. 5   Trophic structure of the dune ecosystem

cover higher inland, the dune slacks nearer the beach have the most available water. Wind, driving sand movement and salt spray as well as distributing litter and seeds, is the most important ecological factor structuring these dune communities.

4. DUNE/BEACH INTERACTIONS

The preceding section endeavoured to synthesise the key features of two discrete but interacting ecosystems from an ecological perspective. The terrestrial ecosystem, comprising the dunefield, has its boundaries at the driftline on the seaward side (the landward limit of marine animals, not only the upper limit of wave upwash) and at the base of the precipitation ridge on the landward side. The marine ecosystem, comprising the beach and surf zone, lies between the dune toe and the outer limit of surf circulation cells.

In the beach/surf zone system sand transport is wave-driven whereas in the dunefield it is wind-driven. Because of the overriding influence of local winds on the wave climate, sand transport in the beach/surf zone and in the dunefield is in similar directions. These two systems comprise the active littoral zone and are intimately connected by the sand exchange between them. In physical terms of sand budgets they may, therefore, be viewed as components of a single system. However, from an ecological perspective, they are two quite discrete systems and no resident species of animals or plants are shared. The exchanges between them are not limited to the movement of sand; salt spray, groundwater and organic materials are also transported across the interface. These exchanges will now be examined with special reference to the Alexandria coast.

4.1. Sand exchange

The sand of the Alexandria coast consists predominantly of quartz (61%) with the remainder calcite in the form of molluscan shell fragments (38%) and a small amount of heavy minerals (1%). The average grain size is 0.2 mm (2.4 ϕ) in the dunes and 0.25 mm (2.0 ϕ) on the beach and the sand is well sorted, becoming finer inland.

The Alexandria coast is characterised by extensive longshore sand transport: in the surf zone 5 to 9×10^5 m^3 a^{-1} (Swart, 1986) and in the dunefield 4 to 5×10^4 m^3 a^{-1} (Illenberger and Rust, 1988). The average rate of longshore transport in the dunefield is 20 m^3 m^{-1} a^{-1}, 30 m^3 m^{-1} a^{-1} near the beach and 10 m^3 m^{-1} a^{-1} farther inland (Illenberger, 1986). Thus the amount of sand moving past a fixed point in one year is an order of magnitude higher in the surf zone than it is in the dunefield, due to the greater transport capacity of water. There is a strong onshore movement of sand to the dunefield, which is accreting by 375 000 m^3 a^{-1} (Illenberger and Rust, 1988). Because of the configuration of the coast, little of this sand is passed alongshore and the dunefield functions largely as a sand sink. However, some sand is lost to the sea in the east by wave erosion of the Woody Cape cliffs at a rate of 45 000 m^3 a^{-1}. The dunefield thus grows by 330 000 m^3 a^{-1}. Sand storage occurs in two processes: the dunefield creeps landwards at an average rate of 0.25 m a^{-1} along its 45 km length and it slowly thickens by about 1.5 mm a^{-1} (Illenberger and Rust, 1988).

Sand supply from the beach to the dunefield represents an important and obvious exchange between the two systems. Furthermore, the movement of the sand exerts a direct and powerful control on the structure of the dunefield and the development of its vegetation communities (Young, 1987).

The eastward movement of dune ridges smothers vegetation in both slacks and bush pockets, creating marked successional sequences in both cases. In slacks, for example, dunes advance 7 m a^{-1} across the slack floors which are 45 m wide on average. Thus only 6 to 7 years are available for succession and development before the vegetation is overrun (McLachlan, Ascaray and Du Toit, 1987). Slack vegetation therefore attains no more than 10% cover and a diversity not exceeding 16 species.

Bush pockets, typically about 1 km inland, are flanked by higher dunes which advance at an average rate of 4 m a^{-1}. This gives a 13-year succession based on a width of 50 m. Bush pocket vegetation averages 40 to 80% cover and diversity up to 20 species (McLachlan, Sieben and Ascaray, 1982). Sand movement is thus one of the primary factors controlling dune vegetation types and the structure of the dune plant and animal communities.

4.2. Salt spray

The winds are predominantly onshore in this area and much salt spray is transported inland. This occurs when bubbles burst in the surf zone, releasing droplets into the air which are transported by the wind. Because of the relatively large size of the droplets, salt is not transported far inland.

Studies by Young (1987) have shown that large amounts of salt spray are blown into the dunefield by the predominant southwesterly winds throughout the year. Salt load increases exponentially with wind speed and decreases exponentially inland, the bulk of the salt spray being dropped within the first 200 m behind the beach (Fig. 6).

The rapid increase in cover and diversity of vegetation communities moving away from the beach, despite the decreasing amounts of available moisture in these areas, indicates that, like sand movement, salt spray is a major factor structuring the dune vegetation communities and limiting plants close to the beach. Vegetation cover and species diversity increase from 10% and 16 species in slacks near the beach to 50% and 42 species on the precipitation ridge, despite more moisture at the former site (McLachlan, Sieben and Ascaray, 1982).

Studies on the effects of salt on germination of seedlings and growth of a number of foredune plants have confirmed the importance of this limiting factor (Avis and Lubke, 1985; Boyd and Barbour, 1986; Young, 1987; Sykes and Wilson, 1988). In *Juncus*, *Arctotheca* and *Gazania*, germination is markedly reduced in salinities above 2 to 3°/oo (Young, 1987). Salt tolerance is, therefore, an important determinant of the proximity to the sea at which different plants can germinate and grow.

4.3. Groundwater

The dunefield has a strong rainfall gradient ranging from 800 mm a^{-1} at the eastern end to 400 mm a^{-1} at the western end. Groundwater flow rates are thus highly variable. Groundwater release occurs in the form of unconfined aquifers, discharging through the dunes to the beach and surf zone. Flow rates vary over more than one order of magnitude but average about 1 m^3 per metre of beach per day (McLachlan and Illenberger, 1986). Values in excess of 20 m^3 m^{-1} day^{-1} have been recorded in front of blind rivers leading into the dunefield. Where fresh and saline groundwater mix below the backshore, a brackish zone is created. This may be colonised by a unique brackish interstitial fauna.

Groundwater seeps via the beach into the surf zone. Because the volumes involved are relatively small in comparison to the volumes of saltwater being flushed through the intertidal sand by wave and tide action (McLachlan, 1979), dramatic drops in beach water salinity seldom occur.

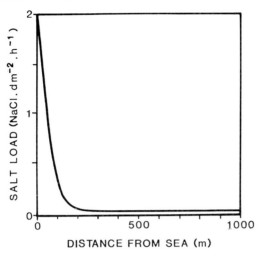

Fig. 6   Relationship between salt load and distance from the sea in the Alexandria dunefield (after Young, 1987)

With an average inorganic nitrogen content of $2.13\,mg\,l^{-1}$, groundwater provides an annual input of $777\,gN\,m^{-1}\,a^{-1}$ or 30 tons along the entire beach. This is sufficient to provide about 5% of the nitrogen requirements of the surf zone phytoplankton, the sole primary producers in the beach/surf zone ecosystem. Thus, while groundwater is not important as a source of freshwater to the beach/surf zone system, it is significant in supplying nutrients.

4.4. Organic materials

A variety of organic materials are exchanged between the beach and dunes (Fig. 7). They include: (i) the stranding of carrion from the sea which is then consumed by terrestrial predators; (ii) predation by land animals on intertidal organisms; (iii) windblown insects carried into the sea which are then eaten by marine animals; and (iv) vegetation litter blown from the dunefield into the sea. All of these exchanges are extremely difficult to quantify because they are highly variable and tend to occur irregularly.

Nearshore islands in Algoa Bay support penguin, gannet and seal breeding colonies and these relatively large vertebrates are regularly cast ashore. Although such carcases are consumed by intertidal invertebrates, the main scavengers are terrestrial mammals in the form of jackals, which regularly patrol the beach. There is no estimate of the quantitative importance of this and it is certainly not highly significant to the beach system, but it must represent an important input to the dunes.

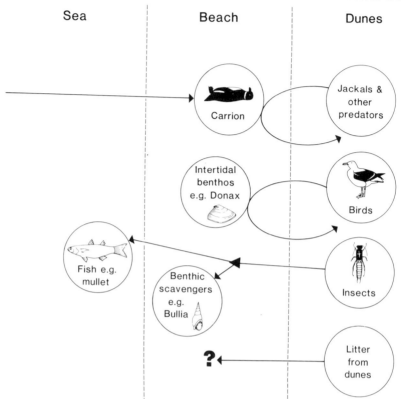

Fig. 7   Organic exchanges across the dune/beach interface

Predation mainly takes the form of birds moving into the intertidal zone to feed
during the low-tide periods. These birds include gulls, sanderlings, oystercatchers
and sandplovers. Their impact on the beach has been quantified by McLachlan
et al. (1980), who showed that they can remove about 20% of the total production
of intertidal macrobenthos. As these birds are mainly residents of the dunefield,
their migration to the beach to feed constitutes an exchange of organic materials
across the dune/beach interface. Supra-littoral crustaceans are nocturnal and move
across the interface to feed on the beach at times and in the dunes at other times.

   The third form of organic exchange, windblown insects, has proved impossible
to quantify. During offshore winds, numbers of insects are blown from the dunes
into the surf zone. When insects are swarming, this can include large numbers
of flying ants as well as moths, butterflies and others. They are often eaten
by birds (McLachlan *et al.*, 1980) and by intertidal scavengers such as crabs
and plough shells. In addition, insects have regularly been recorded in the
stomachs of surf zone fishes (Romer, 1986). Mullet, in particular, pick them
from the water surface when feeding on surf phytoplankton foam.

The final organic exchange between the dune and beach systems concerns litter, vegetation fragments which blow onto the beach and surf zone. Much of this material may be refractory and not available to marine animals; nevertheless, it does represent a potentially significant input.

5. CONCLUSIONS

The beach and surf zone comprise a discrete marine ecosystem containing primary producers, consumers and decomposers and with distinct boundaries determined by surf circulation patterns. The dunefield comprises a discrete terrestrial ecosystem, demarcated by the seaward and landward limits of aeolian sand transport. Like the beach/surf zone system with its marine biota, the dunefield contains a terrestrial flora and fauna which is characteristic.

The beach/surf zone system is primarily wave-controlled; wave-driven water movement transports sand, accumulates phytoplankton, pumps water and organic materials through the interstitial system and moves animals. Wind represents the most important factor impinging on the dunefield, transporting sand, salt spray and litter and burying vegetation. Because of its influence on wave energy, both through initial generation of waves and later local modification of breakers on the shore, wind is therefore ultimately the major parameter controlling both systems.

This chapter has attempted to show that, although sand represents an important exchange material between dune and beach systems, it is not the only form of material exchange. Other exchanges, including salt spray, groundwater flow and organic materials, although quantitatively small, may be highly significant in terms of effects. All four exchanges have direct effects on biological processes: (i) sand transported inland smothers vegetation, and sand transport rates determine successional sequences; (ii) salt spray blown inland limits vegetation development near the beach; (iii) groundwater discharging into the surf zone supplies nutrients for phytoplankton; and (iv) organic exchanges supply food chains on both sides of the dune/beach interface. These two ecosystems should, therefore, be seen as distinct, harbouring quite different faunas, floras and ecological processes, but at the same time interacting through the fluxes and impacts of these key materials.

REFERENCES

Avis, A. M. and Lubke, R. A. (1985) The effect of wind borne sand and salt spray on the growth of *Scirpus nodosus* in a mobile dune system. *S. Afr. J. Bot.*, **51**, 100–110.

Boyd, R. S. and Barbour, M. G. (1986) Relative salt tolerance of *Cakile edentula* (Brassicaceae) from lacustrine and marine beaches. *Amer. J. Bot.*, **73**, 236–241.

Campbell, E. E. (1987) The estimation of phytomass and primary production of a surf zone. PhD thesis, University of Port Elizabeth, 429 pp.

Campbell, E. E. and Bate, G. C. (1988) The estimation of annual primary production in a high energy surf zone. *Bot. Mar.* **31**, 337–343.

Campbell, E. E., Fock, H. P. and Bate, G. C. (1985) Exudation of recently fixed photosynthetic products from surf zone phytoplankton of the Sundays River beach. *Bot. Mar.*, **28**, 399–405.

Davies, J. L. (1972) *Geographical Variation in Coastal Development*. Longman, London, 204 pp.

Davis, R. A. (ed.) (1978) *Coastal Sedimentary Environments*. Springer-Verlag, New York.

Illenberger, W. K. (1986) The Alexandria coastal dunefield: morphology, sand budget and history. MSc thesis, University of Port Elizabeth, 87 pp.

Illenberger, W. K. and Rust, I. C. (1988) A sand budget for the Alexandria coastal dunefield, South Africa. *Sedimentology*, **35**, 513–521.

Lewin, J. and Schaefer, C. T. (1983) The role of phytoplankton in surf ecosystems. In McLachlan, A. and Erasmus, T. (eds), *Sandy Beaches as Ecosystems*. Junk, The Hague, pp. 381–389.

McGwynne, L. E. (1990) Direct measurement of bacterial production and flagellate grazing in the surf zone of a sandy beach. In prep.

McKee, E. D. (ed.) (1979) *A study of global sand seas*. US Geological Survey Professional Paper 1052.

McLachlan, A. (1979) Volumes of sea water filtered through eastern Cape sandy beaches. *S. Afr. J. Sci.*, **75**, 75–79.

McLachlan, A. (1987) *Sandy beach research at the University of Port Elizabeth 1975–1986*. University of Port Elizabeth, Institute for Coastal Research, Report No. 14, 111 pp.

McLachlan, A., Ascaray, C. and Du Toit, P. (1987) Sand movement, vegetation succession and biomass spectrum in a coastal dune slack in Algoa Bay, South Africa. *J. arid. Environ.*, **12**, 9–25.

McLachlan, A. and Bate, G. C. (1985) Carbon budget for a high energy surf zone. *Vie Milieu*, **34**, 67–77.

McLachlan, A. and Erasmus, T. (eds) (1983) *Sandy Beaches as Ecosystems*. Junk, The Hague, 757 pp.

McLachlan, A. and Illenberger, W. (1986) Significance of groundwater nitrogen input to a beach/surf zone ecosystem. *Stygologia*, **2**, 291–296.

McLachlan, A., Sieben, P. R. and Ascaray, C. (1982) *Survey of a major coastal dunefield in the eastern Cape*. University of Port Elizabeth, Zoology Department, Report No. 10, 48 pp.

McLachlan, A., Wooldridge, T., Schramm, M. and Kuhn, M. (1980) Seasonal abundance, biomass and feeding of shorebirds on sandy beaches in the eastern Cape, South Africa. *Ostrich*, **51**, 44–52.

Romer, G. (1986) Faunal assemblages and food chains associated with surf phytoplankton blooms. MSc thesis, University of Port Elizabeth, 194 pp.

Romer, G. (1990) Bacterioplankton, pelagic heterotrophic protozoa and other micro-components of a surf phytoplankton based food chain: standing stocks and demands on phytoplankton production. Submitted for publication.

Short, A. D. and Hesp, P. (1982) Wave, beach and dune interactions in southeastern Australia. *Mar. Geol.*, **48**, 259–284.

Short, A. D. and Wright, L. D. (1983) Physical variability of sandy beaches. In McLachlan, A. and Erasmus, T. (eds) *Sandy Beaches as Ecosystems*. Junk, The Hague, pp. 133–144.

Swart, D. H. (1986) *Physical environment interactions in the Sundays River/Skelmhoek area*. CSIR, Natal, Research Institute for Oceanology.

Sykes, M. T. and Wilson, J. B. (1988) An experimental investigation into the responses of some New Zealand sand dune species to salt spray. *Ann. Bot.* (London), **62**, 159–166.

Talbot, M. M. B. (1986) The distribution of the surf diatom *Anaulus birostratus* in relation to the nearshore circulation in an exposed beach/surf zone ecosystem. PhD thesis, University of Port Elizabeth, 356 pp.

Talbot, M. M. B. and Bate, G. C. (1987a) Distribution patterns of rip frequency and intensity in Algoa Bay, South Africa. *Mar. Geol.*, **76**, 319–324.

Talbot, M. M. B. and Bate, G. C. (1987b) The spatial dynamics of surf diatom patches in a medium energy cuspate beach. *Bot. Mar.*, **30**, 459–465.

Tinley, K. L. (1985) *Coastal Dunes of South Africa*. S. Afr. Natnl. Sci. Progr. Rep. 109, 297 pp.

Young, M. M. (1987) The Alexandria dunefield vegetation. MSc thesis, University of Port Elizabeth, 248 pp.

Chapter Eleven

Erosional landforms in coastal dunes

R. W. G. CARTER,
Department of Environmental Studies, University of Ulster

PATRICK A. HESP,
Carlingford, New South Wales

AND

KARL F. NORDSTROM
Institute of Marine and Coastal Sciences, Rutgers University

1. INTRODUCTION

Set against the extensive literature on depositional landforms in coastal dunes, there are few publications focussing on their erosional counterparts. However, erosional landforms constitute a significant proportion of many dune systems, and form an integral part of many postulated dune morphological cycles (Sokolow, 1894; Aufrère, 1931; Melton, 1940; Davies, 1972). Moreover, erosional forms add a significant aesthetic dimension to dunescapes (Engel, 1981, 1983) and require special management (Ranwell and Boar, 1986; Nordstrom and Lotstein, 1989).

Erosional dune forms include a range of slope failure and deflation structures from small isolated features to more complex terrains. Most dune landscapes tend to be indicative of negative sediment budgets and are likely to include erosional forms. Many other dune systems are mixtures of erosional and depositional morphologies, often with the former controlling the latter and giving rise to a suite of second-phase eolian landforms. The aim of this chapter is to provide a brief review of the erosional landforms encountered in coastal dunes, including wave-eroded features, large-scale deflation surfaces, and blowouts.

Coastal Dunes: Form and Process. Edited by K. F. Nordstrom, N. P. Psuty and R. W. G. Carter
©1990 John Wiley & Sons Ltd

Small-scale erosional and depositional forms associated with these features are also discussed. Examples of dune systems in Ireland, the USA and Australia are used to supplement results of previous investigations. Blowouts are examined in detail to illustrate the feedback between dune shape and wind flow as well as the reversals of net sediment transport caused by alterations in the location of deposition and scour. The complex interaction of processes with dune form leads to a reorganization of topography and creates complex mosaics of topography with vegetation. One of the principal goals of the chapter is to highlight the significance of erosional processes in the interpretation of these seemingly irregular dune landscapes.

2. EROSION OF DUNES BY WAVES

The reworking and remobilization of dunes by wave attack (Fig. 1) is a common process that has been discussed by numerous authors including Bremontier (1833), Dolan (1972), Parker (1975), Leatherman (1979), Vellinga (1983, 1984), van de Graaf (1986), Carter and Stone (1989) and several others in this volume.

Fig. 1   Undercutting and erosion of coastal dunes by storm waves (photograph by John Greer)

Because of the potential dangers, there have been a number of attempts to define conditions leading to dune erosion (Edelman, 1968, 1972; van der Meulen and Gourlay, 1968; van de Graaff, 1977, 1986; Hughes and Chui, 1981; Vellinga, 1983, 1984). These studies, including theoretical, empirical and experimental approaches, have highlighted the critical parameters leading to erosion, including morphological form (beach slope, dune height), and sediment texture (grain size, shape and packing) as well as hydrostatic (water level) and hydrodynamic (wave height, period and type) factors. Van de Graaff's (1986) work would suggest that surge height is by far the most important variable (82.8% variance in his studies), followed by particle size (7.3%) surge duration (2.6%) and initial profile (1.3%). Research indicates that the rate of erosion decays exponentially through a storm as the dune to beach sediment exchange re-establishes an equilibrium to the changed conditions. In fact, Hughes and Chui (1981, p. 186) state that between 70 and 90% of dune erosion is accomplished *before* the surge peak. Both Edelman (1972) and Hughes and Chui (1981) point out that the speed of storms is often the key factor in predicting the amount of erosion at any point on the shore. Van de Graaff (1986) recognizes two types of erosion: (i) gradual, perhaps involving $50 \, m^3 \, m^{-1}$ of shoreline each year (approximately 2 m of retreat) and (ii) storm, when as much as $400 \, m^3 \, m^{-1}$ (equivalent to 15 to 20 m of recession) may occur in 5 to 10 hours.

Erosion is caused by basal undercutting due to periodic wave attack (Fig. 2), although inherent slope stability factors, like soil moisture, increases in overburden and dynamic loading due to vegetation may also play a part. The main focus of attack is the foot of the foredune, which may be directly undercut by waves or trimmed by swash. Water in the sediment interstices usually provides sufficient cohesion to allow a vertical scarp to form. Scarp height (H) is related to repose (residual) angle (ϕ), slope angle (i), cohesion (c), unit weight of sediment (γ) in the form (Lohnes and Handy, 1968):

$$H = \frac{4c}{\gamma} \frac{\sin i \cos \phi}{[1 - \cos(i - \phi)]}$$

If the sediment was cohesionless ($c = 0$) then $H = 0$ and no scarp would form. However, undercutting leads to a state of tension on the upper slopes, often visible as parallel cracks (Carter and Stone, 1989, Fig. 4C), several tens of centimetres deep. Scarping alters the balance of forces on the dune slope, leading eventually to slab wedge-type or steeply inclined rotational failure, which may leave a vertical mid-slope scar and so cause further progressive slumping (Fig. 2). Assuming $\gamma = c. 18 \, kNm^{-3}$ (a reasonable figure for moist, loose sand of mixed grain size) and $\phi = 34°$, then a vertical scarp ($i = 90$) of between 0.2 m and 2 m will be maintained as cohesion, c, increases from 0.5 kPa (kilo Pascals) to 10 kPa.

Rapid oversteepening of the lower slope leads to the collapse of the vertical scarp and eventually secondary failure of the entire cliff. Following undercutting, the time to failure of the dune face is associated with the cohesiveness of the

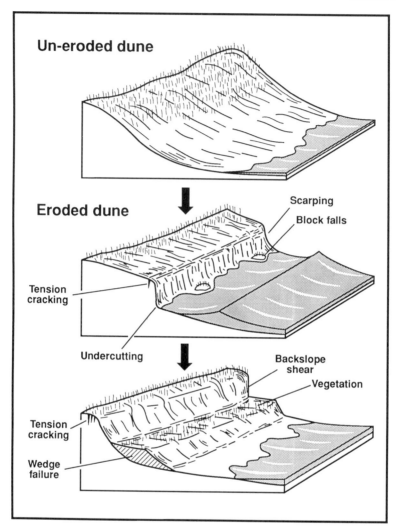

Fig. 2 Sequence of undercutting, failure and post-failure slope development on coastal dunes

dune, which in turn is related to the density of vegetation roots and rhizomes, soil moisture and chemical cementation. Slope failure may take place continuously, intermittently or only once. A single failure may protect the dune foot from further collapse, especially where a vegetation and soil mat slumps to form a protective sheath at the base of the dune. On granular, non-cohesive dune slopes above the vertical scarp, failure following undercutting is usually continuous and occurs as shallow surface slides or avalanches with slope angles

Fig. 3A   Retrogressive slope failure of a dune scarp following basal undercutting

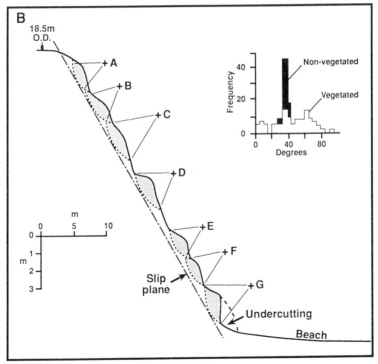

Fig. 3B   Slope profile of same site illustrating the relationship between the shallow rotational slumps and the overall failure surface. The inset on the upper right shows the distribution of slope angles immediately after failure. The vegetation slopes (white) show a complex distribution of slope angles while the unvegetated (black) display only one mode correlated with the angle of repose remaining around 32° to 34°, the repose angle for loose sand grains. Presence of vegetation roots increases cohesion from zero (cohesionless) to 5 to 15 kPa, and allows steeper (40° to 43°), stable slopes (Carter, 1980; Greenway, 1987). Mode of failure also changes to larger, less frequent collapses. Vegetated or loosely cemented slopes fail intermittently, perhaps two or three times during each undercutting episode, usually as discrete slumps or slides (Fig. 3A). A wave-undercut dune slope may also fail as a series of shallow retrogressive translational slides with rotational elements. Fig. 3B illustrates a dune slope failure with seven small rotational slumps, apparently moving on a shallow (*c.* 1.8 m) slip plane coincident with the root vegetation depth. Higher slumps overlie lower ones (E over F, B over C, A over B on Fig. 3) indicating concatenated failure.

The slope failures associated with marine undercutting may be considered as first-phase adjustments. Second-phase activity includes both erosion and deposition. Drying-out of the slope often triggers small avalanches and block falls, occasionally associated with gravity tunnelling, where dry sand underlies a moist cohesive layer. Desiccation and slumping may form breccias (Vortisch and Lindstrom, 1980) and earth and root falls, which may be rolled by swash action into balls (Foss, 1985; Carter and Stone, 1989). In some cases, whole slump blocks, held together by vegetation (commonly a single species), litter

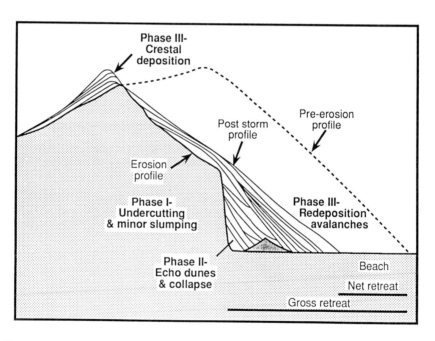

Fig. 4   Three phases associated with the undercutting, slumping and reforming of the dune slope

the avalanche slope. Where the vegetation has multiple roots (e.g. *Ammophila* spp.), it is not uncommon for the plant to survive and establish new roots within the surrounding and underlying avalanche material. Scarp slopes are revegetated by this means. In the examples shown in Fig. 3, the Phase I (Fig. 4) was followed by a number of small avalanches and chutes, which together with fresh accumulations of blown sand, formed a rectilinear facet at 38° within 3 months (Phase II—Fig. 4) (Carter, 1980).

Sediment removed from the dunes during scarping is usually returned to the slope face as part of the beach/dune recovery cycle (Phase III on Fig. 4). The initial accumulations are commonly echo dunes (Tsoar, 1983) forming between the HWM and the scarp (Fig. 5A). Later, windblown material may accumulate at the crest (Fig. 5B) and/or on the mid-slope (Fig. 5C), leading to overloading, instability and further failure, with the material sliding down to mask the lower slope and upper beach (Fig. 5D). Once the entire scarp is covered by aeolian sediment, material may accrete at or beyond the crest (Hesp, 1988), often in distinct lobes. The scarp may be filled by marine sediments thus 'armouring' the dunes and facilitating wave overtopping (Orford and Carter, 1985). Once the scarp slope is filled, and lying below the angle of repose, vegetation growth by vegetative expansion from slump blocks, propagation of seed material, or colonization by rhizomatous species extending from the scarp crest gradually results in partial or full recovery. The internal structure of dunes subjected to scarp/recovery cycles is distinctive, with dominant large-scale, seaward-dipping cross-beds (Fig. 6), often overlain by landward-dipping sets representing an advancing sand-sheet. Some shoreline dunes show a strongly bipolar, shore-normal azimuth structure (Goldsmith, 1973). As the dune retreats, the rebuilding of seaward faces and crests tends to maintain the dune form (Figs 4 and 5D), particularly on well-vegetated slopes.

3. DEFLATION SURFACES

Eolian processes remove the finer material on flat or gently inclined surfaces, leaving an immobile, often coarser, residue (Fig. 7A) (Segerstrom, 1962; Cooke, 1970; Carter, 1976).

The formation of an immobile surface marks a limit to eolian entrainment and thus sediment supply, and introduces a distinctive aerodynamic domain through the creation of a new boundary layer. Limits to deflation may arise via the aggregation of coarse particles at the surface (Carter and Rihan, 1978), the formation of chemical crusts (Carter, 1978; Pye, 1983), presence of algal mats (Van der Ancker, Jungerius and Mur, 1985); the exhumation of buried soils (Cooper, 1958), or on encountering the water table (Ahlbrandt and Fryberger, 1981; Wiedemann, 1984). Where a salt crust forms or there are temporary fluctuations in the water table deflation may cease for a period before resuming. The presence of an immobile shell or gravel pavement may form a semi-permanent deflation limit, which is occasionally re-exposed (Carter and Rihan, 1978).

Fig. 5A–D   A scarp fill sequence on eroding dunes from Fens Embayment, Myall Lakes
National Park, Australia. A: Post-scarp slumping. *Spinifex sericeus* rhizomes are being
undermined and exposed (Phase I in Fig. 4). B and C: Scarp filling takes place via
slumping and later eolian ramp formation (Phases II and III in Fig. 4). Sand blown into

the original scarp crest vegetation forms asymmetric ridges and lobes. In C small-scale
dry sand flows are visible. These are being fed from within and above the vegetation.
 D: Subsequent regrowth of *Spinifex* results in stabilization of the scarp fill

226

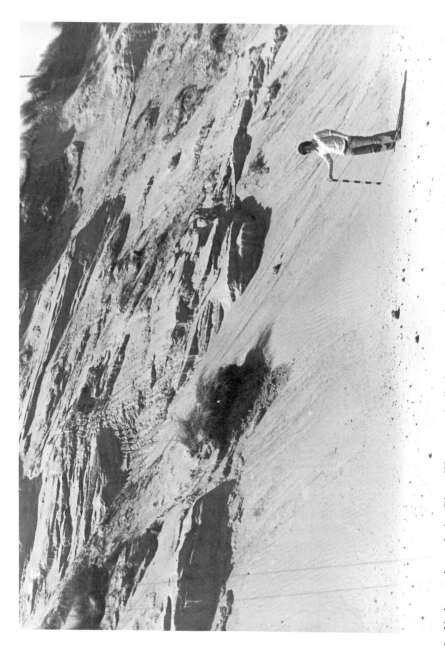

Fig. 6   Undercut slopes during Phase III commonly display a range of erosional and depositional forms. In this photograph seaward-dipping cross-beds indicating an earlier erosion/accretion cycle have been exposed. The lower slopes are being covered by dry sand avalanches, tongues often fed by upper slope rills and chutes. Vegetation block falls are also visible as are soil 'balls' at the foot of the slope

Formation of a deflation surface is often rapid. Carter (1976) describes shell lags 'emerging' in only a few days when conditions are favourable. During emergence, a variety of minor structures may appear including adhesion ripples and tears (Berry, 1973), saltation impact depressions, pedestals (Carter, 1978; Vortisch and Lindstrom, 1980) and tilted stones (Mattson, 1976). Most of these structures are destroyed as deflation proceeds. For example, laboratory experiments on 5 mm thick crusts suggest that the time taken both for pitting (impact craters) and penetration (crust collapse) is inversely related to an exponent of wind velocity above 4.4 m s^{-1} (Franzen, 1989). In a wind speed of 14 m s^{-1}, crusts were destroyed within 40 minutes. In contrast to erosional forms, Ahlbrandt and Fryberger (1981) note that where water tables are rising, aggradation may take place, and adhesion ripples can provide the basis for depositional features as high as a metre.

The formation of a deflation surface radically alters the near-surface air flow (Carter, 1976). Coarse lag deposits often allow particle overpassing (Everts, 1973) as single grains or as discrete bedforms ranging in size from ripples to small dunes (Fig. 7B). Transport is often within strong turbulent airflow, characterized by streaking, bursting and kinematic waves, so that movement is spatially and temporally intermittent (Allen, 1985). Some deflation surfaces may accumulate sediment in interparticle voids during periods of light winds. As wind velocity increases, this stored material is rapidly re-entrained and transported downwind. The ability of coarse-grained deflation surfaces to trap and then release material enables a transport flux to be maintained.

Deflation surfaces linked to water tables may become occupied by dune slacks, ponds (lagoons) or sabkhas (evaporite interdunes) and they may attain considerable size. The deflation plain of the Oregon dunes on the Pacific northwest coast of the USA is up to 2 km wide, and some deflation plain surfaces have increased 0.8 km in width the past 50 years. The rapid expansion is attributed to the elimination of sediment input from the beach caused by high, linear foredunes (Fig. 8) that built up following the spread of exotic European beach grass, *Ammophila arenaria* (Pinto *et al*, 1972; Wiedemann, 1984).

Some lagoons that form on deflation surfaces are large enough to form distinct shorelines, so that significant reworking of marginal dunes may occur (Ahlbrandt and Fryberger, 1981). On occasions, almost complete deflation may occur, leaving broad terrace-like structures (Segerstrom, 1962). The coastal dunes may reform downwind as coherent bedforms, such as transverse ridges, parabolics or sand seas (Hesp *et al*., 1989). Alternatively the sand may disperse into cover sands (e.g. Kocurek and Nielson, 1986) engulfing wide areas. One especially advanced deflation form is the 'machair' (Fig. 9A), a widespread, vegetated dune plain found in western Scotland and northwest Ireland (Ritchie, 1976; Bassett and Curtis, 1985). This flat or slightly landward-dipping feature probably originates from a mixture of factors including slow sea-level change, strong onshore winds, failing sediment supply and anthropogenic intervention. Ritchie

Fig. 7A   A shell and gravel lag deposit forming an extensive interdune surface

Fig. 7B   Small barchan dunes moving across a shell deflation surface. The foreground barchan is 0.8 m high and moving right and away from the camera exposing the internal structure. Note the scoured deflation surface with numerous 'tear' marks and shell pedestals

229

Fig. 8   The deflation plain in Oregon Dunes National Recreation Area showing the high foredune upwind (foreground), the vegetated deflation plain and the active migrating dunes downwind (background)

230

Fig. 9A Dune machair in northwest Ireland. Machair appears to be a strongly erosional landform, dominated by deflation to the water table

Fig. 9B The dune machair at Trawenagh Bay in Co. Donegal was formed in the late eighteenth century following destabilization of high dunes by agricultural impact. Blowing sand engulfed numerous buildings and farms leaving an extensive machair plain at the water table

(1977) proposes two models of machair development, both involving erosion and subsequent landward dispersal of a coastal foredune and deflation to a base imposed by the water table. Many machairs in Ireland rise steeply at the seaward margin, which is often cut by semi-circular blowouts with rim dunes, marking active landward sediment movement (Carter, 1990). In Co. Donegal, Ireland, nineteenth-century destabilization of the dunes at Trawenagh Bay (Fig. 9B) produced a modern machair plain. The sediment deflated from this site was blown over a wide area, much of it infilling and closing a tidal channel.

4. BLOWOUTS

4.1. Terminology

The generic term blowout is usually employed to describe an erosional hollow, depression, trough, or swale within a dune complex. These landforms were identified by early workers as important morphological elements (the 'trough-shaped wind sweeps' of Cowles, 1898). The actual word 'blowout' appears to have gained scientific acceptance in Melton's (1940) paper on the semi-arid dunelands of the southern High Plains and Bagnold's (1941) monograph on desert landforms, although the term was well known before (Kurz, 1942; Oosting and Billings, 1942). Melton used the term to describe parabolic dunes arising from the deflation of sand surfaces, whereas Bagnold's blowouts referred to wind-scoured gaps in an otherwise continuous transverse dune. Despite Brothers' (1954) and Landsberg's (1956) use of Melton's definition to explain parabolic coastal landforms in New Zealand and Scotland, Bagnold's definition has gained common currency, through the works of Laing (1954), Cooper (1958), Olson (1958) and Ranwell (1972).

4.2. Initiation of blowouts

Blowouts form readily in vegetated dunes (Fig. 10), where stable and unstable morphologies may co-exist. There are a number of ways that blowouts are initiated. Most of these involve the acceleration of wind where deflation potential has increased due to shoreline erosion and/or washover (Laing, 1954; Godfrey, Leatherman and Zaremba, 1979), vegetation die-back and soil nutrient deficiency (Jungerius, Verheggen and Wiggers, 1981), destruction of vegetation by animals (rabbits—Ritchie, 1972, Ranwell and Boar, 1986; bears—Martini, 1981), overland flow (Jungerius and van der Meulen, 1988) and diverse human activities including recreation (Mather and Ritchie, 1978) and fencing and house-building (Nordstrom and McCluskey, 1984). Perhaps the most intriguing of these initiating mechanisms is the cyclic model of Jungerius, Verheggen and Wiggers (1981). These authors suggest that periodic, spatially impersistent die-back of vegetation due to natural nutrient depletion leads to disintegration of the soil

Fig. 10A   Small coastal blowouts forming along the shoreline. Note the 'healed' blowout
to the right, and the 'closed' blowout topography farther inland

Fig. 10B   Elongate trough blowout comprising a deep deflation basin, lateral erosion walls,
trailing ridges and a depositional lobe. The blowout has been migrating at 18 m a^{-1} and
the houses are in imminent danger of engulfment (photograph by S. Chape)

Fig. 10C Shallow saucer blowout developing on the stoss face of an established foredune. A coarse deflation lag has formed at the base of the blowout. Steep, low erosion rims mark the lateral margins of the blowout, and sand is deposited in small asymmetric bedforms immediately downwind

Fig. 10D A trough blowout dissecting stable dune grassland. The cross-sectional form of the blowout is highly asymmetrical with erosion (farthest from camera) and accretion (nearest)

surface and eventually deflation, forming shallow blowouts. In almost all the other cited cases initiation of blowout activity is due to external influences. Jungerius and his co-workers also remark that blowout-forming winds are not necessarily the strongest.

Blowout topography need not arise from erosional processes. 'Blowouts' may develop as areas of non-deposition between mobile dune ridges (Gares and Nordstrom, 1988) or as gaps in incipient foredunes that remain open as the dune grows around them (Hesp, 1984; Carter and Wilson, 1988). Blowouts of non-erosional origin often assume the incised side wall characteristics of their erosional counterparts, and discrimination is not always straightforward. Blowout orientation appears to depend on antecedent topography and patterns of external disturbance as well as prevailing winds. Most authors describe a preferred orientation broadly commensurate with prevailing winds (Landsberg, 1956; Jungerius, Verheggen and Wiggers, 1981), but Gares and Nordstrom (1988) note three clear orientations on the New Jersey coast associated with storm wind direction (44% total blowouts), dominant winds (18%) and pedestrians (33%). Many blowout orientations on the Irish coast are shore-normal (Carter, 1990) (Fig. 10A), and probably reflect local micro-climates with wind blowing parallel or sub-parallel to the beach, and veering obliquely into the dunes.

Although there is a large variety of blowout morphologies (see Ritchie, 1972), two basic types have often been identified (Cooper, 1958), the saucer blowout (Figs 10A and C) and the trough blowout (Figs 10B and D). Saucer blowouts are shallow, ovoid, dish-shaped hollows with a steep marginal rim and commonly a flat-to-convex downwind depositional lobe. Trough blowouts are relatively deep, narrow, steep-sided topographies with more pronounced downwind depositional lobes, and marked deflation basins. Along the Australian east coast, saucer blowouts tend to develop on low gradient slopes on the windward faces of large foredunes, and on low, rolling dune topography where the vegetation cover has been locally removed. Trough blowouts are particularly well-developed where they cut through high dunes (e.g. foredunes), and they commonly evolve into parabolic dunes.

4.3. Blowout dynamics

Once initiated, the dune blowout will enlarge through a combination of deflation and slope or side wall failure. Most active blowouts enlarge laterally by wind scour that oversteepens side walls and leads to slumping and avalanching. They enlarge vertically by deflation of the blowout floor and extend downwind by deflation of the original sand surface and migration of the depositional lobe.

The wind flow in blowouts is topographically accelerated and altered as it moves through the landform. Fig. 11 illustrates wind velocity profiles through a narrow, relatively deep (8 m) trough blowout (Fig. 12). Fig. 11 indicates that jet flow is common within the blowout. Wind speeds are significantly accelerated

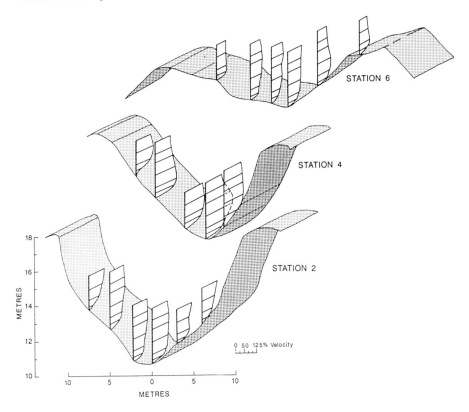

Fig. 11   Wind velocity profiles in a narrow blowout from South Australia. The profiles are expressed as a function of a permanent station sited in the middle of the blowout. The three-dimensional one metre topographic slices are taken from the throat entrance region (Station 2), mid-deflation zone (Station 4) and lower stoss slope of the depositional lobe (Station 6)

up the axis of the deflation basin (see centre profile, station 4). The steepness of the lateral erosion walls accelerates the wind flow and high jet velocities occur along the walls. Maximum shear stresses and sediment transport occur along the base of the trough and along the steepest part of the wall. As the blowout trough expands, for example, where it has cut through to the lee side of a large foredune (Fig. 11), jet flows expand and decelerate. This deceleration occurs in a radial concentric pattern across the blowout depositional lobe, whenever winds blow directly up the blowout.

   When winds approach the blowout directly, these flows maximize sand transport and erosion along the deflation basin and the side walls. Flows are generally strongest up the centreline axis and decrease away from it. Shallow basins result. Erosion of the deflation basin removes support for the sides of

Fig 12A   Entrance/throat region of the studied blowout illustrating narrow deflation floor (partially revegetated in upwind region), steep lateral walls and skewed orientation

Fig 12B   Upper blowout illustrating upwind face of the depositional lobe, narrow deflation floor and erosional walls (with instruments)

Fig. 12C   Similar to B except for inch-view of lateral wall 'overtop' deposition in foreground on left

Fig. 12D   Southwestern lateral wall indicating steep upper slope held by vegetation and roots, mid-lower slope dominated by avalanche cones, and deflation floor (right-hand bottom). The original lee slope of the foredune may be seen as indicated by the soil profile in the mid right-hand portion of the avalanche material

238

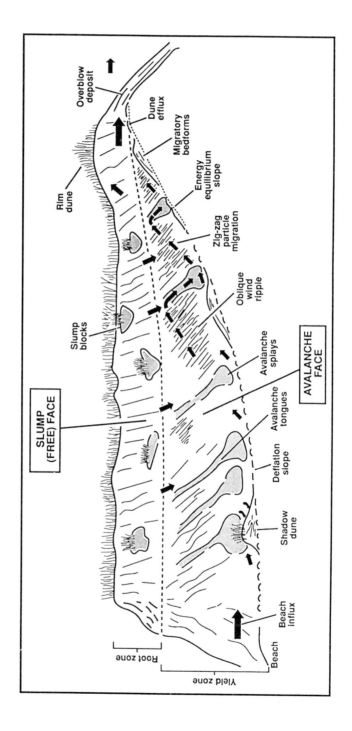

Fig. 13 Schematic long profile of a dune blowout indicating the slope facets and deposition forms. Arrows indicate the likely transport direction if wind is blowing from the beach

the blowout. Side walls are generally composite slopes with an upper free face (>40°) formed in the vegetation root zone overlying a lower, loose avalanche incline at or near the angle of initial yield (30 to 34°) (Fig. 12D). The transition between the free face and the avalanche slope is often sharp. Sediment tends to accumulate along this transition before falling back under gravity to the blowout floor, from where it is removed. Sediment is also transported directly along the blowout walls by oblique ripple migration, often across very steep slopes.

Regular exchanges of sediment between side wall and basin often occur in a zig-zag migration pattern combining ripple patches and small avalanches (Fig. 13). The blowout floor is often inclined, with the lower, flatter segment provided by a hard deflation surface (often shell or gravel) and the upper by an energy equilibrium facet reflecting the increasing wind velocity through the blowout. Spiralling helicoidal flows moving along the side walls transport sediment from the upper wall over the crest and into marginal vegetation forming rim dunes along the lateral blowout margins. Rapidly decelerating flows result in maximum sediment transport up the blowout axis with decreasing movement towards the lateral margins (Fig. 11), producing a parabolic-shaped depositional lobe. This is formed of two elements: (i) a delta-like 'lobe', often semi-circular in shape dominated by largely unvegetated foreset deposition; and (ii) beyond the lobe, a semi-circular or elliptical zone of slow deposition, perhaps marked by more vigorous sand-binding vegetation.

Where the primary wind approaches a blowout obliquely it may be directed by the topography. Many blowouts are highly asymmetrical with one side wall displaying under cutting while the other is dominated by accretion or is in the process of revegetating. Such occurrences are common on the east Australian coast where many blowouts are initiated by occasional SE storms but experience frequent NE to S prevailing winds. Here the wind enters the blowouts obliquely, attacking the facing (exposed) lateral wall, but not the sheltered wall. Blowout depositional lobes are also skewed obliquely as the wind is blowing preferentially out one side of the blowout.

4.4. Evolution of blowouts

The blowout enlarges as the side walls recede and the deflation area extends downwind. Almost all blowouts are limited in terms of depth, with erosion either arrested through the formation of a lag deposit, or controlled by the presence of a fluctuating water table (Ritchie, 1972).

There appears to be a fundamental distinction between 'open' and 'closed' blowouts, with the former having clearly defined wind gaps into and/or out of the hollow. Sediment flux is greater through open blowouts, and many act as transport corridors.

Jungerius, Verheggen and Wiggers (1981) point to a relative constancy in width/depth ratios (between 3 and 6), although Wilcock (1976), working in a more open dune system, suggested that blowout width eventually became independent of blowout depth once downward erosion ceased.

The length of blowouts usually depends on the available relief. The potential length of the small blowouts studied by Jungerius, Verheggen and Wiggers (1981) was largely unrestricted, yet most failed to develop beyond 30 m along the axis

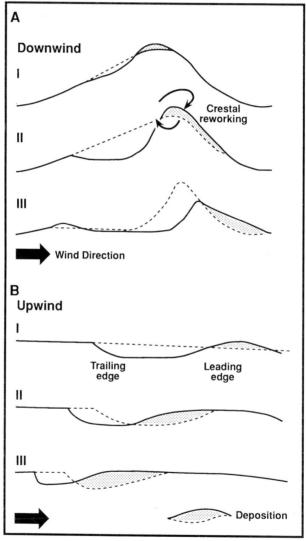

Fig. 14   Schematic views of downwind (A) and upwind (B) migration of blowouts. In A sediment efflux exceeds blowout storage and the form enlarges. In B upwind erosion is more than balanced by downwind infilling, with the result that the form appears to move into the prevailing wind direction

of the prevailing wind, perhaps because the wind-run over a 30 m length scale is accelerated insufficiently to continue sand transport downwind, so that the blowout becomes dormant and eventually revegetates. Blowouts may grow large enough to breach their host dune (Landsberg,1956; Ritchie, 1972; Gares and Nordstrom, 1987). The resulting breach may function more as a transport corridor, moving sand through the dune ridge, rather than directly from it. Breaching usually involves the flattening of the axial slope, presumably indicating a relative deceleration of airflow within the blowout, commensurate with a reduction, if not a cessation, of erosive activity.

Blowouts may migrate downwind or upwind. The downwind migration (Fig. 14) has the leading erosional edge advancing slowly with the wind, often infilling leeward of the deflation zone (Landsberg, 1956; Ritchie, 1972; Martini, 1981). This type of blowout leads to extensive reworking of the dune sediments. Blowouts may migrate upwind, eroding at the upwind edge and accumulating downwind (Jungerius, Verheggen and Wiggers, 1981; Jungerius and van der Meulen, 1989) (Fig. 14).

Significant topographic changes can often be measured over periods of a few years (Ritchie, 1972; Wilcock, 1976; Gares and Nordstrom, 1987). The rate of development varies between sites, depending on the direction, frequency and force of sand-transporting winds, exposure and pre-existing relief, the degree of vegetation cover and the characteristics of both the sand and the sand body.

Fig. 15 shows the typical arrangement of sediments within a single blowout at Portrush, Northern Ireland. Whereas the total deflation from the blowout is about 7200 m^3, almost twice this amount (13 600 m^3) is deposited in the rim dunes and the depositional plumes, showing the utility of the blowout to transfer beach material inland. Carter (1980) estimated that almost 0.25 m of an annual shoreline erosion rate of 0.3 m might be ascribed to this process.

Gares and Nordstrom (1987) measured 3.03 m of deflation in 5 years at the location where a gap formed in the foredune (Fig. 16A) while 2.02 m of deposition occurred adjacent to this location, in the direction of the resultant of the northeast storm winds and dominant northwest winds. Sediment trap data gathered at this blowout over a 5-month period in 1981 prior to breaching revealed that the highest transport rates were at the side wall on the south, which was eroded by the northwesterly winds, and at the saddle south of the blowout. The rate of transport through the saddle was higher than at any other location, including the open beach, and documents the effects of topographic channelling.

Sediment entering these oblique blowouts often collects in the throat (Fig. 16A and B) or on the outer side wall. The height of the (shadow) dune landward of the gap in Fig. 16B grew 1.8 m between 1982 and 1988. The most rapid growth followed creation of the gap by the dominant northwest winds. Sediment transported by northeast and southeast storm winds created the new zone of deposition northwest of the gap. More sediment accumulated south of the gap,

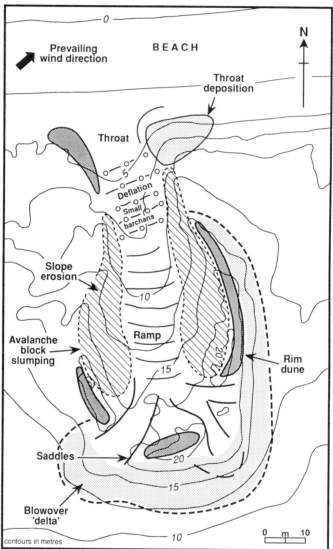

Fig. 15   Landforms and sedimentary features in a small shore-normal blowout at Portrush, Northern Ireland. The blowout comprises a series of slope facets and depositional zones commensurate with its role in the erosion and transfer of coastal sediment landward

where sand delivered both from the beach and from the blowout (by northwest winds) was blown over the crest by northeast winds. The gap thus provides a conduit for delivery of sand to the dune crest and to former low ground within the blowout as well as a conduit facilitating removal of sand from other portions of the blowout.

Fig. 16   Two views of deflation (A) (1982) and subsequent accretion (B) (1989) in a blowout at Island Beach State Park, New Jersey

5. EROSIONAL DUNE COMPLEXES

Complex blowouts may arise where there is great initial topographic variability (e.g. massive dunefields) or where local zones of deposition and scour alter wind and sedimentation through complex feedback mechanisms. Laing (1954) identifies small blowouts forming within large ones, while David (1977) and Filion and Morisset (1983) record a variety of dune blowout forms (imbricate, *en échelon*, digitate and hemicycle). A common complex form found in Ireland is a double or stacked blowout (Fig. 17), in which the formation of a blowout on the lower dune slopes initiates another on the upper slope. These blowouts have two erosional/depositional slope facets separated by an intermediate crest. Visual observations indicate that deposition on the intermediate crest disturbs airflow and vegetation on the upper slope, creating another erosional slope. The bottom part of Fig. 17 indicates the sediment transport pathways up this

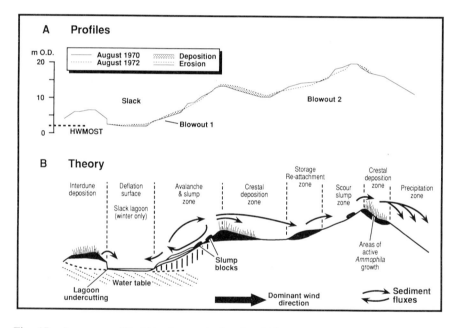

Fig. 17   A cross-profile (A) of a stacked pair of blowouts at Portstewart, Northern Ireland. Erosion of the lower blowout (1) leads to the formation of the upper blowout (2). An anatomical description of this situation is given in B

Fig. 18 (opposite)   The erosional dune complex at Inch, Co. Kerry, Ireland. This unstable dunefield has been hugely dissected by onshore winds. The photograph shows numerous linear trough blowouts allowing sediment to move landward. Much of the material is stored in depositional lobes several hundred square metres in area, before being transported onto the estuarine flats and rejoining the marine transport system via the tidal channel to the south

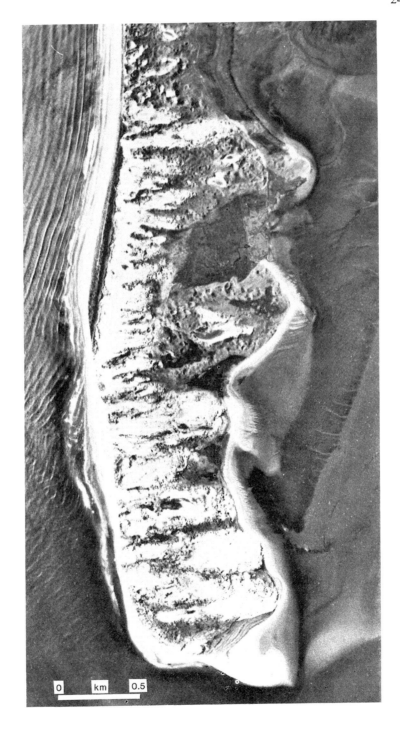

slope. The sequence of events at the blowout portrayed in Fig. 15 also demonstrates how topography redirects wind and creates new locations of accretion and scour, resulting in a dynamic and complex topography. Constant reworking (or cannibalization) of depositional landforms often leads to complete topographic reorganization.

A dominantly beach-aligned system can become dissected by secondary forms, breaking up the simple dune-interdune pattern, and creating a more complex aerodynamic environment, which leads to more irregular topography. In time, the dune complex may become almost divorced from its associated beach system, with internal reorganization subsuming any external forcing factors. Garcia-Nova, Ramirez Diaz and Torres Martinez (1975) provide an example of this from the Donaña dunes in southern Spain.

Erosional complexes are competitive (Cooper, 1958). Major forms subsume minor forms; minor forms are superimposed on major forms; and locations of accretion and scour are transposed (Fig. 18). The resulting diverse and dynamic geomorphological landscape supports a varied vegetation succession with steep ecological gradients and species richness. The varied vegetation, in turn, contributes to further differences in sedimentation rates. Such a landscape is almost fractal in nature as process-form interactions are replicated at many scales, and the gradual evolution into chaotic form is often evident. The interpretation of these landscapes is a challenging task for geomorphologists.

6. CONCLUSION

Development of erosional landforms (especially blowouts) through time requires more extensive investigation. A considerable proportion of the literature on dune management is aimed at prescribing remedies for erosional landforms, although there is a paucity of information to determine the extent to which such forms are self-healing. Long-term studies of the origin and history of blowouts would be particularly valuable given the probability of natural cycles of stability and instability associated with both internal (nutrient cycles, plant successions, animal population fluctuations) and external (climatic changes, human alterations) factors. The net result may be a relatively balanced system, with no long-term losses or gains of sediment, species or productivity. However, the role of blowouts in dune system dynamics has yet to be defined sharply.

Given the great ecological value of erosional dunescapes, effective collaborative research between geomorphologists and ecologists would seem to be an essential prerequisite for future progress. The sheer complexity of erosional dunes offers a major research challenge to coastal scientists.

ACKNOWLEDGEMENTS

Many thanks to Paul Gares, Norb Psuty and Peter Wilson for finding time to comment on this chapter, and to Mary McCamphill, Kilian McDaid and Nigel McDowell for typing, drafting and photography.

REFERENCES

Ahlbrandt, T. S. and Fryberger, S. G. (1981) Sedimentary features and the significance of interdune deposits. In Ethridge, R. G. and Flores, R. M. (eds), *Recent and Nonmarine Depositional Environments: Models for Exploration*. SEPM, Tulsa, Okla., pp. 293–314.

Allen, J. R. L. (1985) *The Principles of Sedimentology*. Allen & Unwin, London, 272pp.

Aufrère, L. (1931) Le cycle morphologique des dunes. *Ann. Geogr.*, **40**, 34–49, 362–385.

Bagnold, R. A. (1941) *The Physics of Blown Sand and Desert Dunes*. Methuen, London, 256pp.

Bassett, J. A. and Curtis, T. G. F. (1985) The nature and occurrence of sand-dune machair in Ireland. *Proc. R. Ir. Acad.*, **85B**, 1–20.

Berry, R. W. (1973) A note on asymetrical structures caused by differential wind erosion of a damp, sandy forebeach. *J. sediment. Petrol.*, **42**, 205–206.

Bremontier, N. T. (1833) Mémoire sur les dunes. *Ann. Pont. Chaussé.* **69**, 145–224.

Brothers, R. N. (1954) A physiographic study of the recent sand dunes on the Auckland west coast. *N. Z. Geogr.*, **10**, 47–59.

Carter, R. W. G. (1976) Formation, maintenance and geomorphological significance of an eolian shell pavement. *J. sediment. Petrol.*, **46**, 418–429.

Carter, R. W. G. (1978) Ephemeral sedimentary structures formed during Aeolian deflation of beaches. *Geol. Mag.*, **115**, 379–382.

Carter, R. W. G. (1980) Vegetation stabilisation and slope failure on eroding sand dunes. *Biol. Conserv.*, **18**, 117–122.

Carter, R. W. G. (1990) Geomorphology of the Irish coastal dunes. *Catena Suppl.* **18**, 31–39.

Carter, R. W. G. and Rihan, C. L. (1978) Shell and pebble pavements on beaches; examples from the north coast of Ireland. *Catena*, **5**, 365–374.

Carter, R. W. G. and Stone, G. W. (1989) Mechanisms associated with failure of eroding sand dunes, Magilligan, Northern Ireland, *Earth Surf. Proc. Landf.*, **14**, 1–10.

Carter, R. W. G. and Wilson, P. (1988) Geomorphological, sedimentological and pedological influences on coastal dune development in Ireland. In Psuty, N. P. (ed) *Dune/Beach Interaction*, *J. Coast. Res. Spec. Issue No. 3*, pp. 27–31.

Cooke, R. U. (1970) Stone pavements in deserts. *Ann. Assoc. Amer. Geogr.*, **60**, 560–577.

Cooper, W. S. (1958) *The Coastal Sand Dunes of Oregon and Washington*. Geol. Soc. Am. Mem. **72**, 169pp.

Cowles, H. C. (1898) The ecological relations of the vegetation on the sand dunes of Lake Michigan. *Bot. Gaz.*, **27**, 97–117.

David, P. P. (1977) *Sand dune occurrences of Canada*. Dept. Indian Affairs/Geol. Surv. Canada Rept. 183pp.

Davies, J. L. (1972) *Geographical Variation in Coastal Development*. Oliver & Boyd, Edinburgh, 204pp.

Dolan, R. (1972) Barrier dune systems along the Outer Banks of North Carolina: a reappraisal. *Science*, **176**, 286–288.

Edelman, T. (1968) Dune erosion during storm conditions. *Proc. 11th Conf. Coast. Eng.*, 719–722.

Edelman, T. (1972) Dune erosion during storm conditions. *Proc. 13th Conf. Coast. Eng.*, 1305–1311.

Engel, J. R. (1981) Sacred sands: the civil religion of the Indiana Dunes. *Landscape*, **25**, 1–10.

Engel, J. R. (1983) *Sacred Sands*. Wesleyan University Press, Middleton, Conn., 352pp.

Everts, C. H. (1973) Particle overpassing on flat granular boundaries. *J. Waterway Harb. Coast. Eng. Div. ASCE*, **99**, 425–438.

Filion, L. and Morisset, P. (1983) Eolian landforms along the eastern coast of Hudson Bay, Northern Quebec. *Nordicana*, **47**, 73–94.

Foss, P. J. (1985) Some observations on 'sea balls' discovered in West Donegal. *Ir. Nat. J.*, **21**, 526–528.

Franzen, L. G. (1989) Experimental studies of eolian erosion on a dune sand surface, protected by an artificial crust. *Zeit. Geomorph. NF*, **33**, 355–360.

Garcia Nova, F., Ramirez Diaz, L. and Torres Martinez, A. (1975) *El sistema de duñas de Doñana*. Publ. No. 5 ICONA, Ministerio di Agricultura, Madrid.

Gares, P. A. and Nordstrom, K. F. (1987) Dynamics of a coastal foredune blowout at Island Beach State Park, N. J. *Proc. Coast. Sed.*, '87, *ASCE*, 213–221.

Gares, P. A. and Nordstrom, K. F. (1988) Creation of dune depressions by foredune accretion. *Geogr. Rev.*, **78**, 194–204.

Godfrey, P. J., Leatherman, S. P. and Zaremba, R. (1979) A geobotanical approach to classification of barrier beach systems. In Leatherman, S. P. (ed.), *Barrier Islands*. Academic Press, New York, pp. 99–126.

Goldsmith, V. (1973) Internal geometry and origin of vegetated coastal sand dunes. *J. sediment. Petrol.*, **43**, 1128–1142.

Greenway, D. R. (1987) Vegetation and slope stability. In Anderson, M. G. and Richards, K. S. (eds), *Slope Stability*, Wiley, Chichester, pp. 187–230.

⁕ Hesp, P. A. (1984) Foredune formation in southeast Australia. In Thom, B. G. (ed.), *Coastal Geomorphology in Australia*. Academic Press, Sydney, pp. 69–77.

Hesp, P. A. (1988) Morphology, dynamics and internal stratification of some established foredunes in southeast Australia. *Sediment. Geol.*, **55**, 17–41.

Hesp, P. A., Illenberger, W., Rust, I., McLachlan, A. and Hyde, R. (1989) Some aspects of transgressive dunefield and transverse dune geomorphology and dynamics, south coast South Africa. *Zeit. Geomorph. Suppl.-Bd*, **73**, 111–123.

Hughes, S. A. and Chui, T.-Y. (1981) *Beach and dune erosion during severe storms*. University of Florida, Dept. Coastal and Oceanographic Engineering, Report UFL/COEL-TR/043, 290pp.

Jungerius, P. D. and van der Meulen, F. (1988) Erosion processes in a dune landscape along the Dutch coast. *Catena*, **15**, 217–228.

Jungerius, P. D. and van der Meulen, F. (1989) The development of dune blowouts, as measured with erosion pins and sequential air photos. *Catena*, **16**, 369–376.

Jungerius, P. D., Verheggen, J. T. and Wiggers, A. J. (1981) The development of blowouts in 'de Blink' a coastal dune area near Noordwijkerhout, The Netherlands. *Earth Surf. Proc. Landf.*, **6**, 375–396.

Kocurek, G. and J. Nielson (1986) Conditions favourable for the formation of warm-climate eolian sand sheets. *Sedimentol.*, **33**, 795–816.

Kurz, H. (1942) *Florida dunes and scrub*. State of Florida Dept. Conservation, Geol. Bull. No. 23, 117pp.

Laing, C. C. (1954) The ecological life-history of the *Ammophila breviligulata* community on the Lake Michigan Dunes. Unpublished PhD thesis, Univ. Chicago, 108pp.

Landsberg, S. Y. (1956) The orientation of dunes in Britain and Denmark in relation to the wind. *Geogr. J.*, **122**, 176–189.

Leatherman, S. P. (1979) Barrier dunes—a reassessment. *Sediment. Geol.*, **24**, 1–16.

Lohnes, R. A. and Handy, R. L. (1968) Slope angles in friable loess. *J. Geol.*, **76**, 247–258.

Martini, I. P. (1981) Coastal dunes of Ontario: distribution and geomorphology. *Geogr. Phys. Quat.*, **35**, 219–229.

Mather, A. S. and Ritchie, W. (1978) The Beaches of the Highlands and Islands of Scotland. Countryside Commission for Scotland, Redgorton, Perth.

Mattson, J. O. (1976) Wind tilted pebbles in sand—some field observations and simple experiments. *Nordic Hydrol.*, **7**, 181–208.

Melton, F. A. (1940) A tentative classification of sand dunes: its application to dune history in the southern High Plains. *J. Geol.*, **48**, 113–174.

Nordstrom, K. F. and Lotstein, E. L. (1989) Perspectives on resource use of dynamic coastal dunes. *Geogr. Rev.*, **79**, 1–12.

Nordstrom, K. F. and McCluskey, J. M. (1984) Considerations for control of house construction in coastal dunes. *Coast. Zone Mngmt J.*, **12**, 385–402.

Olson, J. S. (1958) Lake Michigan dune development. *J. Geol.*, **66**, 345–351.

Oosting, H. H. and Billings, W. D. (1942) Factors affecting vegetational zonation on coastal dunes. *Ecology*, **23**, 137–139.

Orford, J. D. and Carter, R. W. G. (1985) Storm generated dune armouring on a sand gravel barrier system, southeastern Ireland. *Sediment. Geol.*, **42**, 55–82.

Parker, W. R. (1975) Sediment mobility and erosion on a multibarred foreshore (Southwest Lancashire, UK). In Hails, J. and Carr, A. P. (eds), *Nearshore Sediment Dynamics and Sedimentation*. Wiley-Interscience, London, pp. 151–179.

Pinto, C., Silovsky, E., Henley, F., Rich, L., Parcell, J., and Boyer, D. (1972) *The Oregon Dunes NRA Resource Inventory*. US Dept. Agriculture, Forest Serv., Pacific Northwest Region, Portland, Oregon, 294p.

Pye, K. (1983) Coastal dunes. *Prog. Phys. Geogr.*, **7**, 531–557.

Ranwell, D. (1972) *Ecology of Salt Marshes and Sand Dunes*. Chapman & Hall, London, 258pp.

Ranwell, D. and Boar, J. (1986) *Coast Dune Management Guide*. Inst. Terrestrial Ecol./NERC, 105pp.

Ritchie, W. (1972) The evolution of coastal sand dunes. *Scott. Geogr. Mag.*, **88**, 19–35.

Ritchie, W. (1976) The meaning and definition of machair. *Trans. Proc. Bot. Soc. Edinb.*, **42**, 431–440.

Ritchie, W. (1977) Machair development and chronology in the Uists and adjacent islands. *Proc. R. Soc. Edinb.*, **77B**, 107–122.

Segerstrom, K. (1962) *Deflated marine terrace as a source of dune chains, Atacama Province, Chile*. US Geol. Surv. Prof. Paper 450C, C91–C93.

Sokolow, N. A. (1894) *Die Dünen, Bildung, Entwickelung und Innerer Bau*. Berlin.

Tsoar, H. (1983) Wind tunnel modelling of echo and climbing dunes. In Brookfield, M. E. and Ahlbrandt, T. S. (eds), *Eolian Processes and Sediments*. Elsevier, Amsterdam, pp. 247–260.

Van de Graaff, J. (1977) Dune erosion during a storm surge. *Coast. Eng.*, **1**, 99–134.

Van de Graaff, J. (1986) Probalistic design of dunes; an example from the Netherlands. *Coast. Eng.* **9**, 479–500.

Van der Ancker, J., Jungerius, P. and Mur, L. (1985) The role of algae in the stabilization of coastal dune blowouts, *Earth Surf. Proc. Landf.*, **10**, 189–192.

Van der Meulen, T. and Gourlay, M. R. (1968) Beach and dune erosion tests. *Proc. 11th Conf. Coast. Eng.*, 701–707.

Vellinga, P. (1983) Predictive computational model for beach and dune erosion during storm surges. *Proc. Coast. Zone '83, ASCE*, 806–819.

Vellinga, P. (1984) *Movable-bed modelling law for coastal dune erosion.* Water Port Coast Ocean Eng. Div. *ASCE*, **110**, pp. 495–504.

Vortisch, W. and Lindstrom, M. (1980) Surface structures formed by wind activity on a sandy beach. *Geol. Mag.*, **117**, 491–496.

Wiedemann, A. M. (1984) *The Ecology of Pacific Northwest Coastal Sand Dunes: A Community Profile.* US Dept. Interior, Fish and Wildlife Serv. Washington, DC, 130p.

Wilcock, F. A. (1976) Dune physiography and the Impact of Recreation on the North Coast of Ireland. Unpublished DPhil thesis, The New University of Ulster, Coleraine, 169pp.

Section III

SECONDARY DUNES AND DUNEFIELDS

This section contains the results of research conducted in locations where conditions have favored development of large-scale coastal dunes. The three chapters provide perspective on the kind of dune that will develop where sand supply is high and there is no effective means of stabilizing surface sediments. In preceding chapters, McLachlan alludes to the interaction between the beach, foredune, and the more massive dunefield behind it, and Carter, Hesp and Nordstrom identify mechanisms for sediment transport between them. The chapters in this section provide a closer look at the relationship between these components of the eolian environment. The beach is the inevitable source of sediment to both the foredune and the dunefield behind it. The foredune is the active dune ridge above the beach. It becomes a secondary dune if it is stranded by progradation or it is dissected by deflation so as to contribute sand to inland locations. The accelerated movement of sand inland from the foredune is usually associated with destruction of the vegetation cover by natural or human causes, but once initiated, these interior dune forms can undergo alterations that are independent of changes on the beach and foredunes. Their movement can be virtually unstoppable, and they can overwhelm forests, small hills and human settlements and infrastructure (Carey and Oliver, 1918; Besch and Kaminske, 1980; Cullen and Bird, 1980; Wiedemann, 1984). In the process of forming and migrating, a myriad of forms develops. The sequence may be repeated several times, resulting in complex patterns of dune activity that represent a long history of change. Interpretation of these landforms requires investigation at increased time and spatial scales.

The three chapters in this section deal with dunefields in different stages of development. In Chapter Twelve, Hesp and Thom identify the characteristics of dunefields at a stage when there is a conspicuous genetic relationship between the foredunes and the dunefields landward of them. The following chapter by Borówka concentrates on the dynamics of active dune forms that change independently of the interaction between the foredune and the beach. In

Chapter Fourteen, Orme provides a historical perspective, and highlights the geological persistence of dunefields and the cyclic nature of change. The three studies point out that there is no clear latitudinal constraint to the formation of massive, independent dune systems. The dunes studied by Borowka are located on a glaciated coast in high mid-latitudes with abundant rainfall; the dunes studied by Hesp and Thom and by Orme are in non-glaciated areas in low mid-latitudes characterized by hot or dry summers. Migrating sand sheets also occur in more temperate mid-latitude locations (Cooper, 1958), and there is evidence of more widespread occurrence of coastal and inland sand sheets in the past (Castel, Koster and Slotboom, 1989). Changes in climate and associated sea level changes may result in the initiation of new sand drifts or the reactivation of currently inactive dunefields. The investigations in this section thus provide insight into the genesis, form, and duration of presently active areas of sand transport, and they provide a glimpse of conditions that have been more widespread in the past and may be more active in the future.

REFERENCES

Besch, H. W. and Kaminske, V. (1980) Die Oekologie einer Ferienregion—Beispiel Sylt. Fragenkreise 23543, Ferdinand Schoningh, Paderborn, 32 pp.

Carey, A. E. and Oliver, F. W. (1918) *Tidal Lands*. Blackie, London.

Castel, I., Koster, E. and Slotboom, R. (1989) Morphogenetic aspects and age of late Holocene eolian drift sands in northwest Europe. *Zeit. Geomorph. NF.*, **33**, 1–26.

Cooper, W. S. (1958) *The Coastal Sand Dunes of Oregon and Washington*. Geol. Soc. Am. Mem. 72, 169 pp.

Cullen, P. and E. Bird. (1980) *The Management of Coastal Sand Dunes in South Australia*. Geostudies, Black Rock, Victoria, 83 pp.

Wiedemann, A. M. (1984) *The Ecology of Pacific Northwest Coastal Sand Dunes: A Community Profile*. US Dept. Interior, Fish and Wildlife Serv. Washington, DC, 130 pp.

Chapter Twelve

Geomorphology and evolution of active transgressive dunefields

PATRICK A. HESP
Carlingford, New South Wales

AND

BRUCE G. THOM
Department of Geography, University of Sydney

1. INTRODUCTION

Transgressive dunefields are aeolian sand deposits formed by the downwind movement of sand over vegetated to semi-vegetated terrain. In most cases, the dunefields are largely unvegetated when active, and may vary in size from small sheets ($\sim$ 100s m^2) to small sand, seas (several km^2). They occur frequently and most dramatically in temperate humid areas where there is an abundant supply of coastal sand coupled with powerful onshore wind energy (Davies, 1980). However, they are not uncommon in semi-arid coastal areas (Short, 1988) and have been observed in the humid tropics (Pye, 1982).

The term 'transgressive dunes' was coined by Gardner (1955) in Australia to identify sand deposits which were actively migrating downwind and advancing over, or 'transgressing' prior terrain. The term can be applied both to active and to stabilized dunes which can be shown to have moved inland over surfaces composed of forest, swamp, marsh, scrub, or bedrock, or into lagoons. In a well-vegetated coastal region there is a strong contrast between moving sand and the types of surface being buried. Such contrasts are not as distinctive in semi-arid or arid environments, so the generic term transgressive dunes is appropriate in humid regions. Buried soils, peat or stumps exposed in sections or on deflated windward slopes indicate the occurrence of dune transgression. The general term embraces features referred to as blowouts, parabolic dunes,

Coastal Dunes: Form and Process. Edited by K. F. Nordstrom, N. P. Psuty and R. W. G. Carter
© 1990 John Wiley & Sons Ltd

long-walled transgressive dunes, and cliff-top dunes, and thus can be used in a generic sense. The term has received widespread acceptance in Australia and has been applied worldwide in synthesis studies of Davies (1980) and Pye (1983).

Coastal dune classifications which are based on morphology usually distinguish between dunes which develop in the presence of vegetation ('fixed' or 'impeded') and those which involve free-movement of sand in a downwind direction. Goldsmith (1978) devised a classification of four basic dune types: vegetated, artificially induced, medaño and parabolic to which he felt could be added aeolianites and lunettes. In the context of this paper medaños and parabolic dunes would form distinctive types of transgressive dunes although the term medaño may not be correct in this context (see Cornish, 1914). Davies (1972, 1980) classified coastal dunes into primary (derived from the beach) and secondary (derived from erosion of primary dunes) types. In this scheme free dunes, including transverse, barchans, oblique ridges and precipitation ridges, were classed as primary and separated from transgressive dunes (blowouts, parabolics, transgressive sand sheets). He noted that this scheme is not completely satisfactory, if only because the categories are not always as fully separate as it suggests (Davies, 1972, p. 151).

We see the need to combine free and transgressive dune types as did Pye (1983) in his attempt to separate impeded dunes from transgressive dunes. He recognized five main types of transgressive dune: transverse ridges (of which precipitation ridges as defined by Cooper, 1958, are a sub-type), barchans, transgressive sand sheets, parabolic dunes and oblique dunes.

Here we use the term transgressive dunefield more exclusively than the term transgressive dunes described above. Transgressive dunefield is specifically utilized to define a broad, active (free-moving) sand surface migrating landwards or alongshore. Such dunefields have been variously referred to as free dunes, mobile sands, sand drifts and sand sheets (Mort, 1949; Salisbury, 1952; Cooper, 1958; Tinley, 1985).

In this paper, active transgressive dunefield geomorphology, processes and evolution will be reviewed. In particular, the morphology and evolution of the many and various landforms that jointly go to make up the often very complex active transgressive dunefield will be discussed.

2. GEOMORPHOLOGY OF TRANSGRESSIVE DUNEFIELDS

2.1. Gross morphology of active transgressive dunefields

At the broadest scale, two principal sub-types of active transgressive dunefields may be distinguished, namely, tabular fields and sheets, and buttress fields (Hesp *et al.*, 1989). Tabular dunefields are broad, flat to hummocky sand bodies (Fig. 1A) which tend to have a pronounced, more-or-less continuous slipface along the landward margin. Buttress dunefields are more triangular, landward ascending ramps of sand (Fig. 1B; Tinley, 1985). The term is derived from the comparative similarity between buttress dunefields and the buttresses of rainforest trees (Tinley, 1985). Buttress dunefields are more common in regions dominated by sub-tropical or warm-temperature forest and in areas where mobile dunes are climbing up relatively steep seaward sloping terrain. Buttress dunefields may be vegetated on the up-slope margin and commonly display a pronounced vegetated crest and lee slope (Fig. 1B). The marginal, seaward-facing vegetation on the upper slopes of the buttress dunefields appears to act at times as pioneer and intermediate successional species in this environment. Both types of dunefields may occur in a variety of environmental settings ranging from beach parallel bodies, climbing and falling dunes, to headland-bypass dunes (Bigarella, 1975; Cowie, 1963; Pickard, 1972; Pain, 1976; Ritchie and Walton, 1972; Ritchie, Smith and Rose, 1978; Mather and Ritchie, 1977; Thom *et al.*, in press; Tinley, 1985).

2.2. Transverse and oblique dunes

Tabular and buttress dunefields contain a variety of dune features ranging from broad, undulating sheets and domes of sand (e.g. the medaño of Goldsmith, 1978), to the more common transverse and oblique ridges. These ridges are asymmetric sand ridges which lie within 15° of normal, or oblique to the long-term resultant sand-transport direction (Rubin and Hunter, 1985). Although transverse ridges were recognized by Bagnold (1954), he claimed they were unstable forms and were quickly absorbed into other dune types. The point was disputed by Cooper (1958, p. 27) in his classic treatise on coastal dunes.

The transverse dune, as defined by Cooper (1958) is a ridge essentially normal to the wind and moving with it; it is asymmetric in profile, with a gentle windward slope and a steep leeward slope (slipface). Fig. 1A clearly demonstrates the wave-like pattern of transverse ridges as viewed towards the slipface. This pattern was produced by westerly winds during the late autumn and winter of 1960 near Newcastle, NSW, Australia. Here ridge heights on the lee side may exceed 10 m, but the crest line is usually undulating. The ridge is slightly sinuous in plan and attains lengths varying from less than 100 to over 600 m. Lee projections (Cooper, 1958, p. 44) frequently develop in the transverse dunefields. Spacing between ridges is often regular, varying between 100 and 200 m in the example shown in Fig. 1A.

Transverse dunes may vary from simple, straight-crested ridges to more crescentic forms and to more complex axlé forms. Barchans may also occur. These dunes vary considerably in height from small-scale forms 0.5 to 3.0 m high to extremely large dunes 30 to 100 m in height (slipface crest to base). They display characteristic dynamic and geometric relationships in that wavelengths increase as dune heights increase (Matsukura, 1977), morphologic asymmetry increases as regional annual wind speeds become stronger (Wipperman and

Fig. 1A   A broad, relatively flat transgressive dunefield dominated by sinuous transverse dunes, Newcastle Bight, NSW

Fig. 1B   Sinuous and aklé-type transverse dunes on a buttress type transgressive dunefield near Port Elizabeth, South Africa (photograph: W. Illenberger)

Gross, 1986), and annual rates of downwind movement decrease as dune height increases (Long and Sharp, 1964; Hastenrath, 1967). In the case of barchans, there is a linear relationship between the average width between the horns and dune height (Finkel, 1959).

Simple transverse dunes display characteristic morphologies, having steep slipfaces commonly lying at the angle of repose (~ 30 to 34°) and lower angle stoss faces lying around 10 to 15° (Sharp, 1966). In relatively uni-directional wind fields, the dunes tend to migrate in one direction, maintaining spacings related to dune height and flow separation vortice length; rates of movement are related to dune height and flow velocities.

A regular seasonal construction and destruction of transverse dunes was observed by Cooper (1958) along the Oregon coast, USA. In coastal NSW the details of the dune pattern depend on the incidence of high-velocity winds from a particular direction. Here there is a marked seasonal reversal of slipface orientation. The ridges are aligned obliquely to effective east and west winds. In winter, westerly winds prevail and the steep slipface on the lee side of the ridge is oriented to the east. Under the influence of east to northeast sea breezes the slipface reverses (Fig. 2).

Recently, the simple classification of many dunes on transgressive dunefields as transverse dunes has come into question, as many of these dunes act

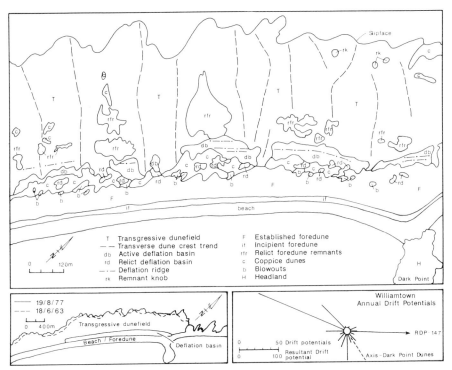

Fig. 2 Geomorphic map of the active transgressive dunefield, northern Fens Embayment, Dark Point, Myall Lakes National Park. The area displays many landforms typical of NSW transgressive dunefields. The dunefield is migrating landwards at 1 to 3 m yr^{-1} (see left-hand box). The reversing transverse dunes do not migrate towards the east as expected from the annual drift potential analysis (right box), but extend to the NNW under the influence of oblique and along-dune winds

dynamically in transverse, oblique or linear modes (Hunter, Richmond and Alpha, 1983; Carson and MacLean, 1986; Rubin and Hunter, 1987). The so-called transverse dunes of the NSW dunefields not only reverse seasonally but also act as linear dunes when onshore storm winds blow parallel or nearly so to the axis of dune crestlines. They are thus dynamically oblique dunes in a similar sense to the Oregon oblique dunes (Cooper, 1958; Hunter, Richmond and Alpha, 1983; Alpha, Hunter and Richmond, 1980). However, in contrast to the Oregon dunes, the NSW dunes show minimal net transverse movement, and maximum landward extension.

Whether transverse and other dunes occur on sand bodies or not seems to depend on the sand volume available and wind fetch. Hesp and Short (1980) found that on the 170 km long Younghusband Peninsula (South Australia), transverse dunes (up to 25 m high) were best developed in the northernmost and highest wave energy region. Here the longest upwind, unvegetated fetch is available (average 410 m; range 250 to 500 m) and the largest sand volumes exist. As the length of upwind fetch decreased so did the scale and magnitude of the dune form. This trend was intimately tied to the volume of transportable sand available. For example, a few small transgressive dunes were found downwind of narrow (< 400 m) largely sand-free unvegetated deflation basins. As the deflation basins widened (400 to 900 m) and the sand sheets narrowed, there were no transverse dunes, even though the basin was almost completely unvegetated.

The migration and movement of dunes on transgressive dunefields can be substantially modified by the presence of wet cores within the dunes (Hyde and Wasson, 1983). Wet cores consist of moist sand sitting a few centimetres to several metres below the surface. This moist sand is stored rainwater. It is not associated with the water table, and it appears to arise because solar drying of the dune surface produces a thermal blanket reducing the evaporation of subsurface moisture. When a sand transport event occurs, the dry surface sand is removed and eventually the moist wet core sand is exposed. This more compact sand is eroded and sand is transported at a much lower rate than the dry sand. In the case of the NSW reversing transverse dunes discussed above, winter westerly winds are much stronger on average than summer winds. A calculation of the potential sand transport according to the standard formulae (e.g. Bagnold, 1954) indicates that the dunes should be migrating obliquely into the sea under the influence of the oblique offshore winter winds. In fact, the dunes are virtually stationary. This is so because, in winter, the dunes experience slightly higher rainfall and significantly less solar insolation than in summer. In the hotter months, the dunes may receive less rainfall and a significantly greater amount of direct solar insolation. The net result is that the dunes erode and migrate more slowly in winter than one would predict. Even though the winter winds are stronger, the rate of surface drying of the exposed wet core is much lower than in summer, and sand is transported at a much lower rate. Thus, the dunes have a morphology primarily dictated by the winter winds (Fig. 1), but remain stationary overall due to the equal and opposite effect of summer winds (Fig. 2). This wet core effect is not an isolated phenomenon, but occurs in many dunes, both coastal and desert. For example, wet cores have been found at several metres depth in the hyper-arid, large (up to 250 m high) longitudinal dunes of the Namib Desert (J. Ward, *pers. comm*).

2.3. Precipitation and long-walled transgressive ridges

Most active transgressive dunefields display a straight to highly sinuous slipface along their landward margins (Fig. 3) (Bigarella, 1975). This marginal slipface, in effect, represents the advancing edge of the dunefield, and may vary in size and two-dimensional shape from a low convex tongue to a very high (30 to 40 m +) steep (32 to 40°) asymmetric ridge.

Where the margin of the dunefield forms a distinct ridge, the feature can be distinguished from parabolic or parabola ridges by their more elongate form, tending more shore-parallel than shore-normal (Fig. 4; see also Fig. 61 in Bird, 1976 for a map of such features on Stradbroke Island, southern Queensland). Vegetation does not always play such an active part in trapping sand to form the ridge as is the case with parabolics, but rather serves as one of the obstacles against

Fig. 3   Sinuous, long-walled slipface forming the landward margin of a tabular transgressive dunefield (photograph: W. Illenberger)

260

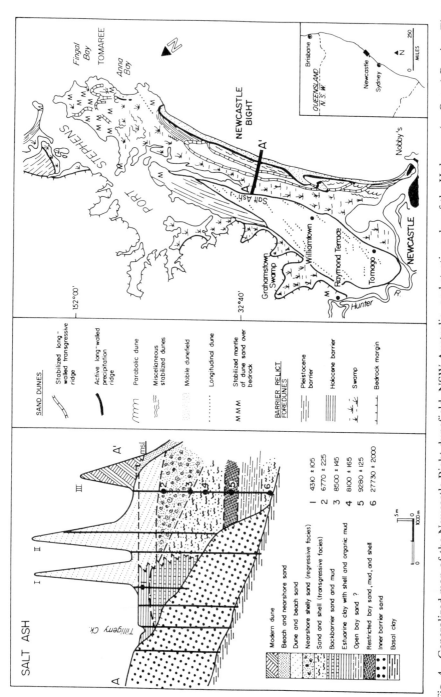

Fig. 4   Generalized map of the Newcastle Bight dunefield, NSW Australia, and stratigraphy of the Holocene (or Outer) Barrier. Three phases of transgressive dune formation are recognized: I, mid-Holocene (>4000 yr BP); II, late Holocene (<2500 yr BP); and III, modern (modified from Thom *et al.*, in press)

which sand may pile as it transgresses downwind. Vegetation may play a significant role in the evolution of the final landform in a stabilized state. Along the coast of NSW it has been possible to distinguish three phases of Holocene transgressive dune formation, each separated by a landward-facing ridge termed a long-walled transgressive ridge (Fig. 4; see Thom, 1978; Thom, Bowman and Roy, 1981; Thom *et al.*, in press, for details on chronology). Each phase appears to have resulted from the complete destruction of vegetated dunes behind the beach.

Active transgressive ridges of the long-walled type are sometimes termed wandering (Wanderdüne) or migrating dunes. Along the Oregon coast Cooper (1958) describes such features invading tall coniferous forests. He refers to them as precipitation ridges. However, similar features transgress marsh, scrub and water as well as forests without any apparent difference in the general morphology of the advancing ridge. The term transgressive frontal dune, as employed by Bird (1976, Fig. 61) for dune ridges of the long-walled type on Stradbroke Island, Queensland, cannot be invoked here because of confusion with the term foredune (see Davies, 1972, Fig. 109).

Various studies have shown that where the rate of advance of such ridges is slow (less than 1 m yr^{-1}), they build up to considerable heights (up to 100 m or more) and if isolated by the growth of vegetation on the upwind deflation plain or basin, may eventually occur as discrete, narrow, asymmetric ridges slowly advancing inland. Examples of these isolated single ridges (now stabilized) may be found along the Australian west coast (Hesp and Chape, 1984; Hesp, 1986b), Lake Huron, Ontario (Davidson-Arnott and Pyskir, 1988), and the Oregon coast (Cooper, 1958, pp. 69–74).

Rates of advance of slipfaces vary considerably and may range from almost stationary to very rapid rates (10 to 20 m yr^{-1}). The rates of advance depend principally on wind energy, slipface height, sand volume and the type of vegetation cover that exists downwind of the advancing slipface. Shepherd (1987a) quotes rates of 17 m yr^{-1} for a dunefield with a 6 m high slipface migrating across pasture and low scrub, and 1.5 m yr^{-1} for an adjacent 15 m high dune precipitating into forest. Where various types of vegetation (e.g. forest vs scrub) exist alongside one another, these variable rates of advance mean that the slipface morphology can be highly crenulate alongshore.

Stabilized vegetated ridges of the long-walled type vary from 10 to 30 km in length and from 10 to around 100 m in height (Fig. 4). They may trend parallel or slightly oblique to the shoreline. These long-walled ridges, which have been interpreted as the vegetated outer margins of former dunefields, differ morphologically in four respects from the transgressive ridge fronts of active dunefields.

(I) The windward slope is steeper in the case of the stabilized type, which frequently ranges between 15° and 25°. The comparable windward side of an active ridge is often not clearly distinguished from other features on the mobile sand surface.

(II) Lee slopes of the fixed ridge vary from 23° to 29° compared with the slipface angle of repose (30° to 35°) for the active ridge, although occasionally stabilized ridge slopes may be in the 35° to 46° range where vegetation has acted to produce steeper angles.

(III) Currently mobile ridges in NSW seldom exceed 20 m height on their leeward side. Stabilized ridges in NSW may exceed 50 m.

(IV) Micro-relief of the stabilized ridge is more intricate and diversified than that of the active dune. The former is composed on conical-shaped hillocks sharply separated by cols 5 to 15 m below hillock summits.

In some areas pre-existing vegetated or partially vegetated dune hillocks and remnant knobs provide a prominence around which mobile sand is wrapped. In central NSW such features form steep-sided dune ridges which are generally symmetrical in section. Enclosed depressions separate the hillock from the wrap-around dune (Thom *et al.*, in press). A remarkable pattern of these dunes may be observed at the eastern end of a long-walled ridge of the Newcastle Bight embayments where three hillocks on the windward side of the ridge are almost completely encircled by sharp crested ridges, 2 to 4 m high. Their morphological character suggests a climbing series of three or four ridges, which have advanced from the south and embraced a stabilized or semi-stabilized hillock at successively lower elevations. These appear similar to the echo dunes of Tsoar (1983).

3. EROSIONAL MORPHOLOGIES

Although transgressive dunefields are chiefly composed of moderate to large-scale unvegetated to partially vegetated sands often arranged into various types of landforms (e.g. transverse dunes), many sub-environments of erosional origin also occur. These include deflation plains, slacks, deflation ridges, and remnant knobs.

3.1. Deflation plains and basins, and slacks

Deflation plains (or basins) and slacks are commonly found along the seaward margin of landward migrating transgressive dunefields. Deflation basins are wind-eroded deflated hollows, semi-circular troughs and depressions. Deflation plains are more elongate and form relatively extensive flat surfaces. Slacks usually evolve in the same manner, but are generally distinguished by the occurrence of groundwater at, or near the surface (Tansley, 1949; Salisbury, 1952). Slacks may also occur as inter-dune troughs or depressions between transverse dunes (Fig. 1A; see also McLachlan, Ascaray and du Toit, 1987).

Along the east coast of Australia, deflation basins and slacks commonly develop via a temporal sequence (Fig. 5). Initially, large established foredunes are wave-eroded and scarped. Wind erosion of the devegetated steep stoss face results in gradual erosion and backwearing of the scarp face (Fig. 5A). Localized blowout development may result in the accelerated development of this process in places, as may be seen in the foreground of Fig. 5A.

As the foredune scarp erodes, the toe of the scarp recedes landward, and an incipient deflation surface begins to form in the seaward region. Figure 5B illustrates an advanced stage in this evolutionary trend. The original foredune (I in the distance) has all but been removed by aeolian erosion, the leeward upper dune slope, partially protected by a soil capping, representing the only remaining remnant (II). As the foredune has eroded, the deflation basin (III) has formed and become increasingly wider. Sands transported from the deflation basin and eroded from the foredune are blown landward to form a tabular, undulating transgressive dune sheet which mantles the former lee slope of the foredune and older, more landward dunes. This process may continue until the entire foredune is completely removed, and a deflation plain or basin and attendant slacks occupy the foredune's former position (Fig. 5C). Deflation generally proceeds until a base level is reached. This is usually the water table or a calcrete layer. The latter is more common in the carbonate-dominated coastal environments of South and Western Australia. Fig. 5D shows that in some instances deflation plains may become extensive (see also the Coos Bay dunefield, Alpha, Hunter and Richmond, 1980; Hunter, Richmond and Alpha, 1983).

In the more semi-arid dunefields where foredunes are negligible, the deflation basin or plain develops immediately landward of the beach and may extend downwind for a considerable distance.

3.2. Deflation ridges

Low, elongate, discrete, coast-parallel narrow ridges which are covered by pumice pebbles are sometimes found within deflation basins around the Australian coast (foreground of Fig. 5C). The ridges are typically 0.5 to 1.0 m wide, around 0.5 m high and vary in length alongshore from a few metres to about 15 m. They trend from being relatively straight to being slightly sinuous. These ridges result directly from deflation of storm and incident swash-deposited sediments. This material is available for reworking because deflation commonly occurs to a level below the maximum height of deposition of storm-deposited beach sediments. Thus, if shoreline progradation has taken place prior to deflation basin development, it is likely that relict beach sediments will be exposed in the base of deflation basins.

The deflation ridges are formed due to differential rates of removal of different sized sediments. Former swash lines composed of larger quantities of pumice and pebbles act as lag deposits, protecting underlying find sands from erosion. Adjacent sands are eroded and the ridges form as the surrounding surface is deflated. If a series of such relict discrete pumice or cobble-rich swash lines are exposed, a suite of deflation ridges may form.

Fig. 5A   Initiation of a deflation basin. The large established foredune has been scarped by storm waves and the stoss face is backwearing (I). Areas cut by blowouts show a more advanced development (II) and greater degree of deflation basin creation

Fig. 5B   The established foredune (I) has been largely removed by aeolian erosion, creating a concave deflation basin (III) to seaward. The original leeward crestline appears as soil-capped yardang-like remnants (II). Sands from the foredune have blown landwards to form a narrow undulating dunefield

Fig. 5C   A wide, water-filled deflation basin has formed to seaward of a transgressive dunefield and in the position of a former large established foredune (same area as Fig. 2). As deflation proceeds (right side) revegetation takes place on the upwind side (left) and a low hummocky dune terrain is formed. Deflation ridges (immediate foreground) are formed as former swash lenses are exposed. In each case (A, B, C) incipient foredunes have reformed to seawards of the dunes

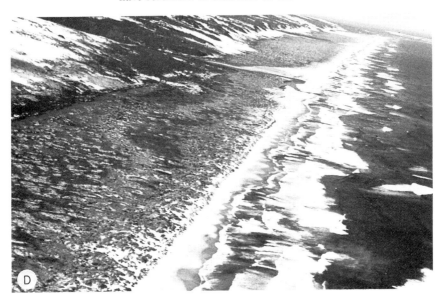

Fig. 5D   Extensive deflation basin (1.5 km wide) fronted by an established foredune and backed by vegetated and active transgressive dunefields on the Warren–Donnelly rivers coast, southwestern Western Australia

Fig. 6A   Remnant knobs. Residual hummocks colonized by *Sonchus megalocarpus*, formed by aeolian erosion of the adjacent unvegetated surface

Fig. 6B   A 4 m remnant knob, formerly erosional and now being colonized and protected from further erosion by the highly adventitious introduced Bitou Bush (*Chrysanthemoides*). An upwind scour bowl can still be seen

Fig. 6C   Large remnant knob formed in partially cemented aeolian calcarenite, blocks of which litter the slopes and reduce the rate of aeolian erosion

3.3. Remnant knobs

The process of deflation and erosion of a pre-existing dune terrain is never uniform. Topographic highs, zones where wind flow is locally accelerated, and poorly vegetated areas are eroded more rapidly than some other terrain/vegetation units. This non-uniformity in the rate of erosion leads to the development of such landforms as remnant knobs (Fig. 6). Remnant knobs are sandy mounds or knobs, partially to fully vegetated, that once formed part of a particular dune type or a more continuous vegetated dunefield. These dune forms are also termed turret dunes (Olson, 1958) and remainié dunes (Guilcher, 1958).

Large-scale remnant knobs commonly develop via the sequence illustrated in Fig. 5 (see also Fig. 1 in Hesp, 1988a). In Fig. 5A the formation of two blowouts has resulted in the removal of a significant portion of the seaward and lateral parts of the established foredune, leaving a large remnant knob behind. In Fig. 5B the foredune has largely been removed in the foreground and central portions, leaving soil-capped and partially vegetation-protected, yardang-like knobs.

The species and habit of colonizing vegetation can be a critical factor in determining knob morphology and evolution. Small remnant knobs may evolve around a single plant. Plants which do not have stolons and rhizomes seem to be particularly susceptible to erosion during high wind velocity events. Examples in the east Australian environment include *Stackhousia spathulata*, *Arctotheca nivea*, and *Sonchus megalocarpus*. These species grow in small clumps. Observations indicate that once the bed surrounding the clumps is eroded, the sand underneath the clumps is gradually removed, leaving the plants perched on small pedestals above bed level (Fig. 6A). Once this occurs, the bed may be lowered rapidly, and the plants removed from their substrate. Mat plants such as *Scaevola*, *Hibbertia* and *Carpobrotus* appear more able to cope with erosional conditions during high-velocity events because of their stoloniferous or rhizomatous nature. When the surface surrounding the plants is eroded, the plant rhizomes and stolons can maintain contact with the bed. This maintains their aerodynamic roughness and reduces substantially the local erosion around and within the plant body (Fig. 6B). These vegetated mounds may thus gain in relative height through adjacent surface erosion. According to Pidgeon (1940), Salisbury (1952) and Ranwell (1972) the mounds provide a means for re-establishment of foredunes and revegetation of mobile dunes following severe wind erosion.

The extent and nature of root development can also be an important factor in influencing knob evolution. Knobs covered by vegetation with close multiple root systems display a high degree of sub-surface binding which can substantially reduce the rate of lateral erosion. Fig. 6A shows plant species which display minimal root development or lateral spread; these species tend to have less sub-surface binding capacity and rates of knob erosion may be greater (Esler, 1970; Hesp, 1988a).

In the more carbonate-rich environments, the formation of calcarenite and calcrete deposits within dunes is common. Where deflation takes place, remnant knobs may be formed and survive for some time due to the presence of calcarenite, which reduces the rate of erosion. In these cases, plants do not need to be present, and the knobs may be covered by a litter of calcarenite blocks (Fig. 6C).

Remnants knobs tend to be preferentially eroded on the windward side. The operation of winds from various directions produces a semi-circular shape (Fig. 6C). If the knob is not exposed to certain wind directions, the unexposed areas may be better vegetated and less erosional. Once formed, the remnant knobs tend to influence the surrounding topography. Brighton (1978) has shown that airflow approaches a knob horizontally, but it accelerates and separates around the sides, forming a downstream wake. The accelerated near-surface horizontal flow around the knob is responsible for concave scouring at the knob base, in a fashion similar to the scouring that occurs on riverbeds in the vicinity of bridge piers (Thwaites, 1960). Echo dunes may form around the upwind margins. Brighton's (1978) observations have been verified in the field by Scorer

(1955) and Mason and Sykes (1979). Wind velocities also increase at the knob crest; this acceleration would induce erosion on the sparsely vegetated crests of knobs (see Cooper, 1958, Plate 20).

The growth of some plant species, particularly stoloniferous ones, during low wind velocity periods or the colonization by more adventitious species, may eventually lead to the covering and stabilization of remnant knobs. In east Australia such species include *Scaevola calendulacea, Chrysanthemoides moniliferum* and *Hibbertia scandens. Chrysanthemoides* (Bitou bush) is particularly adept at stabilizing erosion knobs, because of its long woody root system which, once exposed, provides a high degree of cover and its multi-branching promotes rapid growth (Fig. 6B).

4. DEPOSITION MORPHOLOGIES

A variety of ecological and geomorphological changes may take place as deflation basins and slacks develop. In regions where aggressive pioneer vegetation species are present (e.g. *Spinifex* sp. on the Australian coast), incipient foredunes may rapidly reform on the seaward margin of deflation basins (Fig. 5B). From this position the vegetation may colonize, partially or fully, the deflation basin. One such example occurs on the Oregon coast, where the introduced *Ammophila* sp. has colonized the upper beach and seaward deflation zone of the Coos Bay dunefield, bringing widespread stability to a formerly, and naturally active deflation basin and near-vegetation-free beach. In NSW, deflation basins may be revegetated in a successional sequence partially illustrated in Fig. 5C. Here, the upwind margins of water-filled slacks are colonized by moisture-loving species such as *Isolepis nodosa* and the grass *Acidanthera brevicollis.* As these plants spread by rhizome and seed production, they trap sand and, eventually, low, undulating, irregular dunes are formed, particularly around the upwind margins of the slacks. Sometimes these dunes may develop as discrete ridges if the slack margin shoreline remains stationary for some time. These dunes are probably similar to the Gegenwalle ridges of Paul (1944), and their morphological development is similar to the so-called ground patterns of Pye (1982). As the height of the low ridges and hummocky aeolian flats increases due to sand deposition, other pioneer plants and eventually shrubs colonize the area in a well-defined succession. As deflation basins and slacks advance downwind and revegetate on the upwind margins, a confused vegetated hummocky dunal and relict slack basin topography may be left behind (see Figs 5C, 5D).

Various types of dunes, including coppice dunes and shadow dunes and a variety of other dunal topographies (largely unnamed), may also develop in deflation basins and on active transgressive dunefields.

Fig. 7   Depositional mounds. (A) A variety of Marram (*Ammophila*) mounds formed
by the deposition of sand within Marram seedlings. The mounds grow to 2 to 3 m high,
and (B) as the surrounding deflation plain is vegetated, the Marram dies and is replaced
by other herbs and shrubs. Warren–Donnelly coastline (see Fig. 3D)

4.1. Coppice dunes

Figs 5 and 6 indicate that once a deflation basement is reached, vegetation may again colonize the surface, trap sand and form dunes. Sometimes this dune formation is fairly regular, sometimes chaotic. In the former case coppice dunes similar in mode of formation and morphology to Type 1 incipient foredunes (Hesp, 1984) may be formed by the growth of discrete clumps of pioneer plants and other vegetation species. Fig. 7A illustrates an example where Marram grass (*Ammophila arenaria*) has naturally colonized a wet deflation plain (as depicted in the foreground of Fig. 5D). Sand blown across the plain is trapped by the grass, forming a semi-circular conical mound. Aeolian deposition encourages shoot growth. The clumps spread laterally, trap more sand and continue to increase in height and width (Fig . 7A background). Over time, usually several years, the intervening unvegetated plain is colonized by a suite of low grasses and herbs, particularly water-tolerant plants such as *Isolepsis* sp. and aeolian sand is no longer available to the Marram coppice dunes. The Marram dies out and is replaced by shrub species (Fig. 7B).

In more harsh environments where the water table is not so accessible or the deflation surface is a coarse pavement (often nodular calcrete), more typical desert-type coppice dunes develop (Fig. 8). Here, strong winds transporting sand and salt spray form dunes which slowly migrate downwind. The seawardmost exposed leaves of plants are pruned and die and the landward, more protected leaves and shoots preferentially grow downwind, resulting in the gradual

Fig. 8   Coppice dunes formed within a single shrub species, Cocklebiddy region, Western Australia. The dunes are gradually migrating downwind across a nodular calcrete deflation plain. A transgressive dunefield occurs to landward

Fig. 9   Shadow dunes, around 1 m high, formed on a tabular transgressive dunefield,
Ningaloo coast, Exmouth Peninsula, Western Australia

migration of the dune. The actively growing portion of the plant and its
associated dune may eventually be located several metres downwind of the initial
germination point and tap root zone. Such plants provide a reasonably accurate
indication of the direction of net regional winds for the dunefield.

4.2. Shadow dunes

Shadow dunes are commonly formed on transgressive dunefields and may be
depositional types (Fig. 9; see also Hesp, 1981) or erosional types (Clemmensen,
1986). They may form downwind of coppice dunes, but more commonly occur
on more sand-rich substrates. In the latter case, large-scale shadow dunes may
form downwind of remnant knobs and vegetated mounds (Hesp, 1981). Cooper
(1958) has described and illustrated basal scouring of large, forested remnant
dunes with the eroded sand often deposited in the lee of the knobs as pyramidal-
shaped or lobate shadow dunes. If the knobs are large, or the sand supply
is large, these shadow dunes may continue to form landward of the knob,
and they have been known to migrate downwind under certain conditions
(e.g. the salt rush and willow precipitation hillocks and associated tongues
of Oregon; Cooper, 1958). Large shadow dunes also form downwind of
gully heads on Pleistocene cliff tops in several coastal dunefields (e.g.
New Zealand; Brothers, 1954). These have sometimes been classified as
longitudinal dunes in both desert (Melton, 1940) and coastal (Cotton, 1942)
environments.

Fig. 10   An incipient bush pocket formed within the lee of a transverse dune, Alexandria dunefield, Port Elizabeth, South Africa

Fig. 11   Evolutionary sequence of bush pockets formed in the lee of transverse dunes, Alexandria dunefield, South Africa. Small herbs and pioneer species (I) gradually spread and are followed in a defined successional sequence by intermediate shrubs (II) and eventually low heath and forest (III) (photograph: W. Illenberger)

4.3. Bush pockets

Large pockets or areas of shrubs and even forest occur within some active dunefields (Fig. 10). Cooper (1958), Hunter, Richmond and Alpha (1983), and Alpha, Hunter and Richmond (1980) have described and illustrated the discrete forested dunes in the Coos Bay dunefield. McLachlan, Sieben and Ascaray (1982) have termed such environments bush pockets, a term retained here since the genetic origin of such sites is questionable.

A variety of plants may grow within transgressive dunefields. As these spread laterally, other plant species may colonize the site and a successional suite of plants evolve, trapping sand and forming large vegetated depositional mounds (Figs 10 and 11). Windblown detritus or litter often collects at the base of transverse dune slip facies, held there by reversing vortices formed within flow separation envelopes. Once covered with a minimum of sand and provided some moisture is available, plant seeds within the litter germinate and thrive. In the South African example illustrated in Fig. 10, the bush pocket is initiated by the germination of *Myrica cordifolia*. Once this is established, a variety of other species including *Stoebe plumosa*, *Cynanchum natalicum*, *Passerina rigida* and *Rhus crenata* grow (McLachlan, Sieben and Ascaray, 1982; Hesp, 1986a). As the slipface of the transverse dunes advances, the upwind margin of the bush pocket is destroyed, the central and downwind margins trap sand forming mounds, and the downwind edge of vegetation extends outwards. In examples such as that depicted in Fig. 10, the downwind vegetation margin extends up to 3.75 m yr^{-1} (McLachlan, Sieben and Ascaray, 1982). If slipface advance rates are very low, the slipface itself may be colonized (Fig. 10). Figure 11 illustrates bush pockets in various stages of evolution. On the left-hand side, the vegetated pockets are small and consist of 2 to 3 species (I arrowed). In the centre the pocket is more extensive and dominated by intermediate shrubs (II). On the right margin the pocket is extensive and dominated by woodland species (III).

This evidence for the evolution of large-scale bush pockets within active dunefields raises questions about the validity of assuming that all such pockets are remnants. For instance, Cooper (1958) refers to the large forested dunes in the Coos Bay field as remnant (p. 158), yet it is possible that the dunes are depositional, resulting from the processes described above and indicated in Figs 7, 8 and 9.

Just as remnant knobs may be recolonized by a rapidly spreading prostrate vegetation species and revert to an accretional, or a stationary feature, so too may the coppice dunes, shadow dunes, depositional mounds and bush pockets evolve and become so large that they are a significant element within the lower boundary layer and they actually begin to erode. Cooper (1967, his Plate 11, Fig. 2) illustrates examples of this process. Thus, it is fairly difficult to determine at any one time whether a vegetated mound or bush pocket is a remnant feature or a depositional feature. The presence of a successional pattern within the vegetation, the species involved, as well as the presence of buried soils may provide clues to the origin of the feature.

5. MECHANISMS FOR ACTIVE TRANSGRESSIVE DUNE DEVELOPMENT

The initiation of active transgressive dunefields can take place either as the result of a new pulse of beach sand being made available for onshore aeolian transport, or as a result of reactivation of previously stabilized dunes (Davies, 1980). However, the causes of active dune development may be different in both cases. Furthermore, there is often uncertainty as to whether the causes are induced by changes in external boundary conditions or due to internal morphodynamic adjustments in the sense of Wright and Thom (1977; see also Chappell and Thom, 1986). Additional uncertainty is created by the difficulty of determining whether periods of initiation are episodic and regional or are random events triggered by the local impact of external or internal causes. In the following sections we seek to explore further the problems of transgressive dunefield initiation.

5.1. Sea level change and dune initiation

The literature contains many references to the possibility that a rise or a fall in sea level may induce dune mobilization (see Cooper, 1958, for a discussion of earlier literature). There is a school of thought which argues that such dunes form when sea level falls, exposing a wider sand plain to wind energy (e.g. Dominguez, Martin and Bittencourt, 1987). Even slight falls in sea level have been invoked as a cause of major periods of sand blowing in the Holocene (e.g. Schofield, 1975, in New Zealand). It is possible on some coasts that at times of glacially lowered sea level there were active dunes on what are now submerged parts of continental shelves. However, the lack of firm dating and morphostratigraphic evidence from shelf deposits and the role and importance of vegetation (often ignored) makes this hypothesis difficult to test.

It is easier to correlate sand migration onshore with the period of rapid sea level rise which terminated the Postglacial Marine Transgression (PMT) on many coasts of the world. Large-scale erosion of the shoreline combined with a significant supply of sediment during a rising sea level event such as the PMT may lead to the development of transgressive dunefields. Thom, Bowman and Roy (1981) provide stratigraphic evidence which indicates that transgressive dunefields have developed during rising sea levels on the NSW coast. Pye and Bowman's (1984) and Shepherd's (1987a) datings of parabolic and transgressive dunefields indicate that at least some of the Australian and New Zealand coastal aeolian systems were initiated prior to the so called stillstand (the last 6000 to 7000 years in Australia). This work substantially confirms the hypothesis developed by Cooper (1958) from his Oregon work.

5.2. Sediment supply

Sand may be supplied to the coast by various mechanisms and become available for onshore transport by wind. Many barrier systems in Australia and New Zealand were formed during the early and middle parts of the sea level stillstand post-6500 years ago (Thom, 1984; Thom and Roy, 1985). This development indicates that nearshore profiles were in disequilibrium and sediment was transported shorewards for some time after sea levels reached the present height in order to establish equilibrium profiles. Secondly, river supply and longshore drift may contribute significant sediment volumes to the nearshore-beach-dune system. River supply is important for the continuing development of transgressive dunefields on the west coast of New Zealand, and the central Manawatu coast is prograding at 1 to 2 m yr^{-1} (McLean, 1978). Many dunefields on the California-Oregon coast have developed as a result of sediment supply from rivers (Cooper, 1958, 1967). On the Baja California, Mexico coast (Inman, Ewing and Corliss, 1966; Fryberger, Schenk and Krystinik, 1988), and the southern coast of South Africa (Tinley, 1985; Hesp *et al.*, 1989) large transgressive dunefields have developed, in part in response to large fluxes of sediment transported into sediment traps by longshore drift (Fig. 12).

Fig. 12   Sinuous transverse dunes separated by wet (occasionally tidally inundated) slacks on a tabular active transgressive dunefield, Guerro Negro, Baja California, Mexico. Here southerly longshore drift delivers large quantities of sediment to the coast resulting in concurrent beach progradation and transgressive dunefield development

In all these cases, the sand supply to the backshore of the beach is so large that the local pioneer plants are inundated and killed (Marta, 1958; Shepherd, 1987a), or else they do not colonize the newly arriving sand with sufficient speed and cover to prevent onshore wind transport and the formation of a transgressive sand sheet or dunefield.

5.3. Changes in storminess and shoreline erosion

Changes in the duration and intensity of storms or the degree of storminess has been cited as a possible determinant in shoreline destruction and an initiation mechanism for transgressive dunefield development (e.g. Cooper, 1958). For coastal NSW, Thom (1978) proposed a model of climatic change in which the determining factor is a change in the degree of autumn-winter-spring cyclogenesis, caused in part by a change in the intensity of atmospheric blocking action in the Tasman Sea. Such changes may lead to significant increases in wave energy along the southeast Australian coast, producing partial or complete destruction of established foredunes and initiation of blowouts and transgressive dunefields (Thom, 1978). Whilst data available to test such climatological models have improved (e.g. Bryant, 1987), we have little information to test this or other models invoking climatic change involving changes in storminess. The series of storms which affected the NSW coast in 1974 (e.g. Foster, Gordon and Lawson, 1975) resulted in significant scarping of established foredunes, most of which had by 1988 recovered a reasonable level of vegetation. Subsequent blowout development was minimal, and transgressive dunefield development did not occur. Even though these storms were perhaps the second largest the State has experienced in historical times, they were not sufficient to produce the effects required for dunefield development, and it seems that the minimum requirement is a series of such storms over a period of following years.

The relationship of storminess to shoreline erosion and dune stability/instability has been discussed by Chappell and Thom (1986, p. 113; see their Fig. 2). They argue that:

> Dunes are mobilised after a sufficiently damaging storm event(s) and revegetation commences only when deflation in the windward part of the active dune has lowered the surface sufficiently for the ephemeral water table to be adequately tapped by spreading plants. Response time for restabilisation . . . [then] . . . involves both geomorphic and vegetational factors . . . [and they suggest] . . . that vegetation is stable up to a threshold of high storm magnitude, with low frequency of recurrence. According to . . . [their] . . . model, revegetation occurs when random variation leads to the storm index (or number of storms of sufficient magnitude) being low.

Shorelines and dune systems which were once stable can become unstable once significant erosion of the shoreline takes place (Dolan, 1972; Leatherman, 1979). Along the central west coast of Western Australia, where localized shoreline recession is taking place at up to 1 m yr^{-1} (Semeniuk and Meagher,

Fig. 13A   Transgressive dunefields and large-scale parabolic dunes formed on a receding barrier (1 m yr^{-1} recession); Leschenault Peninsula, 100 km south of Perth, Western Australia (photograph: S. Chape)

Fig. 13B   Broad, coalescing parabolic dunes on the Younghusband Peninsula, South Australia. Marginal vegetation has been eroded and buried until the originally discrete parabolic dunes have coalesced to form a transgressive dunefield (photograph: A. Short)

1981), parabolic and transgressive dunefield systems tend to dominate (Fig. 13A; Hesp and Chape, 1984). It is clear from contemporary observations that sea level does not have to be rising for wave and wind erosion to create coastal dune instability. Locally, the diversion of a river or estuary mouth can lead to such instability. The classic example in the literature is the impact of shoreline erosion of the southern shore of the Basin d'Arcachon, Les Landes, in southwest France. Here the southward migration of a spit initiated erosion downdrift, leading to extensive aeolian instability forming the famous Pyla dune (Buffault, 1942; Bird, 1976, Fig. 6.2). In general, foredune erosion and scarping by wave action may result in the removal of the pioneer vegetation community, leaving intermediate and local climax species exposed to higher wind stress, salt spray and sand inundation. These plant communities may die back, exposing the dune surface to wind erosion. The foredune may then be slowly destroyed, and the sand is blown landwards to form a transgressive dunefield.

Even where the vegetation on scarped foredune crests survives, slow aeolian backwearing of scarped stoss faces may take place and the foredune is gradually removed (Hesp, 1988a). Transgressive dunefields commonly occur in the northern ends of embayments along the NSW coast (Langford-Smith and Thom, 1969). Many of these dunefields have developed either via foredune destruction during storms, or by coalescence of blowouts formed within highly erosional foredunes. The latter are often completely removed by aeolian erosion. The occurrence of these dunefields is attributed to increasing exposure to winds, especially storm winds, in the more northerly, south-facing portions of the embayments (Thom, 1974), but may also be related to the increasing potential for sediment delivery, high-energy wave erosion, and foredune destruction (Short and Hesp, 1982; Hesp, 1988b).

5.4. Foredune height/wind shear thresholds

Where foredunes develop on stationary or very slowly prograding shores, they may reach considerable heights. Along the NSW coast, established foredunes are seldom higher than 30 m above MSL (Hesp, 1982, 1988). This appears to be a maximum threshold level above which wind shear is so significant that aeolian erosion dominates over accretion, even where a fairly high vegetative cover exists (Shepherd, 1981; 1987b). Thus, the potential for purely aeolian erosion of foredunes and the development of blowouts increases as foredune height increases, and such erosion may lead to the development of transgressive dunefields.

5.5. Coalescence of blowouts and parabolic dunes

Blowouts and parabolic dunes are common along many coasts. As these dunes evolve, lateral expansion as well as downwind extension and migration take place. Thus formerly vegetated ridges separating discrete blowouts and parabolics are gradually eroded and the dunes coalesce by lateral expansion, forming transgressive dunefields (Fig. 13B; see also Brothers, 1954; Short and Hesp, 1982). In such areas, there is an interaction between the sand-binding capacity of plants and the erosive ability of wind. Any factor which inhibits vegetative growth, such as overgrazing or fire, may create conditions leading to mobilization of sand through the expansion of blowouts. Thom *et al.* (in press) have argued that much of the vegetation stabilization currently observed on previously active dune sheets (i.e. pre-1970 period) may in part be attributed to the removal of cattle. The introduction of new sand-binding species to an area (e.g. Marram in Oregon and the NZ west coast) may have a similar effect.

5.6. Vegetation succession

Along the west coast of the North Island of New Zealand there are areas (e.g. the Manawatu region; Ninety Mile Beach) where well-developed established foredunes have formed within predominantly introduced and pioneer vegetation species. However, whilst the seaward stoss slopes are well-vegetated and reasonably stable, the landward slopes are poorly vegetated and highly unstable. Sand is eroded from the landward slopes and transported downwind forming parabolic dunes and transgressive dunes. It appears that there are few intermediate vegetation species available to colonize the landward foredune slopes, so these slopes are subjected to aeolian erosion. One could speculate from this example that any major hiatus within a vegetation successional sequence may result in unstable dune terrain and lead to transgressive dune development, given an appropriate combination of energy, sand supply and plant species conditions.

5.7. Aeolian erosion of cliffs

Wind acceleration over high sea cliffs can be considerable, and can lead to significant stress for plants growing along cliff edges. Longer than average duration storms, human or animal disturbance, cliff erosion, or climatic stress (e.g. summer droughts) may lead to a weakening of the vegetation cover and the deflation of cover sands. Many cliffs along the southern Australian east and west coasts are composed partly or entirely of aeolian calcarenites overlain by Last Glacial and early Holocene partially- to uncemented cliff-top dunes (Hesp and Chape, 1984; Short, 1988), and in several places these have been reactivated, forming transgressive dunes.

5.8. Summary

Although we can distinguish a variety of mechanisms which could produce the conditions necessary for transgressive dunefield initiation, it is not clear whether dune activity is an expression of a discrete event in response to some common cause, or whether many phases are represented due to a variety of interacting factors, some external and some internal. Campbell (1915) long ago recognized the possibility of localized and occasional breaches in California dunes eventually shifting the material of the whole dune (cited by Pye, 1983, p. 549). We must view with caution any simplistic attempt to correlate transgressive dunes with climatic events. Limited understanding of morphodynamics in coastal environments where transgressive dunes occur makes interpretations of climatic change as a cause of dune activity hazardous. Thus, the link between greenhouse effects and coastal dunes will remain tenuous for the time being.

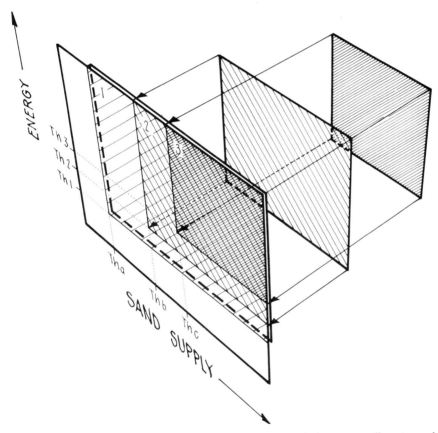

Fig. 14   Conceptual diagram of the relationships between wind energy, sediment supply and the degree and effectiveness of vegetation cover. Three 'windows' are shown, window 1 in which there is little sand-binding vegetation available; window 2 (middle) in which some vegetation is present, and window 3 in which aggressive sand-binding vegetation is present. In order for transgressive dunefields to form at all, minimum thresholds of wind energy and sediment supply must be crossed; these thresholds become increasingly higher as one progresses from window 1 to window 3. (See text for further discussion)

6. DISCUSSION

The initiation and evolution of transgressive dunefields may be conceptually understood by examining the relationship between the three controlling factors noted by Pye (1983) following many authors: (i) rate of sand supply, (ii) wind energy and (iii) effectiveness of sand-binding vegetation. In Fig. 14 we attempt to show three-dimensionally the interrelationships that vegetation may have with sand supply and wind energy. On the first (forwardmost) 'window' there is little or no sand-binding vegetation available. Thus all that is required is a minimum wind energy threshold for sand transport to occur, and a minimum sand supply for any transgressive dunes to form at all. In the left-hand bottom zone of this window may be plotted the minor dunefields of the southern and southwestern Australian coast, USA southwestern coast and South African coast (low to moderate sand supply and low to moderate energy). In the top right-hand corner the major coastal dunefields may be plotted, for instance, the Vizcaino desert (Baja California), the South African dunefields (e.g. the Alexandria), and the semi-arid South Australian fields.

Window 2 indicates the presence of some plant species which are capable of binding sand during periods of high rainfall, periods of high seed productivity, or other favourable environmental conditions. The minimum threshold required to initiate dunefields is a little higher (as indicated on the vertical and horizontal axes) since more energy may be needed to mobilize sand and more sand may be required to overwhelm the vegetation. In this window we could place Coos Bay, Oregon (prior to Marram planting), the NZ west coast dunefields (although the effect of higher rainfall may be important) and the higher rainfall (900 to 1200 mm) west Australian central coast examples.

Window 3 indicates the presence of aggressive sand-binding plant species capable of colonizing a bare sand surface much of the time. The minimum thresholds required to initiate dunefield development are higher than in the previous two windows, as the vegetation presence inhibits creation of transgressive dunes. In this window we can place the more humid coastal dunefields of the Victorian and NSW coast, the semi-arid northwest Australian and sub-tropical northeast Australian dunefields where occasional tropical cyclones provide the energy input, and many of the *Ammophila*-dominated dunefields of the British Isles and Europe.

In essence, whilst each window is seen as distinct or separate from every other window, it is conceptually possible to see that more energy and a greater sand supply would have been needed to initiate transgressive dunefields in humid coastal NSW, where *Spinifex* occurs, compared to humid Oregon prior to the introduction of *Ammophila* sp.

At a second-order level it is possible to utilize the diagram to identify assemblages of dune landforms which express the interaction of sand supply, wind energy and degree of vegetation cover. In this way Fig. 14 may have some value in providing a framework for future studies.

In window 1 where there is little vegetation present, the dominant dune types will encompass: (i) a narrow, relatively flat transgressive sheet or small dunefield and deflation plain where sand supply (SS) and wind energy (WE) are minimal; (ii) moderate-width transgressive dunefields dominated by transverse and oblique dunes and interdune deflation plains and troughs where SS and WE are moderate; and (iii) massive transgressive dunefields dominated by draa with extensive transverse and aklé dunes where SS and WE are high (e.g. Figs 1A and 1B).

In window II, where some vegetation is present, foredunes, blowouts and parabolic dunes are likely to be formed around the threshold regions; parabolics may occur in conjunction with transgressive dunefields so long as SS and WE are not too high. In addition to the dune types found in window I above, where SS and/or WE are low, partial revegetation following dunefield development is possible and foredunes may develop or reform on the seaward margins. The probability of the formation of coppice dunes, shadow dunes, depositional mounds and bush pockets, and partial revegetation of deflation basins is high where SS and WE are moderate to low (Fig. 8). As SS increases, the survival of plants lessens, and the occurrence of the vegetation-induced depositional dune types will decrease. The presence of vegetation will mean that long-walled, marginal slipfaces and ridges develop, dependent on the biogeographic region into which the dunefield migrates (e.g. compare the arid, cacti shrubland of Baja California with the forests of Oregon).

In window III the presence of vegetation ranges from moderate to pronounced. Thus the potential for foredune redevelopment following a dunefield initiation event is high. In regions of low SS and WE, active dunefields will be limited in extent. As sand supply trends from minimal to maximal, there will be a trend from single, narrow isolated transgressive ridges, to moderate-scale broad mounds or a series of ridges separated by deflation basins, to large-scale tabular dunefields and draa with extensive high marginal slipfaces and transverse/oblique dunes (compare Figs 5C, 5D and 13B). As these dunefields advance alongshore and landwards into vegetation, they commonly break into parabolic dunes. Whilst Cooper (1958, p. 65) indicated that there were no vegetated transverse dunes on the Coos Bay dunefield, we find well-defined transverse dunes vegetated with Eucalypt forest in some transgressive dunefields in NSW. Thus, Cooper's Coos Bay example falls within our window II whilst the NSW examples fall within window III.

7. CONCLUSION

Active transgressive dunefields display a wide variety of landforms and landform units. There is considerable variation in the scale of active transgressive dunefields, and to some extent this has led to a plethora of terms to describe the morphologies. A better appreciation of the dynamics of some of the dune types found on transgressive dune sheets now exists as a result of detailed experimental work (e.g. Hunter, Richmond and Alpha, 1983). The link between vegetation cover and dune mobility in different ecological and climatic environments is also gradually being better understood. However, causes of dune initiation and subsequent stabilization appear numerous. Overly simplistic interpretations must be avoided as we enter an era of human-induced climatic change. Evolutionary models must be treated with caution unless supported by independent evidence of environmental change induced by external or internal factors operating on an aeolian sand surface.

ACKNOWLEDGEMENTS

We wish to express our appreciation to Mike Shepherd, Andy Short, Anton McLachlan, Ralph Hunter and Robert Hyde for numerous discussions and field trips on transgressive dunefields, to Mike Shepherd and Peter Roy for fruitful discussions on this paper, to Werner Illenberger and Stuart Chape for provision of photographs, to Janette Brennan and Sue Morsby for typing, to the Australian Research Council (ARC), Macquarie University and CSIR (South Africa) for funding, and to Kathleen and Irene for their patience and support.

REFERENCES

Alpha, T. R., Hunter, R. E. and Richmond, B. R. (1980) Maps showing landforms of the Umpqua South area, Oregon Dunes Nat. Rec. Area. Oregon. Geol. Surv. Misc. Field Studies Map MF 1205.

Bagnold, R. A. (1954) *The Physics of Blown Sand and Desert Dunes*. Chapman & Hall, London 265 pp.

Bigarella, J. J. (1975) Lagoa dunefield, Santa Catarina, Brazil—A model of aeolian and pluvial activity. *Bulletin Para. Geosci.*, 33, 133–167.

Bird, E. C. F. (1976) *Coasts*. ANU Press, Canberra, 282 pp.

Brighton, P. W. M. (1978) Strongly stratified flow past 3-D obstacles. *Q. J. Roy. Met. Soc.*, 104, 289–307.

Brothers, R. N. (1954) A physiographical study of recent sand dunes on the Auckland West Coast. *N.Z. Geogr.*, 10, 47–59.

Bryant, E. (1987) CO_2-warming, rising sea-level and retreating coasts: review and critique, *Austr. Geogr.*, 18, 101–113.

Buffault, P. (1942) *Histoire des dunes maritimes de la Gascogne* Éditions Delmas, Bordeaux, 446 pp.

Campbell, H. (1915) Movement of sand dunes on the California coast. *J. Washington Acad. Sci.*, 5, 328.

Carson, M. A. and MacLean, P. A. (1986) Development of hydrid aeolian dunes: The William River dune field, northwest Saskatchewan, Canada. *Can. J. Earth Sci.*, 23, 1974–1990.

Chappell, J. and Thom, B. G. (1986) Coastal morphodynamics in North Australia. Review and prospect. *Austr. Geogr. Study.*, 24, 110–127.

Clemmensen, L. (1986) Storm-generated eolian sand shadows and their sedimentary structure, Vejers Strand, Denmark. *J. sediment. Pet.*, 56, 520–527.

Cooper, W. S. (1958) *Coastal Sand Dunes of Oregon and Washington.* Geol. Soc. Amer. Mem. 72, 169 pp.

Cooper, W. S. (1967) *Coastal Dunes of California.* Geol. Soc. Amer. Mem. 104, 131 pp.

Cornish, V. (1914) *Waves of Sand and Snow and the Eddies which Make Them.* Unwin, London, 383 pp.

Cotton, C. A. (1942) *Climatic Accidents in Landscape-making.* Whitcombe & Tombs, Christchurch, New Zealand, 354 pp.

Cowie, J. D. (1963) Dune-building phases in the Manawatu district, N. Z. *N. Z. J. Geol Geophys.*, **6**, 268–280.

Davidson-Arnott, R. and Pyskir, N. (1988) Morphology and formation of an Holocene coastal dune field, Bruce Peninsula, Ontario. *Geogr. Phys. Quat.*, **42**, 163–170.

Davies, J. L. (1972) (first edn), (1980) (2nd edn). *Geographical Variation in Coastal Development.* Longman, London, 212 pp.

Dolan, R. (1972) Barrier dune system along the Outer Banks of North Carolina: a reappraisal. *Science*, **176**, 286–288.

Dominguez, J. M. L., Martin, L. and Bittencourt, A. C. S. P. (1987) Sea-level history and Quaternary evolution of rivermouth associated beach-ridge plains along the ESE Brazilian coast: A summary. In Nummendal, D., Pilkey, O. H. and Howard, J. D. (eds), *Sea Level Fluctuation and Coastal Evolution* SEPM Spec. Pub. No. 41 Tulsa, Okla. pp. 115–127.

Esler, A. E. (1970) Manawatu sand dune vegetation. *Proc. N.Z. Ecol. Soc.*, **17**, 14–46.

Finkel, J. H. (1959) The barchans of southern Peru. *J. Geol.*, **67**(6), 614–647.

Foster, D. N., Gordon, A. D. and Lawson, N. V. (1975) The storms of May–June, 1974, Sydney, NSW. *2nd Aust. Conf. Coastal Engr. Inst. Engrs. Austr.*, Pub. No. 75/2, 1–11.

Fryberger, S. G., Schenk, C. J. and Krystinik, L. F. (1988) Stokes surfaces and the effects of near surface groundwater table on aeolian deposition. *Sedimentol.*, **35**, 21–41.

Gardner, D. E. (1955) Beach-sand heavy-mineral deposits of Eastern Australia. *BMR Bull.* No. 28, 103 pp.

Goldsmith, V. (1978) Coastal dunes. In R. A. Davis Jr (ed.), *Coastal Sedimentary Environments.* Springer Verlag, New York, 171–235.

Guilcher, A. (1958) *Coastal and Submarine Morphology.* Methuen, London, 274 pp.

Hastenrath, S. (1967) The barchans of the Arequipa region, southern Peru. *Z. Geomorph. NF*, **11**, 300–331.

Hesp, P. A. (1981) The formation of shadow dunes. *J. sediment. Pet.* **51**(1), 101–111.

Hesp, P. A. (1984) Foredune formation in southeast Australia. In Thom, B. G. (ed.), *Coastal Geomorphology in Australia.* Academic Press, Sydney, pp. 69–97.

Hesp, P. A. (1986a) Notes on the Alexandria coastal dunefield, Algoa Bay. In van der Merwe, D., McLachlan, A. and Hesp, P. (eds), *Structure and Function of Sand Dune Ecosystems.* Inst. Coastal Res., Univ. Port Elizabeth Rept. No. 8, 95–103.

Hesp. P. A. (1986b) Ningaloo Marine Park terrestrial geomorphology and potential development sites. Rept. prepared for the WA Dept. Cons. and Land Management, Perth, WA, 60 pp.

Hesp. P. A. (1988a) Morphology, dynamics and internal stratification of some established foredunes in Southeast Australia. *Sediment. Geol.*, **55**, 17–41.

Hesp, P. A. (1988b) Surfzone, beach and foredune interactions on the Australian southeast Coast. *J. Coastal Res.*, Spec. Issue No. 3, 15–25.

Hesp, P. A. and Chape, S. (1984) 1 : 3 million map of the coastal environment of W. Australia. Central Map Agency, WA Dept. Lands and Survey, Perth, WA.

Hesp, P. A. and Short, A. D. (1980) Dune forms of the Younghusband Peninsula, S. E. South Australia. *Austr. Soc. Soil Sci., Riverina Branch*, 65–66.

Hesp, P. A., Illenberger, W., Rust, I., McLachlan, A. and Hyde, R. (1989) Some aspects of transgressive dunefield and transverse dune geomorphology and dynamics, south coast, South Africa. *Zeit. Geomorph. Suppl.-Bd.*, **73**, 111–123.

Hunter, R. E., Richmond, B. R. and Alpha, T. R. (1983). Storm-controlled oblique dunes of the Oregon coast. *Bull. Geol. Soc. Amer.*, **94**, 1450–1465.

Hyde, R. and Wasson, R. J. (1983) Radiative and meteorological control on the movement of sand at Lake Mungo, NSW, Australia. In Brookfield, M. E. and Ahlbrandt, T. S. (eds), *Aeolian Processes and Sediments.* Elsevier, Amsterdam, pp. 311–323.

Inman, D. L., Ewing, G. C. and Corliss, J. B. (1966) Coastal sand dunes of Guerro Negro, Baja California, Mexico. *Bull. Geol. Soc. Amer.*, **77**, 787–802.

Langford-Smith, T. and Thom, B. G. (1969) Coastal morphology of New South Wales. *J. Geol. Soc. Austr.*, **16**, 572–580.

Leatherman, S. (1979) (ed.) *Barrier Islands from the Gulf of St. Lawrence to the Gulf of Mexico.* Academic Press, New York, 325 pp.

Long, J. T. and Sharp, R. P. (1964) Barchan dune movement in Imperial Valley, California. *Bull. Geol. Soc. Amer.*, **75**, 149–156.

McLachlan, A., Ascaray, C. and du Toit, P. (1987) Sand movement, vegetation succession and biomass spectrum in a coastal dune slack in Algoa Bay, South Africa. *J. Arid Envir.*, **12**, 9–25.

McLachlan, A. S., Sieben, P. and Ascaray, C. (1982) Survey of a major coastal dunefield in the Eastern Cape. Zool. Dept., Univ. Port Elizabeth. Res. Rept. 10, 48 pp.

McLean, R. (1978) Recent coastal progradation in New Zealand. In Davies, J. L. and Williams, M. A. J. (eds), *Landform Evolution in Australasia* ANU Press, pp. 168–196.

Mabbutt, J. A. (1977) *Desert Landforms.* MIT Press, Cambridge, Mass., 340 pp.

Marta, M. D. (1958) Coastal dunes. A study of the dunes at Vera Cruz. *Proc. Coastal Engr. Conf. 6*, 520–530.

Mason, P. J. and Sykes, R. I. (1979) Flow over an isolated hill of moderate slope. *J. Roy. Met. Soc.*, **105**, 383–395.

Mather, A. S. and Ritchie, W. (1977) The Beaches of the Highlands and Islands of Scotland. Rept. Scottish, Countryside Commission. 201 pp.

Matsukura, Y. (1977) A study of the aeolian bedforms in Enshunada Beach, Shizuoka Prefecture, Central Japan—their configurations and criteria of formation. *Geogr. Rev. Japan*, **50**(7), 402–419.

Melton, F. A. (1940) A tentative classification of sand dunes: its application to dune history in the southern high plains. *J. Geol.*, **48**, 113–174.

Mort, G. W. (1949). Vegetation survey of the marine sand drifts of NSW. Some remarks on useful stabilizing species. II. *J. Soil Conserv. Serv. NSW*, **5**, 63–72.

Olson, J. S. (1958) Lake Michigan dune development 2: Plants as agents and tools in geomorphology. *J. Geol.*, **66**, 345–351.

Pain, C. F. (1976) Late Quaternary dune sands and associated deposits near Aotea and Kawhai Harbours, North Island, New Zealand. *N.Z. J. Geol. Geophys.*, **19**, 153–177.

Paul, K. (1944) Morphologie und vegetation der Kurische Nehrung. *Acta Nova Leopoldina Carol.*, NF **13**, 217–378.

Pickard, J. (1972) Rate of movement of transgressive sand dunes at Cronulla, NSW. *J. Geol. Soc. Austr.*, **19**, 213–216.

Pidgeon, I. M. (1940) The ecology of the central coastal area of New South Wales. Types of primary succession. *Proc. Linn. Soc. NSW*, **65**, 221–249.

Pye, K. (1982) Morphological development of coastal dunes in a humid tropical environment, Cape Bedford and Cape Flattery, North Queensland. *Geogr. Annaler*, **64A**, 3–4, 213–227.

Pye, K. (1983) Coastal dunes. *Prog. Phys. Geogr.*, **7**, 531–557.

Pye, K. and Bowman, G. M. (1984) The Holocene Marine Transgression as a forcing function in episodic dune activity on the eastern Australian coast. In Thom, B. G. (ed.), *Coastal Geomorphology in Australia.* Academic Press, Sydney, pp. 179–196.

Ranwell, D. S. (1972) *Ecology of Salt Marshes and Sand Dunes.* Chapman & Hall, London, 258 pp.

Ritchie, W. and Walton, K. (1972) The evolution of the sands of Forvie and the Ythan estuary. In Clapperton, C. M. (ed.), *North East Scotland Geographical Essays.* Dept. Geogr. Univ. Aberdeen, pp. 12–16.

Ritchie, W., Smith, J. S. and Rose, N. (1978) *The Beaches of North East Scotland.* Dept. Geogr., University of Aberdeen Pub., 278 pp.

Roy, P. S. and Thom, B. G. (1981). Late Quaternary marine deposition in NSW and Southern Queensland—an evolutionary model. *Geol. Soc. Austr. J.*, **28**, 4, 471–489.

Rubin, D. and Hunter, R. E. (1985) Why deposits of longitudinal dunes are rarely recognised in the geologic record. *Sedimentol.*, **32**, 147–157.

Rubin, D., and Hunter, R. E. (1987) Bedform alignment in directionally varying flows. *Science*, **237**, 276–278.

Salisbury, E. (1952) *Downs and Dunes: Their Plant Life and Environment.* G. Bell, London, 328 pp.

Schofield, J. C. (1975) Sea level fluctuations cause periodic post-glacial progradation, south Kaipara Barrier, North Island, New Zealand. *N.Z. J. Geol. Geophys.*, **18**, 295–316.

Scorer, R. S. (1955) Theory of airflow over mountains. IV Separation of flow from the mountain surface. *Q.J. Roy. Met. Soc.*, **81**, 340–350.

Semeniuk, V. and Meagher, T. D. (1981) The geomorphology and surface processes of the Australind-Leschenault Inlet coastal area. *J. Roy. Soc., West. Austr.*, **64**, 33–51.

Sharp, R. P. (1966) Kelso Dunes, Mohave Desert, California. *Bull. Geol. Soc. Am.*, **77**, 1045–1074.

Shepherd, M. J. (1981) The Rockingham coastal barrier system of Western Australia. *Western Geographer*, Perth, 67–81.

Shepherd, M. J. (1987a) Holocene alluviation and transgressive dune activity in the lower Mawnawatu Valley, New Zealand. *N.Z. J. Geol. Geophys.*, **39**, 175–187.

Shepherd, M. J. (1987b) Sandy beach ridge system profiles as indicators of changing coastal processes. *Proc. 14th New Zealand Geogr. Conf. and 56th ANZAAS Conf.*, Palmerston North, 106–112.

Short, A. D. (1988) Holocene coastal dune formation in Southern Australia: a case study. *Sediment. Geol.*, **55**, 121–142.

Short, A. D. and Hesp, P. A. (1982) Wave, beach and dune interactions in S. E. Australia. *Mar. Geol.*, **48**, 259–284.

Tansley, A. G. (1949) *Introduction to Plant Ecology, A Guide for Beginners in the Study of Plant Communities.* Allen & Unwin, London, 260 pp.

Thom, B. G. (1974) Coastal erosion in eastern Australia. *Search*, **5**, 198–209.

Thom, B. G. (1978). Coastal sand deposition in southeast Australia during the Holocene. In Davies, J. L. and Williams, M. A. J. (eds), *Landform Evolution in Australasia.* ANU Press, pp. 197–214.

Thom, B. G. (1984) Transgressive and regressive stratigraphies of coastal sand barriers in Southeast Australia. *Mar. Geol.*, **56**, 137–158.

Thom, B. G., Bowman, G. M. and Roy, P. S. (1981) Late Quaternary evolution of coastal sand barriers, Port Stephens-Myall Lakes area, central N.S.W. Australia, *Quat. Res.*, **15**, 345–364.

Thom, B. G. and Roy, P. S. (1985) Relative sea levels and coastal sedimentation in southeast Australia in the Holocene. *J. Sediment. Pet.*, **55**(2), 257–264.

Thom, B. G., Shepherd, M. J., Ly, C., Roy, P. and Hesp, P. (in press). *Quaternary Geology and Geomorphology of the Port Stephens-Myall Lakes Region.* ANU Press.

Thompson, C. H. (1983) Development and weathering of large parabolic dune systems along the subtropical coast of eastern Australia. *Z. Geomorph. Suppl.-Bd.*, **45**, 205–225.

Thwaites, B. (ed.) (1960) *Incompressible Aerodynamics.* Clarendon Press, Oxford, 315 pp.

Tinley, K. L. (1985) *Coastal dunes of South Africa.* S. A. Nat. Sci. Prog. Rept. No. 109, 300 pp.

Tsoar, H. (1983). Wind tunnel modelling of echo and climbing dunes. In Brookfield, M. E. and Ahlbrandt, T. S. (eds), *Eolian Sediments and Processes.* Amsterdam, Elsevier, pp. 247–260.

Wipperman, F. K. and Gross, G. (1986) The wind-induced shaping and migration of an isolated dune: A numerical experiment. *Boundary Layer Met.*, **36**, 319–334.

Wright, L. D. and Thom, B. G. (1977) Coastal depositional landforms: a morphodynamic approach. *Prog. Phys. Geogr.*, **1**, 412–459.

Chapter Thirteen

The Holocene development and present morphology of the Łeba Dunes, Baltic coast of Poland

RYSZARD KRZYSZTOF BORÓWKA
Quaternary Research Institute, Adam Mickiewicz University, Poznań

1. INTRODUCTION

Parts of the southern Baltic coast abound in well-developed coastal dune fields of which two, the Łeba Barrier in Poland and the Kurskiy Barrier in Lithuania, are presently evolving under the influence of intense eolian processes.

Dunes of the Łeba Barrier have for some time been of interest to geomorphologists, geologists and other specialists of natural sciences. The original scientific observations were conducted in the 1920s and 1930s (Hartnack, 1926, 1931; Bülow, 1929, 1930, 1933). Although this work mainly described the geomorphology of the Łeba Barrier, attention was drawn to the number of fossil soil horizons that occurred within the dune system. This early work focussed on the stratigraphy of the dunes, although observations on contemporary eolian processes were also made.

Since 1970, many scientific investigations have been carried out in the Łeba Dunes, mostly aimed at determining the associations between the factors controlling eolian transport and deposition, their frequency of occurrence and the geomorphology and sedimentology of the dunes (Miszalski, 1973; R. K. Borówka, 1979, 1980a,b) and the beach (M. Borówka, 1979a,b, 1986). Palaeogeographic investigations have also been conducted into the Holocene development of the Łeba Dunes (Tobolski, 1975, 1979; 1980; R. K. Borówka, 1975; R. K. Borówka and Tobolski, 1979; Tobolski *et al.*, 1980).

This chapter summarises and discusses the major findings of the research into the contemporary eolian environment of the Łeba Dunes as well as examining their late-Holocene development.

Coastal Dunes: Form and process. Edited by K. F. Nordstrom, N. P. Psuty and R. W. G. Carter
© 1990 John Wiley & Sons Ltd

2. PHYSIOGRAPHY OF THE ŁEBA BARRIER

The Łeba Barrier forms a small strip of the Baltic coast of Poland separating the Gardno and Łebsko Lakes and the adjacent lowland from the Baltic Sea (Fig. 1). The western edge of the barrier zone joins the margin of the youngest end moraines of the Polish Lowland which probably formed during the late-Pleistocene (13.5 Ka). The barrier extends eastward for 75 km to the next coastal cliffs; the most interesting segment of the dunes lies between the towns of Rowy and Łeba (Fig. 1), as the morphology reaches its finest expression in this area, with extensive, mobile sand sheets, barchans and large-scale parabolic forms (Fig. 2).

2.1. Geomorphology

The Łeba Barrier is built primarily of Holocene deposits of marine lagoon and eolian origin, resting on a basement of late-Pleistocene fluvio-glacial sands and gravels (Rosa, 1963; Brodniewicz and Rosa, 1967; R. K. Borówka and Rotnicki 1988). The contact between the barrier and the adjacent fluvio-glacial deposits is erosional, and marked by a residual stone pavement usually at 10 to 12 m below sea level (Rosa, 1963; R. K. Borówka, 1988; R. K. Borówka and Rotnicki, 1988). In some places, early-Holocene peats occur on the surface of the fluvio-glacial deposits, especially under the lagoonal sediments (R. K. Borówka and Rotnicki, 1988).

Marine and lagoon deposits are overlain by eolian sands forming the surface of the barrier. Within the eolian formation, fossil soil horizons occur along with thick, interbedded organic deposits.

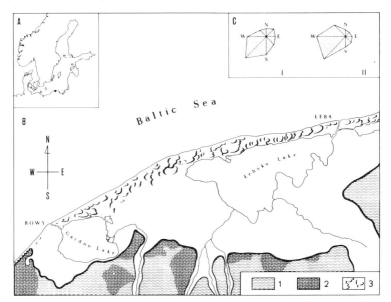

Fig. 1   Location (A), geomorphic setting of the Łeba Barrier (B) and distribution of frequency (C-I) and effectiveness (C-II) of wind blowing from particular directions at Łeba; 1: moraine plateau, 2: terminal moraines, 3: coastal dunes

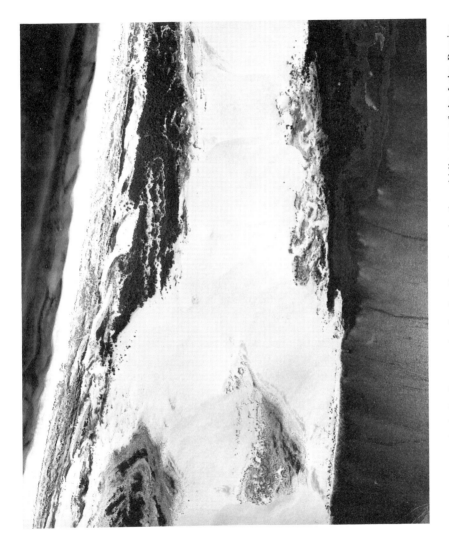

Fig. 2 A vertical air photograph of migrating dunes in the middle part of the Łeba Barrier

After the formation of the Łeba Barrier, during the Littorina Transgression (about 5000 to 6000 years ago) (Bülow, 1933; Rosa, 1963), the surface was covered by eolian sands. Numerous accumulation and deflation structures, both past and present, are the result of eolian processes. At the present, besides dune and slack morphology, sub-fossil forms also occur, associated with former forest and wetland vegetation, especially in old inter-dune troughs.

The Łeba Barrier includes barchans, elliptic dunes and parabolic crescent dunes (Miszalski, 1973). Over a distance of about 35 km, between the western end of the Łeba Barrier and the mouth of the Łeba River, Miszalski (1973) found 78 large dune forms whose relief exceeded several metres and whose maximum height was 50 m. Among these forms, crescent dunes dominate (about 63%) with barchans comprising about 22%. Nearly half of these dunes lack vegetation cover on more than 50% of their surface area and 35% of them are more than 80% bare. Such dunes, partially or completely lacking vegetation, migrate eastward at an average rate of between 1 and $10\,m\,a^{-1}$. Some formerly mobile dunes are wooded. In a narrow zone at the back of the beach, complexes of foredune ridges occur reaching heights of 10 to 12 m above sea level. In some segments of the barrier, the foredunes have been eroded by the sea. Farther inland from the present beach, 'fossil' dune cliffs are often separated from the shoreline by younger foredune ridges. Between specific dunes, it is also possible to find deflation forms, including large inter-dune troughs and smaller basins or deflation hollows as well as deflation remnants, usually in the shape of elongated WNW-ESE orientated ridges. The largest deflation depressions are either damp or occupied by ephemeral lakes which fill during the autumn and winter due to rises in the level of groundwater in the barrier.

The zone directly adjoining the middle part of Łebsko Lake (Fig. 1) is clearly defined as a 5.5 km² area almost entirely devoid of vegetation and consisting of a complex of several 20 to 30 m high barchans migrating at a rate of $10\,m\,a^{-1}$ involving a volume of 0.5 to $2.3 \times 10^6\,m^3$ of sand (Miszalski, 1973).

Eolian forms of the Łeba Barrier comprise quartz sands containing a small proportion (from 0.5 to 1%) of heavy minerals. Among the most common are amphibole, garnet, zircon, magnetite and ilmenite. Dune sands contain about 86% of fine-grained sand, about 13% medium-grained sand and less than 1% of very fine-grained sands dominated by heavy minerals.

2.2. Climate

The climate of the Łeba Barrier area is strongly ameliorated by the Baltic. Low annual variations in average monthly air temperatures (17.9°C) relative to areas farther inland are characteristic. July and August are the hottest months (16.3°C and 16.2°C, respectively) while February, with an average temperature of −1.6°C (recorded over many years), is the coldest month (Fig. 3). February is also the month of the greatest number of cold days (10.4) when the maximum temperature of the 24-hour period remains below 0°C.

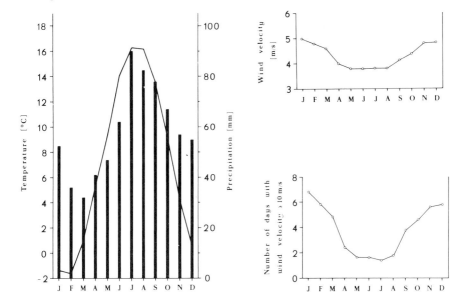

Fig. 3   Mean monthly values of selected meteorological variables at Łeba over the 1951
to 1970 period

The frequency of storms is highest in late autumn and winter. Average
monthly wind velocities are highest in winter, often exceeding $5\,\mathrm{m\,s^{-1}}$ and
occasionally reaching about $7\,\mathrm{m\,s^{-1}}$. The lowest average (about $3.8\,\mathrm{m\,s^{-1}}$) wind
velocities occur in summer (Fig. 3).

An analysis of the annual directional frequency of winds (Taranowska, 1968)
reveals the existence of the following three types of regime on the Łeba Barrier:

(1) The autumn-winter type, from October until March, with south-westerly
and westerly winds dominating but with considerable influence from
north-westerlies.
(2) The spring type, from March until May, with winds of varying directions
due to air mass interaction characterised by frequent southerlies and easterlies.
(3) The summer type, from June until September, with a clear dominance
of westerly winds.

A comparison of wind velocities across the Łeba Barrier with those recorded
by the meteorological station at Łeba reveals that for the unvegetated dunes
wind velocities are 1 to $2\,\mathrm{m\,s^{-1}}$ greater (Rabski, 1986).

Rainfall in the study area is usually associated with westerlies. Annual rainfall
in Łeba is 700 m. The greatest rainfall occurs during summer, from June until
August (about 33.4% of the annual total), with the lowest rainfall in spring
from March until May (17.7%).

2.3. Vegetation

The present vegetation of the Łeba Barrier comprises three separate groups of plant assemblages, among which the largest areas are taken by psamophytic vegetation such as *Elymo-Ammophiletum* and *Helichryso-Jasionetum litoralis*. Among complexes of forest vegetation occurring on stabilised dunes of the Łeba Barrier the *Empetro nigri-Pinetum* complex dominates although considerable areas of *Vaccinio uliginosi-Pinetum* are found in moist inter-dune depressions. Unique dune habitats include those of the *Pino-Quercetum fagetosum* complex. In the southern part of the Łeba Barrier, associations also include *Betuletum pubescentis*, *Carici elongate-Alnetum*, and mires around the lakes.

3. EOLIAN PROCESSES AND TRANSPORT

One object of measuring sediment movement rates on Łeba Barrier dunes was to discover which factors, apart from wind velocity, control eolian transport. A series of measurements were taken between 1973 and 1976 during the dominance of westerly winds, covering a range of wind velocities from 3 to 21 m s^{-1}. All the measurements were taken with the help of sand traps built partly to a design by Riabichin (1969). The sand traps caught material transported both in saltation and in surface creep through a 0.01 m wide section. The trap efficiency was determined in a wind tunnel and was found to be 80 to 90%. All measurements of transport rate were made at the same sites on the windward slope of the selected barchan dune, with local relief of about 20 m and a windward slope facet of 200 m. During simultaneous measurements of transport rates, records were taken of average wind velocity 1 m above the ground level, air temperature, humidity, atmospheric pressure, near-surface moisture of the ground and the degree of the eolian micro-relief.

3.1. Eolian transport and wind velocity

The relation between wind velocity (V) and sand transport rate (Q) in the dune area was determined on the basis of more than 500 measurements (R. K. Borówka, 1980a,b). The character of this empirical relationship can be expressed as a regression curve described by the formula:

$$Q = 0.000301 \ V^{4.68}$$

This formula provides a means of forecasting the magnitude of eolian transport in different directions over a range of wind regimes.

On the basis of the data collected on the Łeba Barrier in 1975, it is feasible to present percentage distributions of frequency and efficiency of winds blowing from different sectors (Fig. 4A,B). A comparison of these figures shows clearly

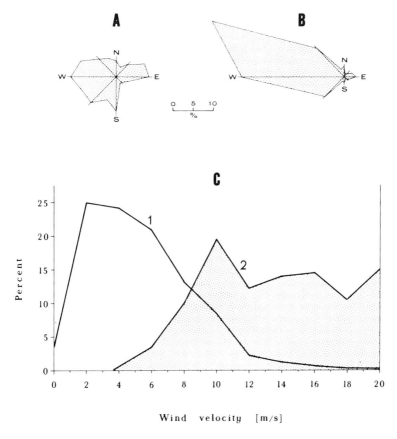

Fig. 4   Frequency (A, C-1) and effectiveness (B, C-2) of wind at Łeba in the year 1975

that, as far as wind frequency is concerned, westerlies were relatively more common in that year (about 44%). If effectiveness is taken into account, westerly winds were clearly dominant (about 86%). This is because relatively weak winds with a velocity of around $6\,\mathrm{m\,s^{-1}}$ dominate in the area of the Łeba Barrier (Fig. 4C). These winds are of little significance for eolian transport, the main work being achieved by storm winds ($>10\,\mathrm{m\,s^{-1}}$) blowing from the west. Although these occur for only a relatively short period of time—about 8% of the entire year—they are responsible for about 80% of the total sand transport (Fig. 4C).

3.2. Effects of morphology on dune transport

The results of transport rate measurements together with profile surveys from the central part of the study site show a marked difference in sand transport at particular locations on the windward dune slope (R. K. Borówka, 1980a).

In the case of measurements taken on a dry rippled surface (Fig. 5, I), transport at the bottom and top of the windward dune slope (points A and B) is almost equal for particular wind velocities. Distal (leeward) transport is 25 to 40% lower (point C), especially when wind velocity exceeds $8\,m\,s^{-1}$. It follows from the above results that on a dry rippled surface only partial accumulation of transported sand takes place on the proximal slope, especially where it faces the wind direction. This conclusion is confirmed by repeated profiling. If such conditions of transport persist, a gradual increase of the height in the zone between the dune crest and the brink of the leeside avalanche slope takes place but at other times crest and brink coincide.

The variability of transport rates is different when the dune surface is moist and devoid of eolian microstructures. Conditions of this type occur most often in the autumn and winter. In this case (Fig. 5, II), especially under higher wind velocities ($>10\,m\,s^{-1}$), a gradual downwind increase in transport rate is observed so that deflation conditions prevail across the entire length of the proximal slope. Under such conditions, given a simultaneous rapid accumulation on the lee slope, the proximal slope shortens while the area between the dune crest and the brink elongates.

3.3. The effect of air temperatures and ground conditions on eolian transport

Based on experience of the Łeba Barrier dunes it was found (R. K. Borówka, 1980a,b) that the relationship between wind velocity and sediment transport rate differs according to moisture, temperature and bed roughness.

A comparison of the relationships between wind velocity and sand transport rate across moist, flat surfaces with those across dry, rippled surfaces revealed that the transport rate across the moist surfaces is much higher than across a dry surface, depending on bed roughness and thermal conditions. Moisture is not an impediment to eolian transport, particularly at wind velocities higher than the velocity required to entrain moist sand, but a clear drop in sediment movement is caused by only very small eolian bedforms such as ripple marks (R. K. Borówka, 1980a,b).

Air temperature also exerts a marked influence on sand transport rates. Air temperature cannot directly affect transport rate, yet it appears that temperature-dependent factors such as the kinematic viscosity and density of air can exert substantial influence on sediment movement, causing changes in bed shear stress and the thickness of the laminar sublayer (see Sherman and Hotta, this volume).

4. METEOROLOGICAL CONDITIONS AND GROUND CONDITIONS IN RELATION TO THE SIZE OF TRANSPORTED GRAINS

In the Łeba dunes environment, the grain size of sand transported in saltation changes according to wind velocity and the nature of the dune surface

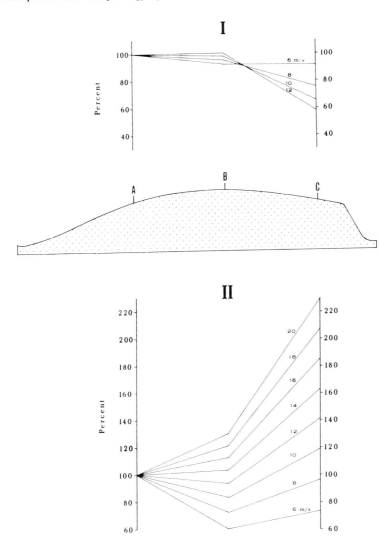

Fig. 5   Changes of transport intensity in particular zones of the windward dune slope (A,B,C) at various wind velocities. I: dry rippled surface, II: smooth wet surface

(R. K. Borówka, 1980a). With respect to sediment transport by saltation-creep flow, an increase in the average size of grains is apparent (a drop in the value of parameter M_z calculated according to Folk and Ward, 1957) as wind velocity increases (Fig. 6). Simultaneously, the standard deviation of the wind velocity increases, leading to poorer sorting (Fig. 6).

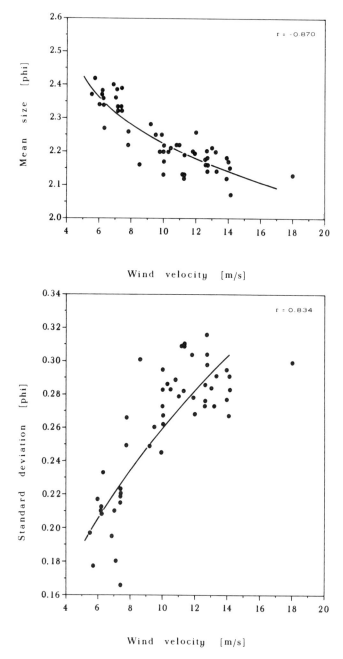

Fig. 6   The relationship between wind velocity and grain size parameters of sand transported in a saltation-creep flow

These grain size trends are subtle, probably stemming from and adjusting to the dune surface form. Experimental investigations conducted in a wind tunnel (R. K. Borówka, in preparation) show that for dry rippled surfaces, fine, well-sorted materials are transported primarily in saltation while the coarser grains remain on the dune surface, moving only slowly by creep. Under such conditions, one is dealing with the selective transport of the fine-grained fraction. In the case of moist surfaces, material exposed on the surface is influenced by wind and transported by saltation. The divergence in the transport mode over the surface types discussed is largely related to varying bed roughnesses which evidently control the wind distribution near the ground (Bagnold, 1960). Hence grains transported over different surface types will have different kinetic energies in the final phase of flight. As a result of this, over a rippled surface one observes not only the selective transport of fine-grained fractions but also a much smaller transport rate and a smaller thickness of the wind-sand stream when compared to a smooth and moist surface.

5. SURFACE MORPHOLOGY AND STRUCTURE OF DUNES UNDER VARYING METEOROLOGICAL CONDITIONS

In those parts of the Łeba Barrier where dunes are migrating, in common with other areas where eolian processes dominate (Wilson, 1972a,b), occasional simultaneous occurrence of three eolian bedform hierarchies can be seen. The first hierarchical level is composed of migrating dunes, coastal barchans, which reach heights of 15 to 45 m and lengths of 250 to 900 m. From the analysis of topographic maps and aerial photographs, Miszalski (1973) demonstrated that the coastal barchans are relatively permanent forms, which have not undergone great changes in size since 1900. The second and third hierarchical levels comprise 'mesobarchans' (with relative heights ranging from 0.3 to 2.0 m and length from 8 to 60 m) and ripple marks (relative height 1 to 16 mm; length 15 to 180 mm). These are more emphemeral elements in the relief, developing periodically on windward slopes of larger barchans (Fig. 7).

Field observations indicate that mesobarchans develop exclusively during dry periods, most frequently in spring when rainfall is low. During humid periods, such forms are dispersed and only the largest mesobarchans survive. The varying morphology of the windward slopes of barchans during dry and wet periods is documented by profiling after periods dominated by either wet or dry conditions. A similar phenomenon from the coast of Oregon is described by Hunter, Richmond and Alpha (1983). It is also worth pointing out that in dry areas of sand deserts second-generation forms are the permanent elements of the relief (Wilson, 1972a,b) where mesobarchans react to short changes of wind directions. During migration, complex, but thin, cross-stratified units are formed (from a few cm to about 2 m) characterised by a large scatter in the dip azimuths of laminae.

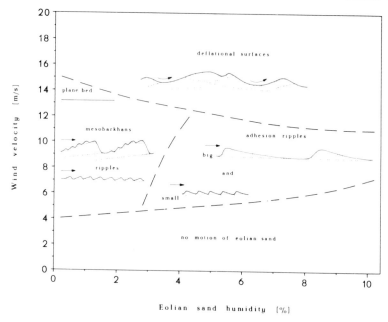

Fig. 7 Classification of ephemeral eolian forms developed on the windward dune slope
in relation to differences in wind velocity and sand moisture

The presence of eolian ripple marks on the surface of dunes and mesobarchans
is also dependent on ground moisture (Stankowski, 1963; Minkevicius, 1968;
R. K. Borówka, 1980a,b) and on wind velocity. Both Hunter's observations
from Padre Island and Borówka's from the Łeba Dunes show that at wind
velocities above 12 to $14\,\mathrm{m\,s^{-1}}$ ripple marks disperse and flat dry surfaces
form.

Various types of stratification, namely sub- and super- critically climbing
translatent stratification, ripple-foreset cross-lamination, and ripple-form
lamination, (Hunter, 1977), are associated with the migration of ripple marks
on surfaces of different slope relative to the dominant wind direction. All these
types of stratification are characterised by thin sedimentation units, usually only
a few mm thick. Plane-bed laminations (Hunter, 1977), usually several cm thick,
are found where smooth dry surfaces have developed during high winds.

During wet periods deflation forms develop on the proximal slope of the dune.
Accumulation forms are represented only by two types of adhesion ripple.
According to observations made by Reineck (1955), Stankowski (1963), Glennie
(1970), Hunter (1973) and R. K. Borówka (1980a), these ripples comprise a short
and steep stoss slope and a long and smooth lee slope. Their development is
connected primarily with the process of a temporary immobilisation of less moist
sands moving over a moister bed.

The larger adhesion ripples usually reach a height of 100 to 250 mm, appearing most often as patches 20 to 30 m apart, perpendicular to the dominant wind direction (Fig. 8A). The height of the smaller adhesion forms ranges between 8 to 15 mm with an average length of 100 to 150 mm (Fig. 8B). During the migration of larger forms, sets of plane-bed lamination several cm thick are formed. The development of smaller forms is sometimes connected with structures defined as pseudo-crosslamination formed by climbing adhesion ripples (Hunter, 1973). A synthesis of the various ephemeral eolian forms found on the windward dune slopes under different environmental constraints is shown in Fig. 7.

The relief variation on the dune surface due to wind velocity and sand moisture content is apparent not only on windward slopes but also on lee slopes. The development of small bedforms, and consequently sedimentation and disturbance structures, is in this case related to gravity. It is known that the amount of sand accumulation varies down the lee slope (Bagnold, 1960; Sharp, 1966; McKee and Douglass, 1971). On the Łeba Dunes it is apparent that the initial volumetric accumulation profiles on the distal slope are exponential

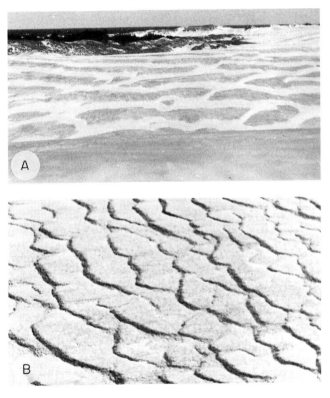

Fig. 8   Large (A) and small (B) adhesion ripple marks

(R. K. Borówka, 1979, 1980a). Hence, primary sand accumulation on this slope is most rapid towards the top. As a result of this phenomenon, the value of the angle of slope inclination steepens, eventually exceeding the value of the natural stability angle of the material. Truncation surfaces are formed which later act as slide surfaces. The value of the critical angle of rest as well as the disposition of the slide surfaces and the mode of gravitational transport are dependent on the degree of apparent sand cohesion, controlled by the surface

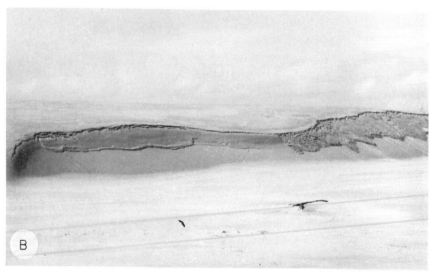

Fig. 9   The lee slope of dunes over a dry (A) and a humid (B) period

tension forces of water. In the case of dry sands, with negligible cohesion, the value of the critical angle of slope stability ranges from 30 to 33°, and downslope movement is in the form of gravity sandflows (Fig. 9). This process begins during eolian accumulation and continues for some time after it has ended. Along the sandflow paths, furrows are formed which are almost immediately obscured by fresh material arriving from the windward slope. When the influx of fresh sand is low the upper part of the lee slope displays a clear micro-relief of furrows and sandflow ribs, whereas at the middle and bottom of the slope, material from individual sandflows often overlaps to form a series of sandflow cones (Fig. 9A).

Cross-stratification from sand flows (R. K. Borówka, 1976, 1979, 1980a; Hunter, 1977; Hunter, Richmond and Alpha, 1983; Clemmensen and Abrahamsen, 1983) is associated with the redeposition of the accumulated sand across the slope. Pseudomorphic scars and sand ribs develop on the upper part of the leeward dune slopes and pseudomorphs of sandflow cones form at the base (Fig. 10A,B).

Sandflow cross-stratification structures are characteristic of leeside deposits formed during spring and summer. However, they arise only when the wind is more or less orthogonal to the distal slope. When the wind blows obliquely, sand is transported across the slope with surface ripple marks forming parallel to the direction of maximum dip. With a constant influx of sediment, tabular cross-stratification develops (R. K. Borówka, 1979, 1980b).

In the case of moist material, which is characterised by a distinct cohesion, leeside gravity displacement occurs much later because the critical angle of rest on the upper slope just exceeds 40°. Grainfall lamination develops until the critical threshold angle is exceeded, after which a slow avalanching takes place (Fig. 9B) forming concave slides with cylindrical cross-sections. The character and shape of the sliding planes are influenced by pre-existing internal dune structure.

Due to a gravitational relocation of slide material, the upper part of the distal slope and its margins display various tensional fissures, often with uneven gash geometrics (Fig. 11A). These fissures are accompanied by slumping. There is a so-called 'separation zone' in which tensional forces dominate. Antithetic faults or gashes, characterised by straight cracks and negligible downthrow (perhaps 10 to 20 mm) also occur in this zone, although less frequently.

A slightly different set of disturbance structures is formed on the bottom and middle parts of the distal slope (Fig. 11B), i.e. in the so-called compression zone, where compression forces are dominant. In the first phase of landslip development, an entire complex of secondary sliding surfaces forms more or less parallel to the bottom of the lee slope. Along the slip planes moist material is forced upwards leading to an *en échelon* system of parallel reverse faults (Fig. 11B). Movement of the distal slope laminae along the fault lines reaches average values of 10 to 40 mm. A further intensification of landslipping in the

304 *Coastal Dunes*

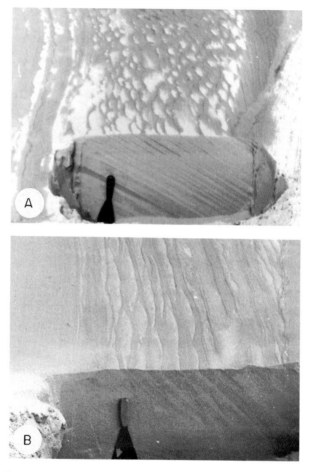

Fig. 10   Sand flow cross-stratification. A: Pseudomorphoses of scars and sand ribs (upper part of lee dune slope stratification). B: Pseudomorphoses of sand flows (lower part of lee dune slope stratification)

compression zone can cause disintegration of the compressed units into individual blocks of moist sandy material, forming the so-called landslip 'breccia' (Fig. 11B). Disturbance structures connected with landslips, such as thrust faults, normal faults or landslip breccia, frequently observed in dune deposits of the Łeba Barrier (Fig. 12), have also been found in some Brazilian coastal dunes (McKee, 1983).

Due to periodic changes in the longitudinal profile of the dunes and to the development of depositional and deflation meso- and micro-bed forms, a characteristic typology of sedimentary structures is formed (Fig. 13). In the higher parts of the dunes, above the winter lee slope, a suite of sedimentary

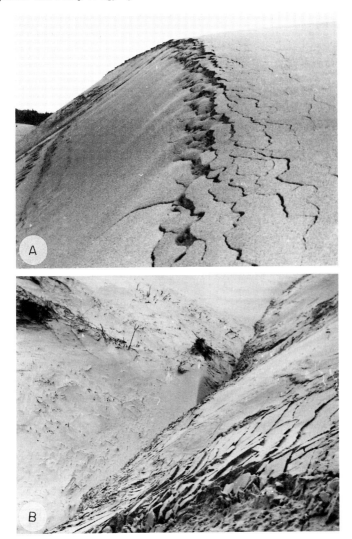

Fig. 11   Microrelief of the lee dune slope associated with redeposition of moist sand.
A: Tension fissures in the upper part of the lee dune slope. B: Compression fissures
and landslide breccia in the lower part of the lee dune slope

structures develops, characterised by extensive lateral and vertical variability of
both the stratification type and the direction and angles of dip of cross-laminae
(Fig. 13). This suite is associated with the migration of ephemeral bedforms as
the wind regime alters.

    In the middle part of the dunes, cross-stratification dominates with thicknesses
broadly corresponding to the average height of the winter leeside slope. Dip

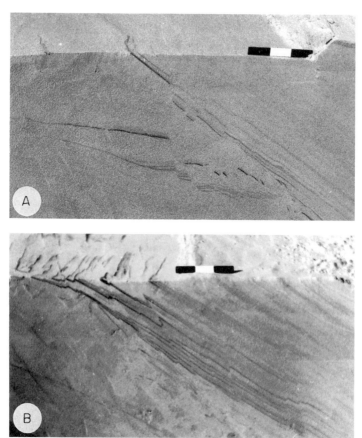

Fig. 12   Disturbance structures in the compression zone. A: Normal faults and landslide
brecia structures. B: Landslide brecia and thrust faults structures

azimuths of cross-laminae are scattered over a range of about 180° (Fig. 13).
Within this layer, both vertical and lateral variability of the types of cross-
stratification is apparent. Sets of sandflow cross-stratification, tabular cross-
stratification and various disturbance structures are common. When a map of
the dune surface structure was made, distinct rhythmic complexes of heavy
mineral laminae were noticed. The genesis of these rhythmic laminae is related
closely to the deflation of the material on the lee slope by occasional easterly
winds. Heavy mineral laminae represent deflation residues forming mostly in
spring, the time of maximum easterly winds. Stratifications containing complexes
of heavy mineral laminae occur every few metres, usually separated by
structureless sand units. Earlier investigations (R. K. Borówka, 1979, 1980a)
showed that mean distances between complexes of heavy mineral laminae,

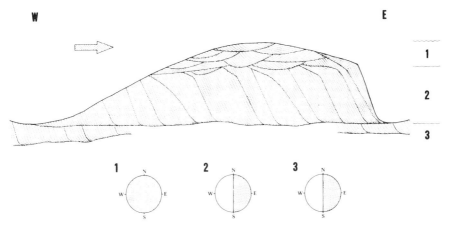

Fig. 13   Schematic view of the vertical internal structures of coastal barchans developed
on the Łeba Barrier

measured in a horizontal plane at right angles to their strike, corresponded to
the mean velocity of dune movement, recorded over many years from aerial
photographs. Thus, the lateral variability and mineralogy of the lamination
within the dunes can, under the conditions of the Łeba Barrier at least, form
the basis of a method for unravelling the palaeogeomorphology of the dune
system.

A few lamina sets with cross-stratification several cm thick occur at the base
of the dune and in the deflation zones of the inter-dune depressions. Dip
directions of laminae in some layers are confined within a range of about 180°.
Some layers are separated by flat or slightly undulating erosional bounding
surfaces. These bounding surfaces are caused by deflation in advance of
migrating. The vertical extent of deflation varies as a result of local conditions,
especially the longer term fluctuations of the groundwater table.

6. HOLOCENE DUNE FORMATION

The history of eolian bedforms on the Łeba Barrier can be elucidated from
'fossil' windblown deposits which, in many places, are separated from
contemporary dune deposits by fossil soils. Several palaeosol and organic
deposits, cropping-out either between or below the eolian sands, have been.[14]C
dated. An analysis of the lithostratigraphic, geomorphological and
chronostratigraphic position of the palaeosols and organic deposits indicates
that four stages of eolian activity can be recognised (Fig. 14).

The oldest known eolian series is found at three sites: (i) adjoining the northwestern limit of Łebsko Lake (R. K. Borówka, 1975; R. K. Borówka and
Tobolski, 1979; (ii) the middle part of the Łeba Barrier (R. K. Borówka and

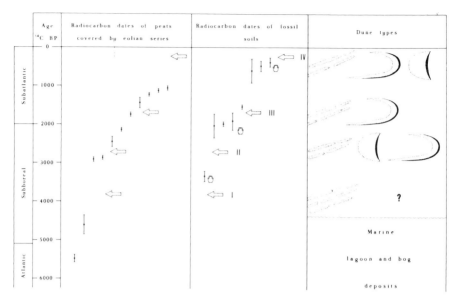

Fig. 14   Stages of eolian activity on the Łeba Barrier during the Holocene

Rotnicki, 1988); and (iii) in the vicinity of the NW edge of Łebsko Lake. Here the oldest eolian series is covered by fossil soil containing numerous charcoals dated to 3340 ± 130 years BP. Numerous fragments of ceramics and well-preserved vessels from the late Bronze Age (3000 to 2600 years ago) were also found in this soil. A structural analysis of the sands forming the oldest eolian series provides little insight into the palaeoenvironmental history of the Łeba Barrier. However, it does appear that there were either low forms of the foredune type or ridge remnants of parabolic dunes. The fossil soils which cover the oldest eolian deposits have poorly-developed illuvial horizons, which is most likely a reflection of the short period over which they developed.

   The second stage of eolian activity between 3300 to 2500 years BP has been recognised at many sites (Tobolski, 1975, 1980; R. K. Borówka and Tobolski, 1979). At one of the sites this later stage is represented by a small area of re-exposed dune sands, which show large-scale cross-stratification with dominant dips from 22 to 36°. This exposure occurs between two outcrops of fossil soil running more or less parallel, about 40 m apart. Both soils dip southwards at an angle of about 30°. The older soil, whose chronological position was determined by archaeological means (R. K. Borówka and Tobolski, 1979), represents a period slightly older than 3000 years BP, while the younger soil, containing numerous charcoals, was dated by ^{14}C to 1920 ± 200 BP (GD-W7, Kobendzina, 1968) and 1940 ± 50 BP (LU-769, Tobolski, 1975). The younger fossil soil, which is also known from other sites, is characterised by a well-developed podzolic profile. It follows from Tobolski's (1975) investigations that it developed under the canopy of an oak forest, as there is a large amount of oak pollen and fragments of oak trees among the charcoals, as well as fossil trunks. In numerous sites in the middle part of the Łeba Barrier, only the presence of an eolian series older than a palaeosol dated about 2000 years BP is documented. In places, where this series was uncovered over a larger area by deflation, detailed structural mapping was undertaken (Fig. 15). These maps show that the deposits originated due both to migration of forms such as barchans, and to a later development of parabolic dunes with crest heights several metres above sea level. Structural indices demonstrate that both barchans and parabolic dunes were formed under the influence of westerly winds.

On the basis of spatial distribution of the palaeosols it was thought at first that these dunes represented old foredune ridges (Marsz, 1966). However, the structural features of eolian sands eliminate this possibility; it now seems certain that the dunes are remains of parabolic dune arms.

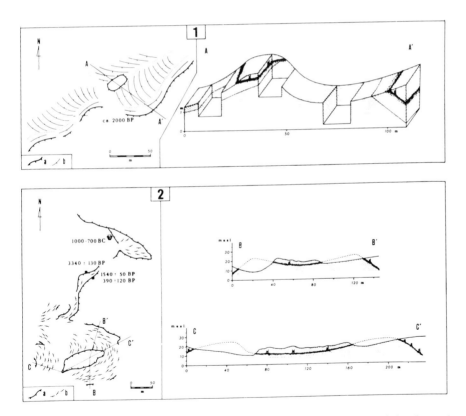

Fig. 15   Structural analysis reveals the stratigraphic position of the buried palaeosol (about 1500 years old) northwest of Łebsko Lake

The third stage of eolian processes is most noticeable in the vicinity of the northwestern margin of Łebsko Lake, although traces may be found in the middle part of Łeba Barrier. Around the northwestern margin of Łebsko Lake it is represented by an eolian series covered by a fossil soil with a ^{14}C age of 1540 ± 50 years BP (Tobolski, 1975). A spatial analysis of the outcrop of this fossil soil and a structural analysis of the eolian series (R. K. Borówka, 1975) indicate that this third phase was represented mainly by crescent and parabolic dunes with heights reaching about 20 to 25 m above sea level (Fig. 15). Soils, which developed on these dunes, are characterised by slightly different morphological formations when compared with the soils dated at about 2000 years BP (Tobolski, 1975). They include a poorly developed illuvial horizon. Pollen analysis shows they were largely formed under the canopy of a pine forest (Tobolski, 1975). In the middle part of the Łeba Barrier, this phase of eolian processes is observed only as thick interbedding of sand within the depositional horizons of fossil podzols. Even here, above the sand interbedding, the pollen spectrum changes decidedly in fossil soils, with the oak forest being replaced by beech and pine (Tobolski, 1975).

Between 1500 and 500 years BP a clear interregnum in eolian activity is observed. A revival of these processes is dated archaeologically and by ^{14}C method to the fifteenth century. Since then a constant and intense development of dune processes has been observed.

Almost all fossil soils found on the Łeba Barrier are associated with traces of human activity. The oldest traces are probably from the later Bronze Age; sites from the Iron Age dated at about 2500 years BP (Marsz, 1972) as well as sites from the late Middle Ages and more recent times are also known. Archival data (Kobendzina, 1976) indicate that economic activity around the town and harbour of Łeba took place in the first half of the fourteenth century and from the second half of the fifteenth century until the sixteenth century. This activity led to a gradual felling of the forests on the Łeba Barrier and consequently triggered a period of intense eolian reworking.

The fact that in the fossil soils, as well as in the peat deposits, charcoal is frequently found, associated with archaeological remains and backed by archival data, suggests that repeated reactivation of eolian processes is closely connected to human activities over the last 3000 years. However, it should be remembered that this period also marks a gradual deterioration of the European climate so that human effects could have become progressively worse. This is confirmed by the extensive alterations in the dune vegetation (Tobolski, 1975, 1980) as well as by simultaneous morphological changes.

REFERENCES

Bagnold, R. A. (1960) *The Physics of Blown Sand and Desert Dunes*. Methuen, London, third edn, 265pp.

Borówka, M. (1979a) Reconstruction of the development of the relief in the beach hinterland in the central part of the Łeba Bar (in Polish). *Badania Fizjograficzne nad Polską Zachodnią*, **32A**, 7–20.

Borówka, M. (1979b) Deflation and accumulation processes on the surface of coastal dune ridges (in Polish). *Badania Fizjograficzne nad Polską Zachodnią*, **32A**, 31–48.

Borówka, M. (1986) Współczesne procesy modelujące powierzchnię plaży piaszczystej morza bezpływowego. Uniwersytet Adama Mickiewicza w Poznaniu, PhD thesis, 133 pp.

Borówka, R. K. (1975) Problem of the morphology of fossil dune forms on the Łeba Bar. *Quaestion. Geogr.*, **2**, 39–51.

Borówka, R. K. (1976) Some types of sedimentary structures and disturbances of coastal dunes and their paleogeographical significance (in Polish). *Poznańskie Towarzystwo Przyjaciół Nauk, Sprawozdania Wydziału Matematyczno-Przyrodniczego*, **94**, 32–34.

Borówka, R. K. (1979) Accumulation and redeposition of eolian sands on the lee slope of dunes and their influence on formation of sedimentary structures. *Quaestion. Geogr.*, **5**, 5–22.

Borówka, R. K. (1980a) Present day dune processes and dune morphology on the Łeba Barrier, Polish coast of the Baltic. *Geograf. Ann.*, **62A**, 75–82.

Borówka, R. K. (1980b) Present day transport and sedimentation processes of aeolian sands—controlling the factors and resulting phenomena on a coastal dune area (in Polish). *Poznańskie Towarzystwo Przyjaciół Nauk, Prace Komisji Geograficzno-Geologicznej*, **20**, 1–126.

Borówka, R. K. (1988) The Quaternary in the vicinity of Łeba (in Polish). *Poznanskie Towarzystwo Przyjaciół Nauk, Sprawozdania Wydziału Matematyczno-Przyrodniczego*, **105**, 40–42.

Borówka, R. K. (in preparation) Wind tunnel studies of grain size composition of sand transported in saltation. *Manuscript*.

Borówka, R. K. and Rotnicki, K. (1988) New data on the geologic structure of the Łeba Barrier (in Polish). *Poznańskie Towarzystwo Przyjaciół Nauk, Sprawozdania Wydziału Matematyczno-Przyrodniczego*, **105**, 26–29.

Borówka, R. K. and Tobolski, K. (1979) New archaeological sites on the Łeba Bar and their significance for paleogeography of this area (in Polish). *Badania Fizjograficzne nad Polską Zachodnią*, **32A**, 21–29.

Brodniewicz, I. and Rosa, B. (1967) The boring hole and the fauna at Czołpino, Poland. *Baltica*, **3**, 61–86.

Bülow, K. (1929) Postglaciale Senkung und Dünenbildung im NO—Hinterpommerschen Küstenbereich. *Jahrb. d Preuss. Geol. Landesants*, **50**, 125–134.

Bülow, K. (1930) Allgemein—Geologische Beobachtungen im Wanderdünengebiet der Lebasse—Nehrung in Ostpommern. *Jahrb. d Preuss. Geol. Landesants*, **50**, 592–606.

Bülow, K. (1933) Ein neuer Fund von Litorina Ablagerungen und der Zeitpunkt der Litorina Transgression in Pommern. *Dohrniana*, **12**, 3–12.

Clemmensen, L. B. and Abrahamsen, K. (1983) Aeolian stratification and facies association in desert sediments, Arran basin (Permian), Scotland. *Sedimentol.*, **30**, 311–339.

Folk, R. L. and Ward, W. C. (1957) Brazos River bar: a study in the significance of grain size parameters. *J. sediment. Petrol*, **27**, 3–26.

Glennie, K. W. (1970) *Desert sedimentary environments*. Developments in Sedimentology, 14, Elsevier, Amsterdam, 222 pp.

Hartnack, W. (1926) *Die Küste Hinterpommerns unter besonderer Berucksichtigune der Morphologie*. Julius Abel, Greifswald, p. 324.

Hartnack, W. (1931) Zur Entstehung und Entwicklung der Wanderdünen an der deutschen Ostseeküste. *Zeit. Geomorph.*, **6**, 174–217.

Hunter, R. E. (1973) Pseudo-crosslamination formed by climbing adhesion ripples. *J. sediment. Petrol.*, **43**, 1125–1127.

Hunter, R. E. (1977) Basic types of stratification in small eolian dunes. *Sedimentol.*, **24**, 361–387.

Hunter, R. E., Richmond, B. M. and Alpha, T. R. (1983) Storm-controlled oblique dunes of the Oregon coast. *Bull. Geol. Soc. Am.*, **94**, 1450–1465.

Kobendzina, J. (1968) Wydmy Słowinskiego Parku Narodowego. *Ziemia*, **1967**, 70–80.

Kobendzina, J. (1976) On the historical geography of Łeba and its surroundings (in Polish). *Przegląd Geograficzny*, **48**, 689–701.

Marsz, A. (1966) Geneza Wydm Łebskich w świetle współczesnych procesów brzegowych. *Poznańskie Towarzystwo Przyjaciół Nauk, Prace Komisji Geograficzno-Geologicznej*, **4**, 1–68.

Marsz, A. (1972) Archaeological findings on the Łeba Bay Bar. *INQUA Subcommission on Shorelines of NW Europe, Guide-Book of the Excursion*, pp. 39–40.

McKee, E. D. (1983) Eolian sand bodies of the world. In Brookfield, M. E. and Ahlbrandt, T. S. (eds) *Eolian Sediments and Processes*. Elsevier, Amsterdam, pp. 1–25.

McKee, E. D. and Bigarella, J. J. (1972) Deformational structures in Brasilian coastal dunes. *J. sediment. Petrol.*, **42**, 670–681.

McKee, E. D. and Douglass, J. R. (1971) Growth and movement of dunes at White Sand National Monument, New Mexico. *US Geol. Surv. Prof. Paper*, 750-D, 108–114.

McKee, E. D., Douglass, J. R. and Rittenhouse, S. (1971) Deformation of lee-side laminae in eolian dunes. *Bull. Geol. Soc. Am.*, **82**, 359–378.

Minkevicius, V. (1968) Die Mikro- und Mesoformen der Wanderdünen aut der Kurischen Nehrung (in Lithuanian). *Geograf. Metra.*, **9**, 97–102.

Miszalski, J. (1973) Present-day eolian processes on the Slovinian Coastline; a study of photointerpretation (in Polish). *Dokumentacja Geograficzna*, **3**, 1–150.

Rabski, K. (1986) Elementy lokalnej cyrkulacji atmosfery oraz wielkosci wybranych parametrow meteorologicznych na obszarze Mierzei Lebskiej. Uniwersytet Adama Mickiewicza w Poznaniu, PhD thesis, 126 pp.

Reineck, H. E. (1955) Haftrippeln und Haftwarzen, Ablagerungs-formen von Flugsand. *Senckenbergiana Lethaea*, **36**, 347–352.

Riabichin, E. L. (1969) Experimental investigations with the sand-traps in field conditions (in Russian). *Problemy osvojenia pustyn*, **4**, 33–37.

Rosa, B. (1963) O rozwoju morfologicznym wybrzeża Polski w świetle dawnych form brzegowych. *Studia Societatis Scientiarum Torunensis, Geographia et Geologia*, **5**, 1–172.

Sharp, R. P. (1966) Kelso Dunes, Mojave Desert, California. *Bull. Geol. Soc. Am.*, **77**, 1045–1074.

Sherman, D. J. and Hotta, S. (1990) Eolian sediment transport: theory and measurement. This volume, pp. 17–37.

Stankowski, W. (1963) Eolian relief of north-west Poland on the ground of chosen regions (in Polish). *Poznańskie Towarzystwo Przyjaciół Nauk, Prace Komisji Geograficzno-Geologicznej*, 4/1, 1–146.

Taranowska, S. (1968) O kierunkach i prędkościach wiatrów dolnych na obszarze środkowego i południowego Bałtyku. *Przegląd Geofizyczny*, **13**, 75–88.

Tobolski, K. (1975) Palynological study of fossil soils of the Łeba Bay Bar in the Słowinski National Park (in Polish). *Poznańskie Towarzystwo Przyjaciół Nauk, Prace Komisji Biologicznej*, **41**, 1–76.

Tobolski, K. (1979) Changes in the local plant cover on the basis of investigations on subfossil biogenic sediments in the beach zone near Łeba (in Polish). *Badania Fizjograficzne nad Polską Zachodnią*, **32A**, 151–168.

Tobolski, K. (1980) The fossil soils of the coastal dunes on the Łeba Bar and their paleogeographical interpretation. *Quaestion. Geogr.*, **6**, 83–97.

Tobolski, K., Pazdur, M. F., Pazdur, A., Awsiuk, R., Bluszcz, A. and Walanus, A. (1980) The radiocarbon dating of subfossil wood from the bars of the Gardno-Łeba Lowland (in Polish). *Badania Fizjograficzne nad Polską Zachodnią*, **33A**, 133–148.

Wilson, I. G. (1972a) Aeolian bedforms—their development and origin. *Sedimentol.*, **19**, 173–210.

Wilson, I. G. (1972b) Universal discontinuities in bedforms produced by the wind. *J. sediment. Petrol.*, **42**, 667–669.

Chapter Fourteen

The instability of Holocene coastal dunes: the case of the Morro dunes, California

ANTONY R. ORME
Department of Geography, University of California, Los Angeles

1. INTRODUCTION

Coastal dune creation and mobility are favored by abundant beach sand and onshore winds above the threshold velocity for sand entrainment. Once formed, coastal dune stability is encouraged by diminished sand supplies or reduced wind velocities, reflecting distance from the beach or environmental change, but along all but arid coasts, the primary stabilizing factor is the effectiveness of sand-binding vegetation. In late Pleistocene and early Holocene times, abundant sand was often available on exposed continental shelves and, in the unstable conditions created by the Flandrian transgression, promoted extensive dune deposition along coasts exposed to strong onshore winds. When the main transgression culminated about 5000 BP, sand supplies declined, existing dunes were stabilized by vegetation, and subsequent dune deposition became less significant. At least that is one theory. In reality, there is much evidence worldwide for the reactivation of Holocene dunes over the past 5000 years, attributable to various causes about which there is little consensus.

The purpose of this chapter is first to sample evidence for coastal dune instability around the world, then to discuss the unstable Holocene dunes at Morro Bay, California, and lastly to evaluate the likely causes of instability at Morro Bay in the context of scenarios invoked elsewhere.

2. HOLOCENE COASTAL DUNE DEVELOPMENT: A GLOBAL SAMPLE

Evidence from around the world supports a generalized model for Holocene coastal dune development whereby (i) onlapping dune sequences are successively

---

Coastal Dunes: Form and Process. Edited by K. F. Nordstrom, N. P. Psuty and R. W. G. Carter

emplaced and destroyed during the Flandrian transgression, (ii) dunes are stabilized sometime following the transgression, and (iii) subsequent destabilization occurs. The following examples reveal some consensus for the first phase, less agreement about the timing of the second phase, and significant differences concerning explanation of subsequent destabilization.

Atop Pleistocene dunes along Australia's east coast, the earlier Holocene dunes were formed as rising sea level and onshore winds forced coastal sands landward across the exposed continental shelf. Thereafter, three main episodes of dune emplacement are recognized (Thompson, 1983). The oldest and most extensive dunes were deposited soon after the close of the transgression, mostly between 4000 and 3000 BP (Thom, Polach and Bowman , 1978). An intermediate series of smaller dunes then developed from the reworking of older dunes, whereas the youngest mobile or recently stabilized dunes originated from blowouts or from inputs of fresh coastal sand. Pye (1983), however, claims that the number and timing of later Holocene episodes are not well understood and, because they vary from place to place, dune mobility must be due to local factors rather than to regional climatic or eustatic causality. Nevertheless, there is broad support for the belief that extensive dune reactivation began in the nineteenth century as European immigrants burned the native vegetation and introduced grazing (Bird, 1974). Along New Zealand's Auckland coast where five Quaternary dune episodes occur, the mobility of the most recent dunes is also attributed to vegetation destruction by fires of either natural or Polynesian origin, but before European settlement (Hicks, 1983). In contrast, local tectonism and related sea-level changes have been invoked to explain the complex stratigraphy of the barrier-dune system on the Leschenault Peninsula, southwest Australia, over the past 8000 years and where the dune front is now eroding at a rate of 1 to 2 m a^{-1} (Semeniuk, 1985).

Along Africa's south coast, episodic Quaternary dune deposition culminated in extensive accumulation during the Flandrian transgression until stabilization and pedogenesis occurred around 7500 BP (Butzer and Helgren, 1972). The dunes were then remobilized and again stabilized after 4200 BP. Between Knysna and Cape St Francis, the dunes remained stable until deforestation and cattle grazing associated with European immigrants exposed the sand in the late eighteenth century. Elsewhere along the south coast, however, shell middens, artefacts and charcoal indicate a much longer human association with Holocene dunes, notably after about 2000 BP when herding was adopted by Khoi (Hottentot) shellfish gatherers. At about the same time, Early Iron Age cultivator-pastoralists began leaving significant imprints in dunes along Africa's southeast coast.

Along the Atlantic and Gulf coasts of the United States, one theory holds that barrier islands and related dunes originated far out on the continental shelf during late Pleistocene times and then migrated erratically upshelf during the Flandrian transgression (Field and Duane, 1976; Leatherman, 1983). As sea-level rise diminished, later Holocene dunes accumulated on both barrier islands and mainland shores. These dunes are mostly destabilizing or eroding at the present time, a fact which many investigators relate to rising sea level although opinions differ as to whether this rise is a eustatic response to renewed melting of global ice, a hydro-isostatic depression of the continental shelf, regional subsidence, or a combination of factors. Other investigators invoke increased storm frequency or magnitude, or reduced sediment budgets, or at a local level the impact of engineering projects and other human activities.

Dune reactivation along the northern and western coasts of Europe confounds simple explanation. The older coastal dunes of the Netherlands, for example, were formed between 5000 and 2000 BP following the Flandrian transgression. Their reactivation and the creation of younger dunes during medieval times reflects a combination of a marine transgression between 500 and 1350 AD, extensive deforestation after 1100 AD, and subsequent overstocking with rabbits after 1300 AD (Maarel, 1979). Dune mobility only began to be arrested significantly after extensive organized use of stabilizing grasses in 1796. Farther east, the same early medieval transgression, most probably related to renewed subsidence of the southern North Sea basin, breached a more or less continuous coastal barrier to form the Frisian Islands. After a late medieval pause, this transgression was renewed in the sixteenth century, from which time human mismanagement of dune vegetation for firewood, grazing and burning for tillage land began to be widely reported along both the North Sea and Baltic Sea coasts (Hartnack, 1931; Schou, 1945). Dune erosion has thus become common in recent centuries along the north coasts of Germany and Poland, and the west coast of Denmark (Rohde, 1978), against which the various protective measures of the past 100 years have had mixed success. Other explanations of dune erosion invoke an increased frequency of storms and strong winds. Hansen (1957) for example, believes that a stronger wind climate during later medieval times caused widespread sand drifting across Denmark. Indeed, there is much documentary evidence from around Europe's more exposed coasts which indicates the burial of lands, villages and castles by drifting sands during the fourteenth and fifteenth centuries, as in the Gower Peninsula of South Wales (Lees, 1982).

The above examples were chosen to illustrate Holocene dune destabilization along a broad front or over a significant period of time. We are not concerned here with local blowouts, which occur in most otherwise stable dunes. Nor is the global scene one of unrelieved dune erosion because locally abundant sediment supplies continue to nourish beach-dune systems, for example downdrift from the Columbia River mouth in the northwest United States and north of the Tugela estuary in southeast Africa. Nevertheless, the global trend towards dune destabilization and erosion is a reality of scientific interest with major coastal management implications, a trend that seeks explanation. The Holocene dunes at Morro Bay, California, will now be examined in this context.

3. THE MORRO DUNE COMPLEX, CALIFORNIA

Dunes occur at frequent intervals along California's Pacific coast in response to dominant northwesterly onshore winds and to the abundance offshore of clastic debris flushed from erodible drainage basins inland. From studies beginning in 1919 and published in 1967, Cooper recognized three episodes of dune formation: active Flandrian dunes, stabilized Flandrian dunes, and subdued pre-Flandrian dunes of uncertain age. Using grain characteristics and discriminant analysis, Orme and Tchakerian (1986) distinguished four main dune phases from mid-Pleistocene times to the present and showed how recurrent dune deposition has been favored by the presence of tectonically subsiding basins. The above studies focussed on the Santa Maria dune complex and made but brief reference to the Morro dunes farther north.

The Morro dune complex occupies a broad triangle, roughly 15 km to a side, at the seaward end of the northwest-trending Los Osos Valley, near San Luis Obispo, 300 km northwest of Los Angeles, California (Fig. 1). Within this area, the Flandrian transgression created Morro Bay, a triangular lagoon 5 to 7 km to a side, that is contained seawards by a later Holocene barrier beach-dune system. Subsequent sedimentation has reduced the lagoon's dimensions so that, whereas 850 ha of water surface and 191 ha of salt marsh are apparent at mean high water, at low water 588 ha of mudflats are exposed and only 262 ha of water remain, mostly as subtidal channels 2 to 4 m deep except where dredged. The Morro barrier varies from 300 to 650 m in width at mean high water and extends 6.5 km northward towards Morro Rock. The drainage area tributary to Morro Bay covers 195 km². The seaward face of the barrier receives sediment from an additional 324 km² drainage area tributary to Estero Bay. The Quaternary dune complex is underlain at depth by Mesozoic metasediments and metavolcanics of the Franciscan mélange which outcrop northward to the Santa Lucia Range. Similar rocks emerge from beneath the dunes south of Los Osos Valley, but are mostly overlain by Neogene marine sediments. Mid-Tertiary hypabyssal dacite plugs form several prominent hills along the northern margin of Los Osos Valley, the most seaward of which is Morro Rock (176 m).

3.1. The late Quaternary dune sequence

The Morro dune complex is exposed along 11 km of ocean front from Islay Creek to Morro Rock, and for a further 3 km north of that rock (Fig. 1). For 2 km north from Islay Creek to Hazard Creek, paleodunes rest upon 1 to 5 m of colluvial, alluvial and beach deposits, which in turn overlie a shore platform 3 to 6 m above mean sea level (MSL). This platform is cut across folded siltstones, claystones and shales of the Miguelito Member of the Pismo Formation (late Miocene to early Pliocene; Hall, 1973). The platform is comparable in height to a shore platform at Cayucos, 10 km north of Morro

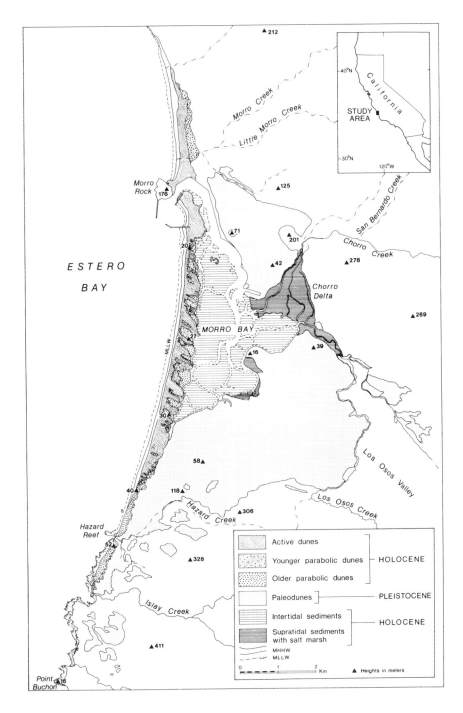

Fig. 1 Late Quaternary dunes around Morro Bay, California

Rock, where fossil corals at the base of overlying deposits have been dated by uranium-series decay to 130 000 to 140 000 ± 30 000 BP (Veeh and Valentine, 1967). With the age uncertainty the Cayucos platform may well correlate with the Last Interglacial sea level around 120 000 to 130 000 BP found at numerous localities worldwide. The shoreline angle of the Cayucos platform was about 6 m above present sea level when the coral formed (Wehmiller *et al.*, 1977). Near Islay Creek, the overlying colluvial and alluvial deposits, whose Miguelito clasts reflect a nearby buried cliff and local streams, have accumulated after the sea's withdrawal from this relatively high interglacial level.

The paleodunes around Morro Bay are 15 to 45 m thick, rising from below present sea level to feather out around 300 m against the Irish Hills. They are presently subject to marine erosion along a broad front north of Islay Creek, and a cover of later dunes increases in thickness northwards (Fig. 2). Erosion has revealed at least four paleosol horizons within the paleodunes, indicating intermittent deposition, but their age is unknown. However, similarity with paleodunes near Point Sal, 48 km SSE, where subjacent organic deposits have yielded ^{14}C ages around 27 000 BP, suggests a major dune transgression during an interstadial of the Last Glaciation when a lower sea level rose to within 10 km of the present coast. The upper portion of the Morro paleodunes, at least, may be related to this dune transgression. At the coast, the paleodunes are capped by a persistent paleosol, 1 to 2 m thick, whose formation probably began during the last cold stade of the Last Glaciation. The paleodunes as a whole have since weathered to a reddish yellow to strong brown color (Munsell 5YR 6/8 to 7.5YR 5/8).

Older parabolic dunes overlie the paleodunes along the cliff top for 3 km northward from Islay Creek, continuing beneath later dunes farther north to form the oldest dunes of the Morro barrier (Fig. 2). In the Montaña de Oro segment, bluff erosion reveals that the basal contact of these older parabolic dunes declines in elevation to present sea level at the south end of the Morro barrier. This indicates that the underlying paleodunes provided a ramp that allowed later dunes to move upslope and inland until stabilized. A similar ramp probably existed to seaward but has since been eroded so that only the landward portions of the parabolas remain. The older parabolic dunes are narrowly apical, sometimes nested within and sometimes truncating one another, rise 10 to 12 m above interior depressions, and reach a maximum elevation of 52 m MSL. The sand is light brown to yellow in color (Munsell 7.5YR 6/4 to 10YR 7/6) and poorly compacted (Fig. 3).

A more precise age for the older parabolic dunes is indicated by a 1 to 2 m thick midden (SLO 1, University of California Archaeological Survey) just north of Islay Creek. This midden overlies the uppermost soil of the paleodune noted above, although some midden material has worked its way into the soil. Near the cliff edge, where the midden underlies the nose of a parabola, a red abalone shell (*Haliotis rufescens*) from near the midden's base has yielded a ^{14}C age

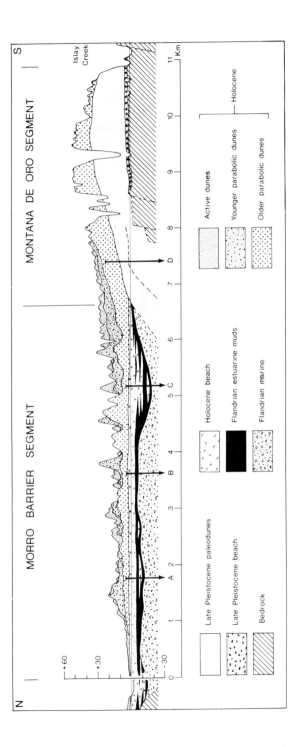

Fig. 2   Cross-section of late Quaternary dunes and other deposits, Morro Bay, California. Sources: dune information from author's field investigations; deposits beneath dunes interpreted by author from well logs (A, B, C, D) obtained by department of Water Resources (1972, 1979). Crest profile (in metres) is along the highest dunes of the Morro barrier and along the highest cliff-top dunes farther south

Fig. 3   Coastal section in late Quaternary dunes, Montaña de Oro segment, 1 km north of Islay Creek. The 25 to 37 m cliff face exposes (1) 1 m of basal alluvial sediments, (3) 20 to 25 m of late Pleistocene paleodunes, and (2) 0 to 12 m of older parabolic dunes whose noses form irregular cliff-top relief. The dark 2 m paleosol complex beneath unit 3 was buried by older parabolic dunes around 3080 ± 90 BP (photo: A. R. Orme)

of 4160 ± 70 BP (Beta 25225). The midden has yielded abundant arrowheads, spear points, hammer stones, knives, bowls, metates, pestles and awls, as well as large quantities of shells, fish bones and animal bones (SLO 1 records). These indicate long-term if intermittent habitation in the lee of advancing dunes, habitation which shifted inland as sand began burying the area but continued long after the older parabolic dunes became stabilized.

Approximately 1 km north of Islay Creek, the uppermost horizon of the persistent paleosol beneath the parabolic dunes contains more scattered shells which have yielded a ^{14}C age of 3080 ± 90 BP (Beta 25991) (Fig. 3).

On the Morro barrier, the older parabolic dunes are associated with several middens. One of these (SLO 978, University of California Archaeological Survey), 2.3 km south of the barrier's northern tip, has yielded a ^{14}C age of 3430 ± 100 BP from a 20 cm deep test pit (Gibson, 1981). Other nearby shell debris, whose stratigraphic relation to the older parabolic dunes is less clear, has yielded ^{14}C ages of 2260 ± 95 and 1150 ± 95 BP, as well as an arrowhead, choppers, worked flakes, asphaltum, and burned or fire-cracked oven rocks (Gibson, 1981). This indicates long if intermittent occupation of the barrier.

Some 3 km north of Islay Creek, the older parabolic dunes are in turn overlain by younger parabolic and lobate dunes (Fig. 2). The contact is marked by a weak paleosol topped by a persistent charcoal seam which has yielded a ^{14}C age of 1730 ± 90 BP (Beta 16024). The younger dunes, comprising yellow sand (Munsell 2.5Y 7/6 to 8/8), form ocean-front bluffs up to 10 m high and pass

Fig. 4   Inner edge of the central Morro barrier showing the main active dunes invading Morro Bay over older parabolic dunes (foreground) and younger parabolic dunes rising to 27 m in the middle distance (photo: A. R. Orme)

Fig. 5   Active foredunes, 1 to 4 m high, forming in a breach in the older dunes near the center of the Morro barrier complex (photo: A. R. Orme)

northward to form stabilized dunes on the Morro barrier, rising there over the older parabolic dunes to 20 to 30 m MSL. It is thus evident that vegetation on the older dunes was destroyed by fire, either naturally or by human agency, thereby initiating a new phase of dune mobility.

The most recent phase of dune activity is represented on the Morro barrier by bare tongues of sand, averaging 6 to 8 m thick, which have moved across the older dunes and invaded Morro Bay in several places (Figs 1, 2 and 4). These active dunes are fed in part from the modern beach and in part by destabilization of older dunes. Beach-dune relations vary along the barrier. Towards the northern end, near the South Breakwater, the beach forms a ramp 100 m wide and medium to fine sand provides a 1° foreshore slope which steepens across an indistinct berm to 5° near the dunes. Farther south, sediments coarsen, coarse sand and small pebbles are more common, the distinct foreshore slopes from 4° to 11°, and the backshore is either flat or slopes 1° landward. Median grain size increases southward from 2.2 phi near the breakwater to between 1.5 and 0.6 phi on the more exposed foreshore of the central and southern barrier, where beach width varies from 100 to 140 m. Where present, foredunes vary from fresh sand hills 1 to 2 m high to older eroded hummocks up to 4 m high (Fig. 5). Where the main dunes have been breached along a broad front, fresh foredunes within the breach are separated from older dunes by corridors or triangular flats 50 to 200 m wide. The median grain size of the dune sands ranges from 1.5 to 2.8 phi depending on exposure and distance from the beach, but there is no significant difference between active and stable dune sands.

Sand found on small islands and intertidal flats within Morro Bay reveals well-sorted distributions mostly in the fine sand range, attesting to its mostly aeolian provenance from the barrier (Krystoff-Jones, 1986). On Grassy Island, 500 m east of the north-central barrier, fine sands (2 to 3 phi) comprise 60 to 78% of all sediments found to depths of 0.6 m, and sand (-1 to 4 phi) constitutes 93 to 98% of all sediments to that depth. Entrainment of such sand by tidal currents within the bay is clearly an active process but the diminishing proportion of typical aeolian sand sizes with depth suggests that the aeolian contribution to these deposits has increased in recent times.

Examining sediment chemistry in the barrier dunes, Brady (1977) was able to distinguish between active and stable dunes, together with the effects of distance from the beach. Calcium, phosphorus and nitrogen all increased with distance from the beach, while soil pH decreased owing to increasing organic matter from dune vegetation. Sodium and potassium revealed distributions characteristic of salt activity in the weathering profile, whereas phosphorus and calcium potentials in the older stabilized dunes were explained by changes in calcium phosphate attributable to weathering.

Clearly, in recent years, aeolian sediment transport across the barrier has exceeded the rate of stabilization by vegetation. Foredune accumulation commonly occurs in association with the herbaceous perennials *Abronia latifolia, Ambrosia chamissonis* and *Carpobrotus aequilaterus* and the annual *Cakile maritima*, but bare wind-swept areas to landward suffer net sand loss rather than a succession to stabilizing vegetation. Farther into the barrier, the younger parabolic dunes are stabilized by a shrubby vegetation of *Haplopappus ericoides, Lupinus albifrons, Baccharis pilularis* and *Eriophyllum staechadifolium,* with herbaceous *Corethrogyne filaginifolia, Croton californicus* and *Dudleya caespitosa.* On the older parabolic dunes coastal sage scrub is well developed, dominated by *Artemisia californica, Salvia mellifera* and *Mimulus aurantiacus.* Farther south and inland, these dunes and the paleodunes exhibit a further succession through dune chaparral, dominated by *Adenostoma fasciculatum, Ceanothus cuneatus* var. *fascicularis, Prunus ilicifolia, Heteromeles arbutifolia* and *Arctostaphylos morroensis,* to dune oak scrub with the dwarfed coast live oak *Quercus agrifolia.*

The Morro barrier dunes overlie beach and nearshore deposits extending from about 2 m above to 12.5 m below MSL (Fig. 2). Based on this author's preliminary interpretation of four well logs obtained in the 1970s during investigations of seawater intrusion (Department of Water Resources, 1972, 1979), the beach component overlies estuarine or lagoonal sandy silt and silty clay, dark greenish blue to dark grey in color, which in turn rests on nearshore marine sediments at depths of 6.5 to 18.5 m below MSL (Fig. 2).

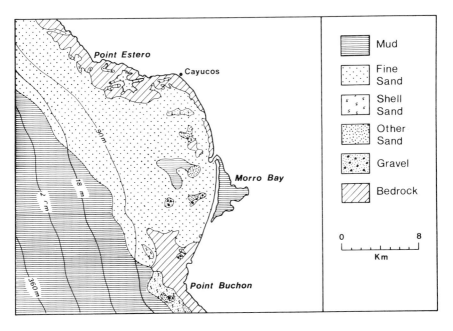

Fig. 6   Offshore surficial deposits of Estero Bay, based on Welday and Williams (1975) and local nautical charts

Table 1   Coastal dune phases at Morro Bay

Phase	Expression	Maximum age
Holocene III	Active dunes	<200 BP
Holocene II	Younger parabolic and lobate dunes	<1730 BP
Holocene I	Older parabolic dunes	≤4160 BP
Late Pleistocene	Youngest paleodunes	<27 000 BP

4. ANALYSIS AND INTERPRETATION

There are at least four main phases of dune activity around Morro Bay, one late Pleistocene and three Holocene in age (Table 1). The morphology and structures of these dunes are oriented ESE, suggesting deposition under conditions similar to today. At present, winds from WNW are dominant for 9 months of the year, with mean hourly velocities from 2.4 to 4.5 m s^{-1}, gusting to 16.1 m s^{-1}. Winds from the NW dominate a further 2 months whereas in December ESE winds are marginally more important.

The immediate source of beach and aeolian sands, now as in the past, is sediment deposited within the Estero Bay coastal cell, a 32 km stretch of shelf between Point Estero and Point Buchon (Fig. 6). Sediment is delivered to this cell from three main sources: from a terrestrial drainage area of 519 km^2, from sea cliff erosion north and south of Morro Bay, and from bottom currents working on the shelf deposits of Estero Bay. The floor of Estero Bay has a gradient of 0.09, comparatively gentle for the California shelf, with the shelf break to deeper water lying at 90 m some 10 km west of Morro Rock. Although bedrock outcrops extensively around Point Estero and Point Buchon, the intervening shelf is widely mantled with fine sand, with coarser sand and gravel locally (Welday and Williams, 1975). Some of this fine sand may be submerged or reworked dune material. Coarser sediments are most likely derived from thick Pleistocene fluvial and marine deposits on the shelf. In any case, bottom currents may activate these shelf deposits and bring sand within the range of wave-related nearshore currents. Quaternary sea-level oscillations would of course increase the mobility of these shelf sediments, and during low sea levels large quantities of sand would be exposed to wind action.

4.1. Pleistocene framework and paleodunes

The Holocene dunes rest upon a framework created by Pleistocene events, of which three elements are relevant to subsequent dune deposition.

First, Pleistocene tectonism played a significant role in shaping the Morro Bay area. Pliocene marine sediments in the Irish Hills southeast of Morro Bay are strongly folded and locally faulted. These rocks are overlain unconformably by Lower Pleistocene fluvial deposits sometimes dipping at 20 to 30°, and these in turn are partly covered with locally deformed Middle and Upper Pleistocene fluvial and marine sediments (Hall *et al.*, 1979). The Los Osos Fault, which brings Franciscan metavolcanics up against Lower Pleistocene sediments in Los Osos Valley, probably continues WNW beneath Morro Bay and the central barrier. The Edna Fault farther south may also cross the coast near Hazard Reef. Substantial faulting would explain why the Miguelito shales, found in the shore zone from Islay Creek northward for 2.5 km, are not encountered 169 m below sea level in Testhole D, 1 km farther north (Fig. 2). Pleistocene faulting probably deepened a complex trough along the ESE-WNW axis of Morro Bay, a trough in which thick sediments then accumulated. Whether continuing tectonism has contributed, at least in part, to Holocene dune instability and bluff erosion by further lowering of the Morro Bay area, cannot presently be ascertained, but it should not be ignored.

Secondly, in the Montaña de Oro segment, Holocene dunes overlie paleodunes which in turn rest on beach deposits most likely correlated with the Cayucos marine terrace formed about 130 000 BP (Veeh and Valentine, 1967). Several paleosols indicate that these dunes accumulated episodically and, based on paleodunes being studied by the author near Point Sal, the latest Pleistocene dunes around Morro Bay probably accumulated during an interstadial interlude shortly after 27 000 BP. Soil formation on these paleodunes then continued well into Holocene times until buried by the older parabolic dunes (Fig. 3).

Thirdly, in the Morro barrier segment, Holocene dunes overlie deposits of the Flandrian transgression and its aftermath, as shown in Testholes A, B and C (Fig. 2). Marine sediments with an Upper Pleistocene and Holocene fauna rise to within 12.4 m, 6.5 m and 18.5 m of present sea level in Testholes A, B and C respectively. These deposits are overlain by blue-green to dark grey silty clays with typical estuarine fauna, which are in turn covered with the sands and gravels of the present beach. This sequence indicates a transgressive barrier environment that transformed an open marine embayment to a sheltered lagoon, whose deposits were progressively overlapped by the Morro barrier as the sea reached its present level. This Flandrian barrier then provided the foundation for later Holocene dune deposition.

4.2. Older parabolic dunes

Near Islay Creek, the older parabolic dunes were transgressing across the paleodune surface 4160 years ago and probably stabilized shortly afterwards. About 1 km farther north, parabolic dunes were still active 3080 years ago but stabilized thereafter. Dunes of this phase were also present on the Morro barrier by 3430 BP. Activation of these older parabolic dunes probably began some time before 4160 BP and continued until around 3000 BP, after which soil began to form on a surface increasingly stabilized by vegetation.

The most probable scenario suggests that these dunes formed from transverse or barchanoid dunes that had accumulated several hundred meters seawards towards and shortly after the close of the Flandrian transgression. As sea level began to stabilize, these transverse dunes would begin to be deprived of plentiful sand supplies from the shelf and in a similar wind environment would likely be breached and their sand transported downwind to form parabolic dunes that climbed over the paleodune ramp until stabilized. There is no surviving evidence of the earlier transverse dunes but the hypothesis is consistent with the belief of McKee (1966) and others that parabolic dunes evolve downwind from transverse or barchanoid dunes owing to decreasing sand supply. Conversely, fresh sand supplies or reactivation through devegetation may cause new transverse dunes to invade older stable parabolic dunes, as is presently occurring at Pismo Beach 30 km southeast of Morro Bay. In the Morro Bay area, the older parabolic dunes were subsequently truncated by bluff erosion along a broad front, so that only their downwind noses now survive (Figs 2 and 3).

4.3. Younger parabolic and lobate dunes

While the shoreline still lay some distance seaward, the older parabolic dunes were destabilized by the burning of their vegetation cover over a broad area, initiating a new phase of dune development after 1730 BP. These younger dunes were also vaguely parabolic in form, although lobate masses also accumulated. Attribution of the fire that produced the charcoal seam beneath these dunes is problematic. Coastal sage and chaparral vegetation are highly flammable, notably in conditions of old growth and low humidity. Lightning could well have ignited the brush, although humidity levels are often higher at the coast than inland. Fire set either deliberately or accidentally by human beings is also possible.

Pursuing the latter theme soon emphasizes the uncertainty. Whereas it is sometimes assumed that the native scrub and woodland of coastal California only gave way to grassland after European settlement began, the diaries of successive Spanish explorers testify to the widespread occurrence of fire during the Indian period. Accompanying Portola's land expedition in 1769, for example, Juan Crespi described how, in travelling north across coastal terraces from Morro Bay to Cayucos on 9 September, 'we set out to the northwest over mesas of good land covered with grass and well supplied with water but without trees' (Crespi, translated by Bolton, 1927). Clearly, the indigenous inhabitants, the Chumash people, had already modified the coastal vegetation, presumably by setting fires, in pursuit of hunting and food collecting. But the Chumash were relative latecomers and if anthropogenic fire is invoked around 1730 BP it is most likely attributable to peoples of the so-called Intermediate or Hunting Cultures (Heizer, 1978). Certainly burned and fire-cracked oven rocks occur in many local middens, as at SLO 978 on the Morro barrier whose use, noted earlier, ranged from 3430 to 1150 BP (Gibson, 1981). Such people may well have set fires to drive game or to stimulate new plant growth, or camp fires may have escaped their control. Destruction of vegetation by fire would have generated widespread sand movement and initiated a new phase of dune accumulation.

4.4. Active dunes

The presently active dunes of the Morro barrier are probably a response to several influences operating over the past 200 years, but most effectively since around 1940. These influences comprise (i) destruction of dune vegetation by fire, grazing, off-road vehicles and military activity, (ii) modification of the Morro Bay entrance channel and relocation of Morro Creek, and (iii) possible relative sea-level rise.

4.5. Destruction of dune vegetation

At the onset of European contact, blowouts doubtless existed as part of the natural dune system, while other stretches of bare sand could probably be

Fig. 7   The Morro barrier, November 1947, showing unstable foredunes and active main dunes moving across older Holocene dunes into Morro Bay. The dune pattern has changed little since 1947. Morro Rock and the adjacent breakwaters appear in the distance (E 13229, Spence Collection, UCLA)

attributed to Chumash activities. The Portola expedition of 1769 encountered both small Chumash villages and semi-nomadic groups around Morro Bay (Crespi, translated by Bolton, 1927). Fish camps existed on the barrier and near creek mouths. The founding of Mission San Luis Obispo de Tolosa in 1772 led to the collapse of native culture and the introduction of widespread grazing which had a major impact on the coastal vegetation. At its zenith, between 1805 and 1820, the mission had as many as 961 Chumash, 8900 cattle, 10 202 sheep, 2258 horses, and raised wheat, vegetables, grapes and other crops (Engelhardt, 1933). The mission was secularized in 1835.

The Mexican and American periods saw a continuation of widespread cattle and sheep raising in the area. In 1842, most land between San Luis Obispo and the coast became Rancho Cañada de los Osos, from which Rancho Pecho y Islay was carved the following year. Although the Morro barrier remained unclaimed unimproved land, there is little doubt that cattle roamed freely across the dunes at this time, especially during the severe droughts of 1828 to 1830 and 1862 to 1864 when pasturage inland was severely limited. Wharf construction at Morro Bay in 1872 and Islay Creek (Spooner's Landing) in 1890 was designed initially for the export of hides and live cattle and later of dairy products and grain, as well as for timber import. The coastal dune cover thus probably suffered from sporadic grazing from 1772 until the 1940s, the vegetation suffering accordingly, as it still does from the black-tailed deer.

During World War II, dune vegetation suffered further from the emplacement of coastal gun batteries and access roads, such that photographs taken in 1947 show a pattern of active dunes similar to today (Fig. 7). In the post-war era, off-road vehicles caused further damage until excluded, following the creation of the state parks in the 1960s. Wildfires ignited by lightning were reportedly common in the Montaña de Oro segment around 1900 but effective fire suppression since the 1920s has limited this agent of change, although the state park system plans to introduce prescribed fires to reduce fuel accumulation (Montaña de Oro State Park, Preliminary General Plan, 1988).

4.6. Modification of Morro Bay entrance channel

Artificial changes around the Morro Bay entrance channel over the past century have had a serious impact on the Morro barrier. Under natural conditions Morro Rock was an island linked to the mainland by intertidal sand, Morro Creek reached the coast east of the rock, and Morro Bay had unstable entrance channels both north and south of the rock (Fig. 8). Describing Morro Rock in 1769, Juan Crespi 'saw a great rock in the form of a round morro, which at high tide is isolated and separated from the coast by little less than a gunshot' (Crespi, translated by Bolton, 1927). The shoaling of one entrance channel was commonly accompanied by the deepening of the other.

As early as 1872, army engineers recommended closing the south channel but

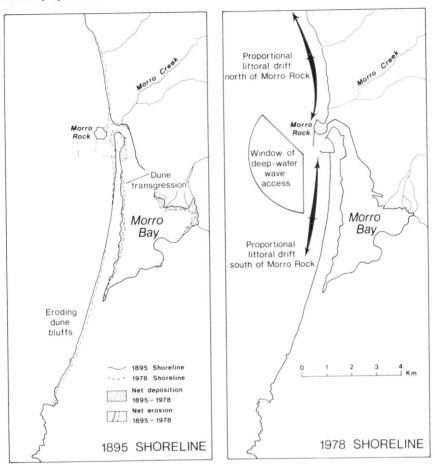

Fig. 8   The 1895 and 1978 shorelines in the Morro Bay area. The left diagram shows
net erosion and deposition between 1895 and 1978 based on comparison of USGS
coordinates. The right diagram shows the window of deep-water wave access off the
present entrance channel and the relative proportions of northward- and southward-
flowing littoral drift computed by Noda and Jen (1975). Note the diversion of Morro
Creek and the placement of the entrance breakwaters

in 1910 it was the north channel that was closed by a partial revetment built
by the San Francisco Bridge Company. This revetment decayed for want of
maintenance after 1913, was rebuilt and extended to 518 m in length and 3.7 m
above MLLW in 1935 to 1936, and was further reinforced in 1943, by which
time accretion against its north side had eliminated the north channel.
Meanwhile, Morro Creek had been diverted directly to the sea north of Morro
Rock. In 1942 the North Breakwater was built southward from Morro Rock
to provide a safe haven for naval patrol boats. As this was a hurried wartime

project, neither wave analysis nor model studies were conducted prior to construction. Wave refraction around this breakwater soon caused shoaling to leeward, and the South Breakwater, extending westward from the barrier, was built to reduce northward littoral drift from moving into the entrance channel (US Army Corps of Engineers, 1986). Shoaling was countered by periodic dredging.

Before these engineering activities, it is likely that the predominant littoral drift moved southward on both sides of Morro Rock, augmented by sediment from Morro Creek. The above changes, however, have yielded three impacts of significance to the stability of the Morro barrier. First, the revetment prevents the southward movement of sediment between Morro Rock and the mainland at high tide, and pushes sediment from the now diverted Morro Creek to the west of the rock. Second, the 575 m long North Breakwater has enhanced Morro Rock's influence on wave refraction and effectively reversed the predominant littoral drift for some 2 km along the barrier. Under present conditions, Noda and Jen (1975) computed a net southward drift of over $14\,000\,m^3\,a^{-1}$ along the beach north of Morro Rock, but south of the rock they estimated a net northward drift of nearly $25\,000\,m^3\,a^{-1}$ (Fig. 8). Because of this the barrier farther south is suffering net erosion and insufficient sand is available to replenish the foredunes. Third, the revetment and North Breakwater have created a large sand trap in the lee of Morro Rock which, to maintain adequate navigation channels, has required periodic dredging since 1949 (Table 2).

As Table 2 shows, large quantities of mostly sand, but with some silt and gravel, have been removed from the entrance channel and discharged in several localities. At first, most dredged spoil was placed updrift on the beaches to either side of Morro Creek north of Morro Rock, and also on the northern end of

Table 2 Dredging history of Morro Bay entrance channel, 1949 to 1987

Year of dredging	Volume dredged (m³)	Disposal area
1949	628 786	
1956	695 786	
1964	536 749	North beach, north barrier
1968	311 109	North beach
1971	145 274	North beach, north barrier
1974	305 840	900 m SW of Morro Rock
1980	445 303	5.5 km south along barrier
1985	420 530	North beach, S of South Breakwater
1987	305 840	4.9 km south of South Breakwater

| Total volume removed | 3 795 217 m³ | |

| Assumed annual rate of deposition in entrance channel, 1944–87 | $88\,261\,m^3\,a^{-1}$ |

Source: US Army Corps of Engineers, Los Angeles District (1988).

the Morro barrier which has widened accordingly (Fig. 8). Some sediment was placed offshore but soon moved back towards the entrance channel. Since 1980, in recognition of the erosion occurring along the barrier, most sediment has been placed on the beach south of the South Breakwater. However, placement of the spoil pipeline along the backshore and through bulldozed foredunes not only accelerated local erosion of the beach and foredunes but provided enough medium to fine sand at sufficiently frequent intervals to maintain unstable dunes. In the longer term this sediment may offset some of the beach erosion caused by construction farther north, but it also ensures continuing dune instability.

4.7. Relative sea level change

In recent years, much evidence has accumulated for a contemporary global rise of sea level with major implications for coastal stability. Using tide gauge data, estimates of eustatic rise in this century vary from 1 to 4 mm a^{-1} (Gornitz, Lebedeff and Hansen, 1982). Some of this rise may be attributable to a redistribution of ocean water without an increase in water volume, specifically as ocean-basin capacity changes in response to isostasy, tectonism and sedimentation. The remainder of the rise is commonly linked to global warming attributable to the 'greenhouse effect'. In California, where measures of sea-level rise are complicated by tectonism, tide gauge data from San Francisco and Los Angeles, about equidistant from Morro Bay, indicate a sea-level rise of 1 to 2 mm a^{-1} during this century. It is perhaps premature to invoke sea-level rise to explain dune instability in the Morro Bay area, especially in view of the human influences discussed above. Nevertheless, the significant erosion of dune bluffs over 4 km northward from Islay Creek may be due, at least in part, to a rising sea level combined with tectonic subsidence along seaward extensions of the Los Osos and Edna faults discussed earlier.

5. CONCLUSIONS

The coastal dunes around Morro Bay reveal an interesting record of deposition and reactivation during late Quaternary times, and the first dated sequence of Holocene aeolian activity along the California coast. The base of the sequence is provided by Pleistocene paleodunes whose lowest member was deposited on an interglacial shore platform some 130 000 years old, and whose uppermost member accumulated during an interstadial interlude shortly after 27 000 BP. In the upland portions of the study area, the Montaña de Oro segment, these uppermost paleodunes with their well-developed soil provided a ramp across which Holocene dunes were subsequently deposited. In the lowland area farther north, Holocene dune development awaited creation of the Morro barrier by the Flandrian transgression. The present barrier and dunes were more or less in place by 3430 BP and probably earlier.

Three Holocene dune phases followed culmination of the main transgression. The older parabolic dunes were transgressing across the paleodune ramp between 4160 and 3080 BP but, as these dates are derived from basal materials close to the landward noses of the parabolas, it is probable that the dune transgression began some distance seaward at an earlier time and ceased around 3000 BP. Younger parabolic and lobate dunes formed after widespread fire destabilized the existing dunes around 1730 BP. The most recent and continuing phase of dune activity began some 200 years ago but has probably accentuated during the present century.

How does this sequence fit the global pattern sampled earlier? Certainly the uppermost stratigraphy of the Morro barrier reveals the nature of the Flandrian transgression and, although dunes specific to the close of the transgression have not yet been recognized, it is probable that the older parabolic dunes were formed from the destruction of transverse or barchanoid dunes developed during and shortly after the transgression. These older parabolic dunes are probably comparable in age to similar forms found along Australia's east coast and to the older Holocene dunes of northwest Europe.

The younger parabolic dunes are probably specific in their age to the Morro Bay area. Whether the widespread fire which led to dune reactivation was due to lightning or to human activity is uncertain, but a case can be made for either hypothesis. Lightning is always possible, as it is today, but we have also seen how fire generated by indigenous peoples has been invoked in Africa and Australasia.

The youngest dunes probably became active after European settlement began some 200 years ago. The causes are complex. Destruction of the vegetation by increased fires, grazing, off-road vehicles and military activity is readily documented, while the modification of the Morro Bay entrance channel and subsequent discharge of dredged sediment have thoroughly upset the stability of the Morro barrier system. Causes such as these are, as noted earlier, common around the world, only the intensity of human activity varying with time and place. Natural forces, notably relative sea-level rise and increased storm frequency and wind intensity, cannot be ruled out but they are difficult to distinguish amid the pervasive impact of recent human activities. Human impacts alone pose major problems for the future management of the local dunes, now in the care of the California state park system. If rising sea level and increased storminess are in fact augmenting human activity, the implications for coastal management, here as elsewhere, are serious indeed.

ACKNOWLEDGEMENTS

A grant from the Academic Senate, University of California, Los Angeles, is gratefully acknowledged. Beta Analytic Inc., Coral Gables, Florida, provided the ^{14}C dates. Amalie Jo Orme assisted in the field. Pamela Castens (UCLA and US Army Corps of Engineers), Robert Gibson (Archaeologist, Paso Robles), Thomas Rice (Soil Science, California State Polytechnic University, San Luis Obispo), Terry Rudolf (University of California Archaeological Survey, Santa Barbara), and LouElla Saul (Los Angeles County Museum of Natural History) provided valuable information incorporated in the text.

REFERENCES

Bird, E. C. F. (1974) Dune stability on Fraser Island. *Queensland Naturalist*, **21**, 15–21.

Bolton, H. E. (1927) *Fray Juan Crespi, Missionary Explorer of the Pacific Coast, 1769–1774*. Univ. Calif. Press, Berkeley.

Brady, K. (1977) Some soil relationships of the Morro Bay sand dunes, Unpublished MS thesis (Agriculture), Calif. State Poly. Univ., San Luis Obispo.

Butzer, K. W. and Helgren, D. M. (1972) Late Cenozoic evolution of the Cape coast between Knysna and Cape St Francis, South Africa. *Quat. Res.*, **2**, 143–169.

Cooper, W. S. (1967) *Coastal dunes of California*. Geol. Soc. Amer. Mem. 104, Boulder.

Department of Water Resources (1972) *Sea water intrusion: Morro Bay area, San Luis Obispo County*. The Resources Agency, State of California.

Department of Water Resources (1979) Morro Bay sand spit investigation. The Resources Agency, State of California.

Engelhardt, Z. (1933) *Mission San Luis Obispo in the Valley of the Bears*. Mission Santa Barbara.

Field, M. E. and Duane, D. B. (1976) Post-Pleistocene history of the United States inner continental shelf: significance to origin of barrier islands. *Bull. Geol. Soc. Amer.*, **87**, 691–702.

Gibson, R. O. (1981) *Cultural resource test program at SLO 978, Morro Bay, San Luis Obispo County, California*. Rept. to US Army Corps of Engineers, Los Angeles District.

Gornitz, V., Lebedeff, S. and Hansen, J. (1982) Global sea level trend in the past century. *Science*, **215**, 1611–1614.

Hall, C. A. (1973) Geologic map of the Morro Bay South and Port San Luis quadrangles, San Luis Obispo County, California. US Geol. Surv., Misc. Field Studies Map, MF 511.

Hall, C. A., Ernst, W. G., Prior, S. W. and Weise, J. W. (1979) Geologic map of the San Luis Obispo-San Simeon region, California. US Geol. Surv., Misc. Invest. Ser., Map I-1097.

Hansen, V. (1957) Sandflugten i Thy og dens indflydelse på Kulturlanskabet. *Geogr. Tidsk.*, **56**, 69–92.

Hartnack, W. (1931) Zur Enstehung und Entwicklung der Wanderdünen an der deutschen Ostseeküste. *Zeit. Geomorph.*, **6**, 174–217.

Heizer, R. F. (1978) *Handbook of North American Indians: Vol. 8*: California. Smithsonian Inst., Washington, DC.

Hicks, D. (1983) Landscape evolution in consolidated coastal dune sands. *Zeit. Geomorph., Suppl.-Bd*, **45**, 245–250.

Krystoff-Jones, J. M. (1986) Sediment analysis at selected sites in the Morro Bay estuary—a measure of habitat suitability for *Callianassa californiensis*. Rept., Biol. Sci. Dept., Calif. State Poly. Univ., San Luis Obispo.

Leatherman, S. P. (1983) Barrier dynamics and landward migration with Holocene sea-level rise. *Nature*, **301**, 415–418.

Lees, D. J. (1982) The sand dunes of Gower as potential indicators of climatic change in historical time. *Cambria*, **9**, 25–35.

Maarel, E. van der (1979) Environmental management of coastal dunes in the Netherlands. In Jefferies, R. L. and Davy, A. J. (eds), *Ecological Processes in Coastal Environments*. Blackwell Scientific Publications, Oxford, pp. 543–570.

McKee, E. D. (1966) Structures of dunes at White Sands National Monument, New Mexico (and a comparison with structures of dunes from other selected areas). *Sedimentol.*, **7**, 3–69.

Noda, E. K. and Jen Y. (1975) *Sand transport analysis, Morro Bay*. Tetra Tech, Pasadena; Rept. to US Army Corps of Engineers, Los Angeles District.

Orme, A. R. and Tchakerian, V. P. (1986) Quaternary dunes of the Pacific coast of the Californias. In Nickling, W. G. (ed.), *Aeolian Geomorphology*. Allen & Unwin, London, pp. 149–175.

Pye, K. (1983) Formation and history of Queensland dunes. *Zeit. Geomorph., Suppl-Bd.*, **45**, 175–204.

Rohde, H. (1978) The history of the German coastal area. *Küste*, **32**, 6–29.

Schou, A. (1945). Det marine foreland. *Folia Geogr. Danica*, **4**, 1–236.

Semeniuk, V. (1985). The age structure of a Holocene barrier dune system and its implications for sea-level history reconstructions in southwestern Australia. *Mar. Geol.*, **67**, 197–212.

Thom, B. G., Polach, H. A. and Bowman, G. M. (1978) *Holocene Age Structures of Coastal Sand Barriers in New South Wales, Australia*. Duntroon Royal Military College, Canberra.

Thompson, C. H. (1983). Development and weathering of large parabolic dune systems along the subtropical coast of eastern Australia. *Zeit. Geomorph, Suppl-Bd.*, **45**, 205–225.

US Army Corps of Engineers (1986) Oral history of coastal engineering activities in southern California, 1930–1981. Los Angeles District.

Veeh, H. H. and Valentine, J. W. (1967) Radiometric ages of Pleistocene fossils from Cayucos, California. *Bull. Geol. Soc. Amer.*, **78**, 547–550.

Wehmiller, J. F., Lajoie, K. R., Kvenvolden, K. A., Peterson, E., Belknap, D. F., Kennedy, G. L., Addicott, W. O., Vedder, J. G. and Wright, R. W. (1977) *Correlation and chronology of Pacific coast marine terrace deposits of continental United States by fossil amino acid and stereochemistry—technique evaluation, relative ages, kinetic model ages, and geological implications*. US Geol. Surv., Open-File Rept. 77–680, Washington, DC.

Welday, E. E. and Williams, J. W. (1975) Offshore surficial geology of California. Calif. Div. Mines & Geol., Map sheet 26.

Section IV
EFFECTS OF
HUMAN DEVELOPMENT

Several of the previous chapters identify complications introduced by human activities. In some cases, human impacts do not appear to result in widespread alteration of dune landscapes (e.g. the effect of artificially introduced exotic species on dune stability in isolated portions of McLachlan's study area or the effect of pedestrian trampling on local diversity of plants and topography identified in Carter and Wilson's study). Other contributors identified impacts at larger scales. Orme, for example, points out the pervasive nature of human-induced change through recent history and the significance of continued human changes on future dune stability. Psuty shows how human efforts to achieve stability are counter to the natural trend of shoreline change, and he demonstrates the rapid failure of structures that introduce disequilibrium into the coastal environment. Both authors show how human efforts complicate interpretation of the landscape and hinder the development of geomorphic models. Borowka calls attention to the importance of humans in the initiation of sand drift in many parts of the world, along with an example of just how dramatic the changes to the landscape can be.

There is a vast literature on coastal dunes in human-altered systems in a management context (Gares, Nordstrom and Psuty, 1979; Cullen and Bird, 1980; Doody, 1985; Chapman, 1989; van der Meulen, Jungerius and Visser, 1989). The bulk of these studies concerns the effects of dune alterations on biota or for achieving shore protection goals. Many of these studies are concerned with strategies to create dunes or to rework the dunes into forms designed to achieve human values, and many are engineering-oriented. It is only recently that scientists have attempted to study the more widespread geomorphic effects of these alterations in a basic research context and to formulate models of change.

Human activities that affect coastal dunes include: (i) alteration of beach processes and sediment budgets; (ii) destruction and destabilization of ground cover by trampling or vehicle use; (iii) alteration of sediment inputs to the dune by bulkheads, seawalls, or sand fences; (iv) introduction of exotic species; and (v) direct manipulation using earth-moving equipment. Some dunes have been so modified that it is questionable whether the term dune is applicable to these features (Nordstrom, 1990). Not surprisingly, these complications have made study of developed dunes somewhat unpopular with geomorphologists and sedimentologists, especially those who are concerned with interpreting natural systems of the past. Although the study of developed landscapes may not be the key to the distant past in such cases, it is almost certainly true that this approach is the key to interpreting the recent past and predicting the future. According to some scientists, it is the only way future changes can be meaningfully anticipated in some areas (Nordstrom, 1987).

Two chapters in this section deal directly with effects of human development. Pye identifies some of the impacts of dredging and the construction of retaining walls on the sediment budget. He also identifies some of the more direct effects on the dunes resulting from the introduction of marram grass, pedestrian pressure, and grading for agricultural use. These effects are incorporated into a model of dune morphology. Gares treats a different suite of human variables, concentrating on sand fences and shorefront buildings. The effects of these human adjustments have been quantified in studies of specific features (Knutson, 1978; Snyder and Pinet, 1981; Nordstrom and McCluskey, 1985), but they have not been placed in a comprehensive geomorphic framework. The contribution by Gares attempts to do this by presenting a comparative model of developed and undeveloped dune systems.

REFERENCES

Chapman, D. M. (1989) *Coastal Dunes of New South Wales: Status and Management.* University of Sydney Coastal Studies Unit, Report CSU 89/3.
Cullen, P. and E. Bird. (1980) *The Management of Coastal Sand Dunes in South Australia.* Geostudies, Black Rock, Victoria, 83 p.
Doody, P. (ed.) (1985) *Sand Dunes and their Management.* Nature Conservancy Council, Peterborough, 262 p.
Gares, P. A., Nordstrom, K. F. and Psuty, N. P. (1979) *Coastal Dunes: Their Function, Delineation and Management.* Rutgers Center for Coastal and Environmental Studies Rept. NJ/RU-DEP-14-12-79.
Knutson, P. L. (1978) Planting guidelines for dune creation and stabilization. *Proc. Coast. Zone '78, ASCE,* 762–779.
Nordstrom, K. F. (1987) Predicting shoreline changes at tidal inlets on a developed coast. *Prof. Geogr.,* **39,** 457–465.
Nordstrom, K. F. (1990) The concept of intrinsic value and depositional coastal landforms. *Geogr. Rev.,* **80,** 68–81.
Nordstrom, K. F. and McCluskey, J. M. (1985) The effects of houses and sand fences on the eolian sediment budget at Fire Island, New York. *J. Coast., Res.,* **1,** 39–46.
Snyder, M. R. and Pinet, P. R. (1981) Dune construction using two multiple sand-fence configurations: implications regarding protection of eastern Long Island's south shore. *Northeast. Geol.,* **3/4,** 225–234.
van der Meulen, F., Jungerius, P. D. and Visser, J. (eds) (1989) *Perspectives in Coastal Dune Management,* SPB Academic Publishing, The Hague.

Chapter Fifteen

Physical and human influences on coastal dune development between the Ribble and Mersey estuaries, northwest England

K. PYE
Postgraduate Research Institute for Sedimentology, University of Reading

1. INTRODUCTION

The importance of coastal dunes as natural sea defences, sites of ecological interest and recreational areas is being increasingly recognized, and both local authorities and national bodies find themselves faced with the need to devise appropriate dune management strategies in the face of competing land-use demands. An understanding of the factors which have influenced the present dune morphology is essential if accurate assessments are to be made about the effects of changing sea level, proposed engineering works, or land-use changes within the dune system itself.

The morphology of coastal dunes, including both the shape of individual dunes and the spatial arrangement of dune complexes, is governed by four main factors:

(a) Beach morphology and shoreline dynamics, which influence the rate of sand supply, the area and size of sand exposed to wind action, and the degree of exposure to storm wave attack on the frontal dunes.
(b) Wind characteristics, including the strength/frequency distribution and the directional variability.
(c) The extent and growth form of natural vegetation cover.
(d) Human activities, which may have either a direct or an indirect impact.

Coastal Dunes: Form and process. Edited by K. F. Nordstrom, N. P. Psuty and R. W. G. Carter
©1990 John Wiley & Sons Ltd

Direct effects of human activities include sand fence construction, planting, trampling, levelling and sand-mining within the dune system itself. Indirect effects include beach sand-mining, dredging, spoil dumping, construction of groynes and training walls, and any other activity which affects the wave climate, nature of tidal currents, and the sand supply. Any change in the beach and nearshore morphology, however caused, is likely in time to have an effect on the neighbouring dunes. As pointed out by Psuty (1988), all variations in beach and dune forms are related to sediment availability, the relative input to the beach and to the dunes, and whether these inputs change over a given time interval.

The purpose of this chapter is to illustrate how the morphological development of coastal dunes in part of northwest England has been influenced both by natural shoreline changes and by human activities. Based partly on field evidence from this area and partly on previous work by the author (Pye, 1982, 1983a,b), a general model is proposed which relates variations in dune morphology to shoreline dynamics, wind energy and vegetation cover.

2. PHYSICAL BACKGROUND TO THE STUDY AREA

The study area is located between Liverpool and Southport in northwest England (Fig. 1). This section of coast is transitional between open coast and estuarine regimes, being influenced by processes both in the eastern Irish Sea and in the Ribble and Mersey estuaries. Much of the coast is sandy, and includes one of the most important coastal dune complexes in the United Kingdom. A considerable part of the area is designated as being of high scientific interest and is protected within reserves managed by the Nature Conservancy Council, the National Trust and Sefton Metropolitan Borough Council.

The dunes have formed in response to large-scale onshore movement of sand during the later Holocene. The coast is macrotidal, with a mean spring range at Formby Point of about 9 m. During storm surges the predicted tidal levels can be raised by up to 1.4 m (Lennon, 1963a,b; Graaf, 1978). Consequently the eastern Irish Sea is characterized by strong tidal currents.

Flood tidal current velocities in Liverpool Bay are higher than the ebb velocities (Admiralty Hydrographic Department, 1960), resulting in a net landward drift of sediment near the bed. The flood tidal streams diverge offshore from Formby Point and flow towards the Mersey and Ribble estuaries. Lower velocity ebb tidal streams diverge offshore from Birkdale and eventually merge with the ebb tidal streams from the Mersey and Ribble.

The alignment of sand waves on the floor of the eastern Irish Sea, and the results of sea bed drifter studies, confirm that net sediment movement takes place in a generally easterly direction (Fig. 1; Sly, 1966; Best *et al.*, 1973; Halliwell, 1973). The main source of sand is provided by fluvio-glacial deposits on the floor of the Irish Sea (Wright *et al.*, 1971). Only a relatively small amount

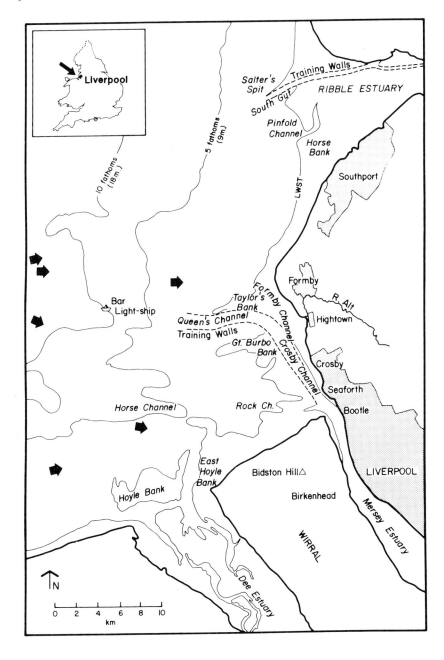

Fig. 1 Location of the study area. The arrows indicate the direction of sand wave movement on the floor of the Irish Sea (after Sly, 1966)

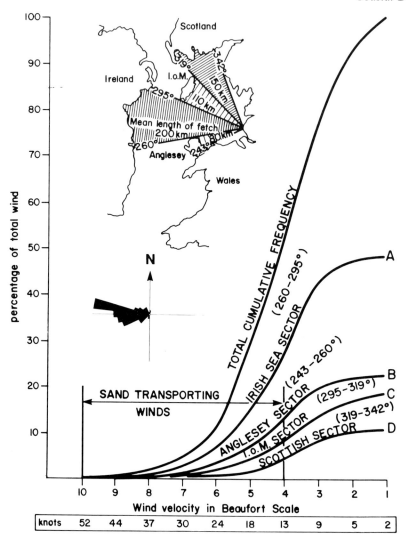

Fig. 2   Wind velocity frequency distribution for different Irish Sea sectors (modified after Gresswell, 1953; Sly, 1966). Also shown are the maximum lengths of wave fetch in each sector and the wind rose for sand-moving winds (>20 km hr^{-1}) recorded at Southport (extracted from Meteorological Office Monthly Weather Reports)

of sand was supplied to the coast by the Rivers Mersey and Ribble during the Holocene (Barron, 1938). Depth soundings and evidence from marine charts indicate rapid shoaling in Liverpool Bay in the last 150 years, amounting to 104×10^6 m^3 of accumulated sediment between 1833 and 1955 (HRS (Hydraulics Research Station), 1958). Since 1955, shoaling has apparently accelerated in some areas. Admiralty charts show that accretion on Taylor's Bank, Jordan's Bank and Zebra Flats, located offshore from Formby Point, resulted in a reduction of the mean water depth from 7 m in 1965 to 5.9 m in 1975, representing an accumulation of 120×10^6 m^3 of sediment. Some of this material is redistributed dredge spoil taken from the Mersey Approaches and Queen's Channel since 1890 (McDowell and O'Connor, 1977; O'Connor, 1987), but much of the sediment has been moved landwards from farther out in the Irish Sea by tidal currents. Sediment accumulation may have been enhanced in the last 50 years by the construction of the Mersey training walls, begun in 1909 and substantially completed by 1939 (Cashin, 1949, O'Connor, 1987). Model studies (HRS, 1958; Price and Kendrick, 1963) suggest that construction of the training walls concentrated the ebb tidal flow in the main channel, leading to an enhanced residual flood transport over the sand banks on either side.

Accretion on a major scale has also taken place in the Ribble estuary since 1850. Construction of training walls to maintain access to the Port of Preston started in 1840, and by 1938 they extended 24 km out into the estuary (Barron, 1938). As in the Mersey, concentration of the ebb flow in the trained channel may have accentuated flood-dominated sediment transport over the sand banks on either side (HRS, 1965, 1968, 1977, 1980). Dredging of the main shipping channel ceased in 1980, since when it has almost completely filled with sand.

The dominant winds and waves come from the west and west-northwest (HRS, 1969a; Bell, Barber and Smith, 1975), corresponding with a maximum fetch of over 200 km in the Irish Sea Sector (Fig. 2). The dominant wave approach angle of 277° (from the north) calculated by Sly (1966) agrees closely with the direction of highest hourly wind speeds recorded at Southport (280°N). Wave energy is moderate; waves higher than 2 m in height occur about 3% of the time (HRS, 1969a), although during severe storms waves in the eastern Irish Sea may reach a height of 5.7 m with a period of 8.7 seconds (HRS, 1977). The eastern Irish Sea is relatively shallow, with the 18 m depth contour located about 15 km west of Formby Point. Approaching waves are refracted as they pass over the offshore sand banks, leading to focusing of wave energy on Formby Point. The wave crests approach approximately normal to the shore near the centre of Formby Point, but diverge at an oblique angle both to the north and the south (Gresswell, 1953). A littoral drift divide is located near Victoria Road, Freshfield (HRS, 1969a,b; Fig. 3), and the beach sand shows a slight longshore fining trend to the north and the south of this point (Pye, unpub. data).

The sediment transport dynamics and bedforms on the shore between Crosby and Southport have been investigated by Gresswell (1937, 1953), Parker (1971, 1974, 1975) and Wright (1976). The most notable features are a series of intertidal ridges and runnels which run obliquely away from the shore to the north and south of Formby Point (Fig. 4). The average beach gradient is relatively steep near the centre of the Point, but becomes much gentler towards the north and south (Fig. 4).

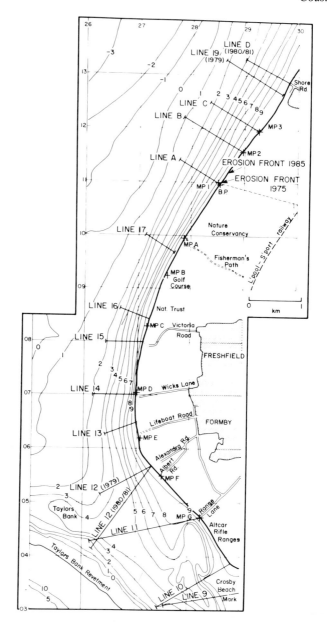

Fig. 3   Map showing the location of marker posts and survey lines where measurements of dune erosion/accretion and changes in the beach profile are made by the Sefton Borough Engineer's department. The contours indicate height in metres relative to Chart Datum (− 4.42 m OD), surveyed in 1979

Evidence from Ordnance Survey maps indicates that the entire shoreline around Formby Point prograded seawards between 1845 and 1906, but after about 1900 erosion began between Victoria Road and Lifeboat Road (Gresswell, 1937, 1953; Pye and Smith, 1988). The underlying reasons for this change are uncertain, but the onset of erosion was closely associated with the establishment of a negative beach sediment budget and reduction in backshore width, thereby increasing the susceptibility of the frontal dunes to storm wave attack. After 1906 the northern limit of erosion advanced slowly northwards, and at present it lies approximately 1 km north of Fisherman's Path (Fig. 3). The southern limit of erosion presently lies between the seaward end of Alexandra Road and Lifeboat Road. Near Victoria Road accretion during the period 1845 to 1906 amounted to about 220 m, but erosion since 1906 has resulted in shoreline retreat of about 400 m (see Fig. 2 in Pye and Smith, 1988). The present coast therefore exposes eroded sections in dunes which were probably formed before the early nineteenth century.

3. AGE AND MORPHOLOGY OF THE DUNES

Much of the shore between Crosby and Southport is backed by dunes which have a maximum inland extent of 4 km at Formby Point. The total area covered by blown sand exceeds 30 km^2. Inland the sand forms a level sand sheet, but near the coast dune forms are well developed and some dunes exceed 20 m in height. The two distinct bodies of blown sand were mapped as the Formby Series and the Dune Sand respectively by the Soil Survey of England and Wales (Hall and Folland, 1967).

The windblown sand overlies blue-grey estuarine or nearshore marine clays (the Downholland Silt) and freshwater peat deposits of early to mid-Holocene age (Wray *et al.*, 1948; Tooley, 1970, 1974, 1976, 1978). A radiocarbon date from peat buried by blown sand at the western edge of Downholland Moss gave an age of 4090 +/− 175 ^{14}C years BP (Tooley, 1977, 1982). A similar date of 4510 +/− 50 ^{14}C years BP was obtained from peat below blown sand at Sniggery Wood, near Little Crosby (Innes, 1982), while a tree trunk in peat exposed on the Formby foreshore yielded an age of 2510 +/− 120 ^{14}C years BP (Pye, unpub). The evidence therefore suggests that accumulation of blown sand began between 2500 and 4500 years ago. The reasons for the change from coastal mud to sand deposition are uncertain, but may reflect a lag period between submergence and effective marine sorting of fluvio-glacial deposits in the eastern Irish Sea. Gresswell (1953) believed that the sand accumulated as a regressive wedge as the shoreline retreated westwards from a maximum landward position attained in the mid-Holocene, but very limited evidence is available to support this hypothesis (Tooley, 1976, 1978). At present little information is available about the stratigraphy and age structure of the blown sand belt.

Fig. 4   Mosaic of air photographs taken in 1979 showing active sand sheets along the eroding shore at Formby Point. The divergence of foreshore ridges away from the Point can also be seen. The distance from north to south represents approximately 5 km. (Reproduced by permission of Sefton Borough Council Engineer and Surveyor's Department)

Although the older blown sands have been levelled further for agriculture and other uses, it is unlikely that these sands ever had well-developed dune topography. Map and documentary evidence suggests that high dunes did not exist at Formby before about 1700 AD (Ashton, 1909). This may have been due to the fact that marram grass (*Ammophila arenaria*), which favours the development of steep, hummocky sandhills (Travis, 1915; Fig. 5), was not common in this area until the beginning of the eighteenth century when strict laws about its planting and cutting were introduced. Marram planting was made a condition of tenure in the Formby area at least as early as 1711, and in 1742 an Act of Parliament was passed, applying especially to the Lancashire and Cheshire coasts, which made uprooting of marram a criminal offence. Planting of marram was obligatory until 1886, when the landowners ceased to enforce it. Prior to marram planting, sand may have been blown inland from the beach almost unhindered, rather than accumulating as steep sandhills near the shore.

The morphology of the sand dune belt near the coast has been influenced by a wide range of human activities. These include excavation of flat-floored plots for agriculture (especially asparagus plantations), sand-mining, waste-dumping, and the creation of recreational facilities such as caravan parks, car parks, roads and golf courses (Fig. 6). In the early part of this century extensive areas at Formby and Ainsdale were planted with conifers (Fig. 7), and many of the foredunes formed in the period 1880 to 1914 owe their form to the erection of brushwood fences and planting along the backshore. Once sufficient sand had been trapped by the fence, further accumulation was encouraged by planting of dune grass species (Gresswell, 1937, 1953). The dunes formed during this

Fig. 5   Marram-covered dunes near Albert Road, Formby

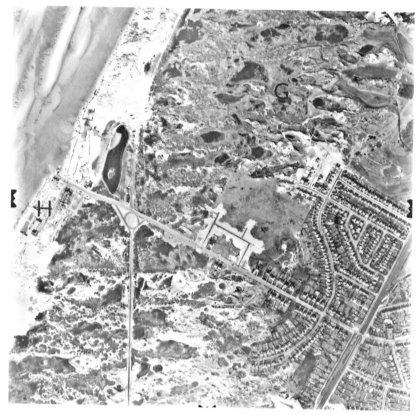

Fig. 6 Air photograph of the dune area around Shore Road, Ainsdale, taken in March 1966. The highly degraded nature of the dunes around the holiday camp (H) can be clearly seen. The dunes on the right have been partly covered by urban development, while those in the top half of the photograph have been turned into a golf course (G). (Reproduced from an Ordnance Survey aerial photograph with the permission of the Controller of Her Majesty's Stationery Office, ©Crown Copyright.) Scale 1:7500

period appear on air photographs as parallel ridges (Fig. 8). Between the sand ridges are linear depressions, known locally as 'slacks', which experience waterlogging when the water table rises above the surface during the winter and spring.

The belt of parallel dune ridges, which in places is more than 200 m wide, was originally continuous around Formby Point. However, the central portion between Lifeboat Road and Fisherman's Path has been gradually removed by erosion since 1906. The truncated ends of the slacks near the northern erosion limit are now subject to marine incursion during major storms.

Behind the belt of parallel dune ridges and slacks is a much more irregular belt of low hummocky dunes and intervening slacks which formed in the period

Fig. 7 Air photograph taken in March 1966 showing a major transgressive sand sheet around the seaward end of Lifeboat Road. The effects of heavy pedestrian pressure are also clearly seen on the frontal dune ridges and in the inland dunes south of Albert Road. The area labelled C is occupied by a caravan park and the area labelled W is a coniferous plantation. (Reproduced from an Ordnance Survey aerial photograph with the permission of the Controller of Her Majesty's Stationery Office, ©Crown Copyright.) Scale 1 : 7500

before 1880 (Fig. 8). At this time brushwood fences were not systematically planted along the backshore, although marram planting was carried out on the frontal dunes, and so sand accumulated in a more haphazard manner. Many of the irregularly-shaped slacks represent natural blowouts, but true parabolic dunes are almost absent, reflecting the patchy nature of the vegetation cover present at the time and the relatively small thickness of sand. The sandhills form

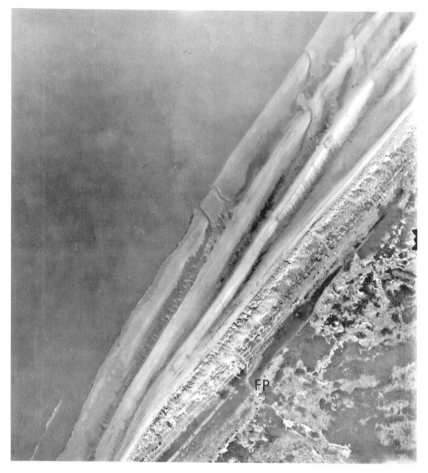

Fig. 8   Air photograph of the frontal dunes in the Fisherman's Path (FP) area, Freshfield North, taken in March 1966. Formation of the parallel dune ridges and intervening swales immediately adjacent to the coast was encouraged by brushwood fence construction and planting between 1880 and 1925. Since the photograph was taken the erosion front has extended beyond Fisherman's Path. (Reproduced from an Ordnance Survey aerial photograph with the permission of the Controller of Her Majesty's Stationery Office, ©Crown Copyright.) Scale 1 : 7500

an irregular assemblage of hummocks which are not clearly related to the position of any former shoreline.

    Close to places where the public can gain access to the beach, recreational pressures have been heavy for almost a century and the dunes have been severely degraded. In the most popular areas, as around Shore Road, Ainsdale (Fig. 6), large bare areas of actively blowing sand existed in the late 1920s, even where the coast showed a natural progradation trend (Gresswell, 1937). However, transgressive sand sheets were especially well developed along a section of coast between Lifeboat Road and Victoria Road which began to erode after 1906 (Figs 4 and 6). Before restoration measures were implemented in the late 1970s, these sand sheets had the form of low sand ramps which sloped seawards at an angle of 5 to 10° (Fig. 9). They were bounded on either side by eroded sandhill remnants and in many places terminated with a 5 to 10 m high slip face on the landward side. The lower part of the sand ramps formed a deflation surface, marked in some places by exhumed soils, while the upper part of the ramp was composed of 1 to 2 m of freshly deposited sand.

The occurrence of transgressive sand sheets at Formby, rather than well-defined parabolic or elongate parabolic dunes, such as those in North Queensland described by Pye (1982), reflects three major factors: (i) the natural dune vegetation, which is dominated by marram, other grasses and annual plants, offers relatively little resistance to blowout development and sand encroachment on a broad front; (ii) the directional variability of the wind is relatively large; and (iii) the development and maintenance of well-defined dune forms is hindered at Formby by the high pedestrian pressure. It is perhaps significant, however, that while bare sandblows also developed on non-eroding sections of the coast which were equally subject to recreational pressure, well-developed transgressive sand sheets with terminal slip faces were restricted to

Fig. 9   A blowout and transgressive sand sheet north of Lifeboat Road, viewed from the beach in April 1977

the eroding section of coast. Periodic wave erosion may play an important part in transgressive sand sheet formation by removing any incipient foredunes which begin to nucleate around backshore debris during periods of fair weather. However, the possibility that slightly higher wind energy plays an important role on the more exposed central part of Formby Point cannot be ruled out, since no field wind data are available.

4. EFFECTS OF DUNE MANAGEMENT POLICY AND CONSERVATION MEASURES

Measures to combat degradation of the dunes and reduce the threat of flooding due to marine incursion were implemented following the establishment of the Sefton Coast Management Scheme in 1978. This scheme was proposed following the reorganization of local government in 1974, which brought the entire coast between Bootle and Southport under common administration. A Coastal Working Party was set up jointly by the Planning Departments of Sefton Borough Council and Merseyside County Council to prepare the framework for the management scheme. An agreement between the two Councils and the Countryside Commission was concluded in November 1978. In the same year dune restoration work began in the most severely eroded areas around Lifeboat Road and on the National Trust Reserve around Victoria Road. This work was subsequently continued in association with the Manpower Services Commission. Eight management objectives were identified by the Coast Management Scheme Steering Group (1983):

(I) Dune conservation—to maintain the dune system as a natural sea defence.

(II) Landscape maintenance and renewal—to maintain the quality of the dune landscape and to restore it in those areas where it had been degraded.

(III) Woodland management—to maintain the continuity and quality of woodland cover, extending it in some areas to screen unattractive developments.

(IV) Nature conservation—to maintain and where appropriate widen the diversity of habitats and the wildlife they contain.

(V) Foreshore management—to improve the appearance of the foreshore through measures such as regular litter collection.

(VI) Visitor management—to deal with recreational demands in such a way as to minimize damage to the dune system and its habitats while continuing to allow public enjoyment of the area.

(VII) Interpretation—to increase people's understanding of the area, their enjoyment of it and care of it.

(VIII) Monitoring—to collect and assess information about changes affecting the area, and to review regularly the effectiveness of management techniques and policies.

In order to meet the dune conservation and landscape renewal objectives, a programme of brushwood fence construction, marram planting and boardwalk construction was put into force. Large brushwood fences were erected across major sandblows, and in many cases have led successfully to sand accumulation and sand sheet stabilization (Fig 10). Farther inland, bare sand patches were fenced off and either covered with loose brushwood or artificially planted with marram (Fig. 11). Attempts have also been made to improve the effectiveness of the frontal dunes as sea defences by artificially encouraging the accumulation of blown sand along parts of the backshore. Where the coast is most seriously threatened by wave erosion, for example to the north of Victoria Road, Freshfield, wooden fences have been erected both parallel and at right angles to the dune front (Fig. 12). However, no dune protection works have been carried out on the rapidly eroding dune frontage of the Nature Conservancy reserve between Massam's Slack and Fisherman's Path. In areas which are less at risk from wave erosion (i.e. between the northern erosion limit and marker post 3 shown on Fig. 3), vehicle access along the beach has been restricted, while brushwood fences have been erected along the shore between Birkdale and Southport which is backed only by low dunes. No fencing has yet been erected between marker posts 1 and 3 at Ainsdale, where pedestrian pressure is not too severe and where embryo dunes are still able to develop naturally (Fig. 13).

Fig. 10   Accumulation of sand behind a brushwood fence constructed across the seaward end of a major blowout, north of Lifeboat Road (photograph taken in 1985)

354   *Coastal Dunes*

Fig. 11   Restoration of degraded dunes by a combination of brushwood fence erection
and marram planting (photograph taken in May 1985)

Fig. 12   Sand accumulation encouraged by construction of wooden fences in front of
the natural foredune ridge, north of Victoria Road (photograph taken in 1987)

Fig. 13   Naturally accreting embryo dunes south of Shore Road, Ainsdale. Vehicle access
to this part of the beach is prevented by a barrier, but no sand fences have been erected
to enhance sand deposition or to prevent trampling by pedestrians

Similarly, no fences have been erected on the naturally accreting coast between
Albert Road and Range Lane.

Monitoring has shown that the rate of coastal erosion has slowed since dune
protection work began, but this trend has been aided by the non-occurrence
of major storm events since 1 February 1983. It is therefore too early to judge
what the long-term effectiveness of the protection works will be.

5. DISCUSSION AND CONCLUSIONS

The dependency of coastal dune morphology on wave energy regime, beach
morphology and littoral vegetation characteristics has been clearly demonstrated
by work in southeastern Australia (Short and Hesp, 1982; Short, 1988; Hesp,
1988). On many European and other developed coasts, human pressure is a
further factor which has a profound effect on coastal dune morphology (e.g.
Carter, 1975, 1980). A general model of dune morphological development, based
partly on evidence from the Sefton coast and partly on evidence from other
areas studied by the author (Pye, 1982, 1983a,b), is shown in Fig. 14. This model
takes into account three factors: (i) rate of littoral sand supply to the shoreline;
(ii) the wind energy available to transport sand inland; and (iii) the effectiveness
of sand-trapping vegetation, which in turn reflects the nature of the natural
flora and the extent of its destruction due to human pressures.

Case (a) in Fig. 14 represents a situation where there is a positive beach sand
budget but wind energy is low. The result is progradation of a beach ridge plain

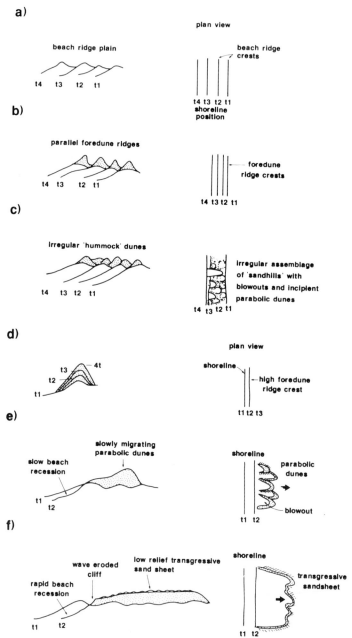

Fig. 14   Schematic model showing the relationship between dune morphology, shoreline dynamics, wind energy, and effectiveness of sand-trapping vegetation. For explanation see text

with minimal dune development over time intervals t1 to t4. In case (b) there is also a positive beach sand budget but higher wind energy which transfers part of the sand supplied to the backshore where it is trapped by effective sand binding vegetation. A series of parallel dune ridges is therefore formed as the coast progrades more slowly than in case (a).

In case (c) there is a positive sand supply to the shore and moderate wind energy but only ineffective or discontinuous vegetation cover. The result is an irregular series of hummock dunes with incipient blowouts and parabolic dunes which prograde seawards over time. In case (d) the rate of sand supply to the beach is balanced by the rate of aeolian transfer to the dunes, where all the arriving sand is trapped by vegetation. The result is a single foredune ridge which grows vertically, with no net change in the position of the shoreline over time. Case (e) indicates a situation where the rate of sand supply to the shore by marine processes is slightly lower than the rate of aeolian transfer from the beach to the dunes. The beach is lowered and the shoreline recedes slowly landwards, causing damage to the littoral vegetation through salt scalding and wind burn. A series of blowouts and small transgressive parabolic dunes is initiated behind the shore. In case (f), little or no sand is supplied to the beach by marine processes, but wind energy is high and sand is blown inland from the beach. The beach is therefore lowered rapidly, increasing the risk of storm wave damage and increasing the rate of shoreline retreat. Littoral vegetation is destroyed and transgressive sand sheets develop on a broad front, especially if trampling pressure is high.

Dune development on the Sefton coast before the beginning of the eighteenth century, when systematic marram planting began, corresponds most closely with case (c) shown in Fig. 14. Between 1880 and 1925, planting of brushwood fences reinforced a natural progradation trend and led to the formation of parallel dune ridges in a manner similar to that shown in case (b). Following the onset of erosion along the central part of the Formby Point coastline after 1906, transgressive sand sheets formed in a way similar to that illustrated in case (f).

ACKNOWLEDGEMENTS

I am grateful to Mr A. J. Smith of Sefton Borough Council Engineer and Surveyor's Department for providing access to air photographs and survey data.

REFERENCES

Admiralty Hydrographic Department (1960) *Atlas of Tides and Tidal Streams, British Isles and Adjacent Waters.* The Admiralty, London, 30 pp.

Ashton, W. (1909) *The Battle of Land and Sea on the Lancashire, Cheshire and North Wales Coast and the Origin of the Lancashire Sandhills.* W. Ashton & Sons, Southport.

Barron, J. (1938) *A History of the Ribble Navigation from Preston to the Sea.* Preston Corporation and Guardian Press, Manchester.

Bell, M. N., Barber, P. C. and Smith, D. G. E. (1975) The Wallasey embankment. *Proc. Inst. Civ. Eng.*, **58**, 569–590.

Best, R., Ainsworth, G., Wood, P. C. and James, J. E. (1973) Effects of sewage sludge on the marine environment, a case study in Liverpool Bay. *Proc. Inst. Civ. Eng.*, **55**, 43–66, 755–765.

Carter, R. W. G. (1975) The effect of human activities on the coastlines of Co. Antrim and Co. Londonderry. *Irish Geogr.*, **8**, 72–85.

Carter, R. W. G. (1980) Human activities and geomorphic processes: the example of recreation pressure on the Northern Ireland coast. *Zeit. Geomorph. Suppl.-Bd.*, **34**, 155–164.

Cashin, J. A. (1949) Engineering works for the improvement of the estuary of the Mersey. *J. Inst. Civ. Eng.*, **32**, 296–355.

Coast Management Steering Group (1983) *Coast Management Scheme. Plan for Coastal Management Between Hightown and Birkdale—Metropolitan Borough of Sefton.* Planning Department, Sefton Metropolitan Borough Council, Bootle.

Graaf, J. (1978) Abnormal sea levels in the northwest of England. *Dock and Harbour Authority*, **58**, 366–371.

Gresswell, R. K. (1937) The geomorphology of the southwest Lancashire coastline. *Geogr. J.*, **90**, 335–349.

Gresswell, R. K. (1953) *Sandy Shores in South Lancashire.* Liverpool University Press, Liverpool.

Hall, B. R. and Folland, C. J. (1967) *Soils of the Southwest Lancashire Coastal Plain.* Mem. Soil Surv. GB, Harpenden.

Halliwell, A. R. (1973) Residual drift near the sea bed in Liverpool Bay; an observational study. *Geophys. J. R. Astronom. Soc.*, **32**, 439–458.

Hesp, P. (1988) Surfzone, beach and foredune interactions on the Australian southeast coast. *J. Coast. Res. Spec. Issue No.*, **3**, 15–25.

HRS (Hydraulics Research Station, Wallingford, UK) (1958) Radioactive Tracers for the Study of Sand Movement: *A Report on an Experiment Carried Out in Liverpool Bay in 1958.* Hydraulics Research Station Report, Wallingford.

HRS (1965) *Investigation of Siltation in the Estuary of the River Ribble, July, 1965.* Hydraulics Research Station Report Ex. 281, Wallingford.

HRS (1968) *Notes on Engineering Works to Reduce Dredging in the Ribble Estuary.* Hydraulics Research Station Report Ex. 391, Wallingford.

HRS (1969a) *The Southwest Lancashire Coastline, a Report of the Sea Defences.* Hydraulics Research Station Report Ex. 450, Wallingford.

HRS (1969b) *Computation of Littoral Drift.* Hydraulics Research Station Report Ex. 449, Wallingford.

HRS (1977) *Sand Winning at Southport.* Hydraulics Research Station Report Ex. 708, Wallingford.

HRS (1980) *River Ribble Cessation of Dredging.* Hydraulics Research Station Report Ex. 948, Wallingford.

Innes, J. B. (1982) Notes on the vegetation history of the Sefton District, Merseyside. In Lewis, J. (ed.), *Archaeological Survey of Merseyside: Sefton Rural Fringes Survey, Appendix 1.* Merseyside County Council, Liverpool, pp. 1–26.

Lennon, G. W. (1963a) Identification of weather conditions associated with generation of major storm surges along the west coast of the British Isles. *Q. J. R. Met. Soc.*, **89**, 381–394.

Lennon, G. W., (1963b) A frequency investigation of abnormally high tide levels at certain west coast ports. *Proc. Inst. Civ. Eng.*, **25**, 451–484.

McDowell, D. M. and O'Connor, B. A. (1977) *The Hydraulic Behaviour of Estuaries.* Macmillan, London.

O'Connor, B. A. (1987) Short and long term changes in estuary capacity. *J. Geol. Soc. Lond.*, **144**, 187–195.

Parker, W. R. (1971) Aspects of the marine environment at Formby Point, Lancashire. PhD thesis, Liverpool University.

Parker, W. R. (1974) Sand transport and coastal stability, Lancashire, U.K. *Proc. 14th Conf. Coast. Eng.*, 828–850.

Parker, W. R. (1975) Sediment mobility and erosion on a multi-barred foreshore (southwest Lancashire, U.K.). In Hails, J. R. and Carr, A. P. (eds), *Nearshore Sediment Dynamics and Sedimentation*. Wiley, London, pp. 151–177.

Price, W. A. and Kendrick, M. P. (1963) Field and model investigations into the reasons for siltation in the Mersey estuary. *J. Inst. Civ. Eng.*, **24**, 473–517.

Psuty, N. P. (1988) Sediment budget and dune/beach interaction. *J. Coast. Res. Spec. Issue No.*, **3**, 1–4.

Pye, K. (1982) Morphological development of coastal sand dunes in a humid tropical environment, Cape Bedford and Cape Flattery, North Queensland. *Geograf. Ann.*, **64A**, 212–227.

Pye, K. (1983a) Formation and history of Queensland coastal dunes. *Zeit. Geomorph. Suppl.-Bd.*, **45**, 175–204.

Pye, K. (1983b) Dune formation on the humid tropical sector of the north Queensland coast, Australia. *Earth Surf. Proc. Landf.*, **8**, 371–381.

Pye, K. and Smith, A.J. (1988) Beach and dune erosion and accretion on the Sefton Coast, Northwest England. *J. Coast. Res. Spec. Issue No.*, **3**, 33–36.

Short, A. D. (1988) Wave, beach, foredune and mobile dune interactions in southeast Australia. *J. Coast. Res. Spec. Issue No.*, **3**, 5–10.

Short, A. D. and Hesp, P. (1982) Wave, beach and dune interactions in southeast Australia. *Mar. Geol.*, **48**, 259–284.

Sly, P. G. (1966) Marine geological studies in the eastern Irish Sea and adjacent estuaries, with special reference to sedimentation in Liverpool Bay and the River Mersey. PhD thesis, Liverpool University.

Tooley, M. J. (1970) The peat beds of the southwest Lancashire coast. *Nature in Lancashire*, **1**, 19–26.

Tooley, M. J. (1974) Sea-level changes during the last 9000 years. *Geogr. J.*, **127**, 18–42.

Tooley, M. J. (1976) Flandrian sea-level changes in west Lancashire and their implications for the 'Hillhouse Coastline'. *Geol. J.*, **11**, 137–152.

Tooley, M. J. (ed.) (1977) *The Isle of Man, Lancashire Coast and Lake District*. Guidebook for Excursion 4, 10th INQUA Congress, Birmingham, Geo. Abstracts, Birmingham, 58 pp.

Tooley, M. J. (1978) *Sea-Level Changes in Northwest England During the Flandrian Stage*. Clarendon Press, Oxford, 232pp.

Tooley, M. J. (1982) Sea-level changes in northern England. *Proc. Geol. Ass.*, **93**, 43–51.

Travis, W. G. (1915) Marram grass and dune formation on the Lancashire Coast. *Lancashire & Cheshire Naturalist*, **8**, 313–320.

Wray, D. A., Cope, F. W., Tonks, L. H. and Jones, R. C. (1948) Geology of Southport and Formby. *Mem. Geol. Surv. GB*, HMSO, London.

Wright, P. (1976) The morphology, sedimentary structures and processes of the foreshore at Ainsdale, Merseyside, England. PhD thesis, Reading University.

Wright, J. E., Hull, J. H., McQuillin, R. and Arnold, S. E. (1971) *Irish Sea Investigations 1969-71*. Inst. Geol. Sci. Rep., 71/19.

Reading University, PRIS Contribution No. 058.

Chapter Sixteen

Eolian processes and dune changes at developed and undeveloped sites, Island Beach, New Jersey

PAUL A. GARES
Department of Geography, Colgate University

1. INTRODUCTION

Much of the east coast of the US has undergone some form of development, resulting in significant changes to the natural system (Peoples and Gregg, 1984; Nordstrom, 1987). Dunes are one component of the coastal system which is severely altered by development. Scientists working on eolian sediment transport in undeveloped dune systems have quantified the relationship between wind velocity, grain size and sediment transport (Bagnold, 1941), demonstrated the effects of vegetation on sediment accumulation (Olson, 1958; Hesp, 1983), and observed alterations on wind flow caused by dune topography (Svasek and Terwindt, 1974; Hsu, 1987). There is also considerable information about air flow around cultural features, such as buildings (Hoxey and Moran, 1983; Nagib and Corke, 1984) or sand fences (Castro, 1971; Phillips and Willetts, 1979). Nordstrom and McCluskey (1985) examined the way in which cultural features affect the eolian sediment budget. Despite the information provided by these studies, the extent to which development and the human activities associated with it affect eolian sediment transport and dune formation still remains to be determined.

One way of evaluating the effects of development is to compare dune systems along developed and undeveloped shorelines, an exercise which has not heretofore been undertaken. This study follows such an approach by comparing data on dune morphology and processes collected at several sites within each

Coastal Dunes: Form and Process. Edited by K. F. Nordstrom, N. P. Psuty and R. W. G. Carter
© 1990 John Wiley & Sons Ltd

type of system. The objective of this chapter is to determine whether these processes and dune characteristics are similar at developed and undeveloped sites. The basic hypothesis is that the systems are different and that the differences can be quantified.

2. GEOGRAPHIC SETTING

This study was conducted along Island Beach, New Jersey, a barrier spit which extends 36 km south from the New Jersey headlands to Barnegat Inlet (Fig. 1). Three general wind directions prevail along Island Beach. West and northwest winds occur most frequently during the winter months, whereas south and southwest winds dominate during the summer. Northeast winds also occur fairly frequently, particularly in the fall and spring. Average annual velocities are $6.7\,\mathrm{m\,s^{-1}}$ for westerly winds, $7.2\,\mathrm{m\,s^{-1}}$ for southerly winds, and $8.3\,\mathrm{m\,s^{-1}}$ for northeast winds (US Army Corps of Engineers, 1981). The northeast winds have high velocities because they are associated with offshore low pressure systems

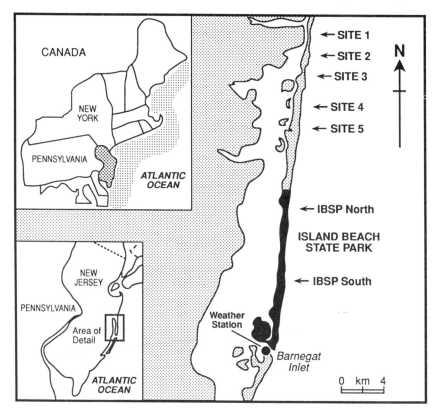

Fig. 1   Location of the Island Beach study sites

which produce coastal storms (Mather, Adams and Yoshioca, 1964). Storm waves also approach this shoreline from the northeast, whereas swell waves come from the southeast. The net direction of longshore current along the spit is to the south, although temporary reversals may occur (Ashley, Halsey and Buteux, 1986). The average wave height along Island Beach is 0.8 m, and the average wave period is 8 s (US Army Corps of Engineers, 1981). During annual storms, wave heights reach 1.2 m, and during the largest storm of recent times, the 1962 northeaster, wave heights of 10 m were estimated to have occurred (US Army Corps of Engineers, 1981).

Island Beach was selected for study because developed (Fig. 2) and undeveloped (Fig. 3) areas exist adjacent to each other. Five sites were located in the developed northern portion of the spit, and two sites were in Island Beach State Park, an undeveloped recreational area at the southern end of the spit (Fig. 1). The proximity of developed and undeveloped areas on Island Beach spit allows the assumption that spatial differences in the wind regime responsible for modifying the dune form are minimal. This permits the study to focus on other factors to explain differences in dune formation processes.

Fig. 2   Normandy Beach (site 3) in the developed portion of Island Beach spit, is typical of New Jersey beach communities. Note the truncated dunes and the abundance of sand fences at the toe of the dunes

Fig. 3   Island Beach spit, New Jersey, with the State Park in the foreground and Seaside Park in the background to the north. Note the old Coast Guard station in the Park in the middle of the photograph. Undeveloped site 1 is located just to the south of the station

3. DATA COLLECTION METHODOLOGY

The objective of the analysis is to explain differences between developed and undeveloped sites in the volume of sediment transported by the wind as measured by sand traps and in elevation changes. Trapped sediment was chosen as one dependent variable because it is indicative of the quantity of sand that is added to or removed from the dune, and thus affects such factors as dune height and width. Elevation change, the second dependent variable, is one morphological

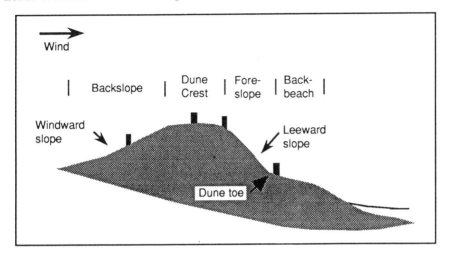

Fig. 4   Dune zones and sediment trap locations

manifestation of eolian sediment transport. The elevation change data are important complements to the trapped sediment data because, although the traps measure how much sand passes a certain point, they do not indicate where or how much sand accumulates.

Sediment transport was sampled with vertical, cylindrical sand traps made from PVC pipe (Leatherman, 1978). The traps were placed in similar beach and dune locations at each study site to permit between-site comparisons (Fig. 4). Other sample stations were selected according to specific characteristics of the site. The traps monitored cross-shore and alongshore sediment transport and were oriented basically north, south, east and west. Dune surface changes were measured with depth of disturbance rods placed at 5 m intervals along survey transects and at significant breaks-in-slope. Single profile lines of these rods were used at the developed sites, but multiple profiles were established at the undeveloped sites.

The field study ran from October 1981 to April 1982. The sediment traps and the depth of disturbance rods were monitored weekly from 1 November, 1981 to April 2, 1982, and at two-day intervals between 1 November to 10 December, 1981. During the weekly sampling period, some traps exposed to high-velocity winds filled completely and the data under-represent the amount of sand trapped. Data from these stations are used in the morphological analysis because they represent a minimal accumulation relative to the other stations.

The differences between shoreline types are analysed by examining four groups of independent variables. The first group comprises process variables: wind velocity and wind direction. Regional wind velocity and direction data recorded every three hours at the Barnegat Light Coast Guard station (see Fig. 1) were aggregated into four directional components corresponding to the trap orientations. The second set of variables pertain to beach characteristics because the beach is the primary source of windblown sand. Beach width and slope were obtained from weekly surveys conducted throughout the study period. Beach sediment size was obtained from a single sample collected in March, 1982.

The third group of variables involves dune characteristics: vegetation density, slope, and elevation. Because vegetation density has a great effect on the movement of windblown sediment (Olson, 1958; Bressolier and Thomas, 1977), the density within 5 m² upwind of each trap was estimated visually. The assumptions here are that only the vegetation upwind of the trap influences sediment that is moved to the trap, and that it is the vegetation closest to the trap which has the greatest effect on sediment trapped. Topographic variables were examined to determine if they could be used as surrogates for wind velocity and direction because of their strong relationship with wind velocity (Taylor and Gent, 1974; Pearse, Lindley and Stevenson, 1981) and with wind direction (Svasek and Terwindt, 1974; Hauf and Neumann-Hauf, 1982). The station elevation and upwind slope were obtained from surveys of the dune zone conducted at each study site.

The final group of variables focuses on cultural variables. Characteristics evaluated were the presence or absence upwind of the trap of: (i) a sand fence; (ii) any other cultural obstruction to air flow (house, boardwalk, road); or (iii) a coarse sediment surface. The first two characteristics alter the velocity or direction of wind flow (Castro, 1971; Nagib and Corke, 1984). The third factor was included because so many houses were partly or entirely surrounded by driveways composed of an imported sand/gravel mix. Despite the unobstructed air flow over these surfaces, the material was packed and coarse enough to affect the availability of sediment for eolian transport. The three variables were quantified nominally; a 1 indicating the presence of the object; a 0 indicating its absence. The distance between the objects and the trap was also measured and analysed.

4. DATA ANALYSIS METHODOLOGY

This study's hypothesis is evaluated in several different ways because no one analysis methodology is strong enough to support or reject the hypothesis. Data collected during the period of more intensive sampling are grouped into developed and undeveloped categories and are further subdivided into beach and dune categories. All data, except for the measures of cultural features, are first subjected to a Student's t-test to compare data from the two shoreline types.

Data from each sub-division are also evaluated using correlation and stepwise multiple regression analyses. The correlation coefficients for the two different shoreline categories are compared to determine if the relationships are similar. The dependent variable in each category is the sand trapped during the sampling interval by each trap deployed. Independent variables analysed for beach stations include elevation, beach slope, beach width, mean grain size, wind velocity, wind direction, presence of fences, and presence of other obstructions. Independent variables included in the analysis of data from dune stations are vegetation density, dune slope, station elevation, wind velocity, wind direction, presence of fences, distance downwind from the fence, presence of other obstructions, distance downwind from the obstruction, and surface condition. Some variables are transformed (see Table 1) to improve the linearity of the relationship between the variables. The variables with the strongest correlation coefficients are used in the regression. The regression equations for data sets from developed and undeveloped sites are examined to identify differences between sites.

The hypothesis that development does explain differences in sediment transport is further tested by combining the developed and undeveloped data sets and submitting the new set to regression analysis along with a nominal dummy independent variable that indicates whether the sample station is at a developed or undeveloped site. The inclusion of the dummy variable in the regression equation would indicate that development plays a significant role in explaining the variation in sediment transport.

A visual analysis of data is also conducted to determine if differences between the two systems are more obvious when the data are evaluated for specific morphologic zones. For this analysis, the beach and foredune system is divided into four morphologic zones: (i) the backbeach; (ii) the dune foreslope; (iii) the

Table 1 Correlation coefficients between sediment transport and independent variables for dune stations at developed and undeveloped sites, according to slope type

Independent variable	Developed windward slope	Developed lee slope	Undeveloped windward slope	Undeveloped lee slope
Wind velocity cubed	−0.060	0.342	0.260	0.037
Wind direction	0.149	0.370^2	0.209	0.217
Vegetation density	−0.406	−0.111	$−0.591^1$	$−0.456^1$
Log vegetation density	$−0.437^2$	−0.100	$−0.643^1$	$−0.620^1$
Elevation	$−0.548^1$	−0.073	0.017	0.086
Log elevation	$−0.549^1$	−0.044	−0.001	0.086
Slope	−0.031	0.006	−0.036	−0.089
Reciprocal slope	−0.084	−0.147	0.342^2	0.296^2
Fence	−0.046	0.384^2		
Distance from fence	−0.038	$−0.410^2$		
Obstacle	0.226	−0.252		
Distance from obstacle	−0.101	0.075		
Coarse surface	−0.218	−0.220		

[1]Significant at 0.01 probability
[2]Significant at 0.05 probability

dune crest; and (iv) the dune backslope (Fig. 4). The foredune is divided
into sub-units because of large variations in elevation change and sediment
transport across the foredune. Trapped sediment and elevation change are
related to variations in vegetation cover and topography. At developed
sites, special characteristics associated with human use of the sites are
examined to explain differences in morphologic change. The location of a trap
relative to a fence, a house, a road or a boardwalk helps to explain differences.
Likewise, the way in which the area is used (beach access only, sunbathing,
sporting activities) may explain elevation change or sediment transport
characteristics.

5. RESULTS OF THE QUANTITATIVE ANALYSIS

5.1. Beach

Beach data are analysed separately from dune data because the beach represents
a relatively flat environment where sediment transport is little affected by
vegetation or topography. Data used in this analysis were obtained from traps
located at the dune toe which faced either offshore or alongshore. Landward-
facing traps are excluded because the presence of the dune immediately upwind
affects the movement of sand in this zone. These traps are included in the analysis
of dune data. Wind velocity and direction have the highest significant
correlations with sediment transport. The only other significant correlation
coefficient is the obstacle variable. The regression equations explain 85.5% of
the variation in sediment transport for undeveloped sites and 83.9% for
developed sites. In both equations, the largest proportion of the variation in
sediment transport is explained by the cube of wind velocity. The only other
variables significant enough to be included in the equations are sediment particle
size for undeveloped sites and beach slope for developed sites. Although the
two R-square values are similar, the equation coefficients and the y-intercepts
are different for each equation, suggesting differences in sediment transport
rates between developed and undeveloped sites.
  That the two shoreline categories are different is also demonstrated when the
data sets are combined and analysed with a dummy development variable. This
dummy variable becomes the second variable entered into the equation after
wind velocity, explaining 19.9% of the variation in sediment transport. This
inclusion indicates that the type of land use in existence along the shoreline
has an effect on eolian sediment transport. The difference is further reinforced
by the results of a difference of the means test on the original data. Sediment
transport, elevation and sediment size are found to be statistically different
at developed and undeveloped sites, whereas slope and beach width are
similar.

5.2. Dune

The dune data are divided into two subsets based on whether the slope upwind of the trap is a windward slope or a lee slope (see Fig. 4). Air flow behaves differently in each of the cases (Taylor and Gent, 1974; Pearse, Lindley and Stevenson, 1981), and sediment transport should also be different for traps with windward or lee slopes. In the case of undeveloped dune stations with windward slopes, only vegetation density and the reciprocal of slope have significant correlations with sediment transport, whereas at developed stations with similar slopes, vegetation density and elevation have significant correlation coefficients with sediment transport (Table 1). For dune stations with lee slopes, vegetation density and the reciprocal of slope are the only significant variables for undeveloped sites, whereas the cube of wind velocity, wind direction, the presence of a fence and the fence distance have significant correlation coefficients for developed sites. None of these variables have particularly strong relationships with the dependent variable.

The multiple R-squares associated with the regressions suggest that there are small differences in the case of windward slopes ($R^2 = 50.3\%$ for developed sites; $R^2 = 52.8\%$ for undeveloped sites). The differences are more substantial for lee slopes ($R^2 = 36.8\%$ for developed sites; $R^2 = 48.2\%$ for undeveloped sites). The variables included in each equation are different and the coefficients vary greatly when the same variables are entered (Table 2).

6. DISCUSSION OF THE RESULTS OF THE QUANTITATIVE ANALYSIS

The correlation and regression analyses for beach data provide no evidence to support the hypothesis that the two systems are different, but the inclusion of the dummy development variable in the second regression analysis and the results of the t-test suggest that significant differences exist between the two systems. The t-test results provide the strongest evidence because the amounts of sediment

Table 2   Regression equations for dune stations at developed and undeveloped sites, according to slope type

---

Developed sites with windward slopes
$q = 41460.67 - 6524.67$ elevation $- 9550.97$ coarse surface

Developed sites with lee slopes
$q = 4870.073 - 49.99$ fence distance $+ 183.40$ wind frequency $- 6527.11$ obstacle

Undeveloped sites with windward slopes
$q = 42401.99 - 12504.55$ log vegetation density $+ 512.17$ wind frequency

Undeveloped sites with lee slopes
$q = (44002.66/\text{vegetation density}) + 5641.57$ elevation $- 29071.30$

---

q: Sediment trapped in $cm^3 \, m^{-1}$

trapped are statistically different. The results of the analyses for dune stations also support the hypothesis that developed and undeveloped systems are different. The first piece of evidence is the inclusion of the dummy variable in the equation for the combined data set. Second, the multiple R-square values for the equations for stations with lee slopes are visibly different whereas those with windward slopes are slightly different. Finally, the results of the t-test indicate that several of the variables, including the sediment trapped, are significantly different.

There are indications in the equations that specific development characteristics explain differences in the amount of sediment trapped in the dune zone. First, the surface variable is included in the equation for windward slopes and it explains 20.23% of the variation in sediment transport. This result reflects the use of coarse sand and gravel fill for driveways in the rear of the house lots which provides little sediment for transport in a seaward direction. In the case of stations with lee slopes, both the fence distance and the obstruction variables are entered into the equation, and they explain 16.83% and 11.69% of the variation in sediment trapped. The negative equation coefficient for fence distance suggests decreasing sediment transport with increased distance from the fence. This relationship appears to be counter-intuitive. However, sediment already in motion upwind is carried through the fence and deposited just to its lee. Sediment transport will only begin again a considerable distance downwind as the distance over which the fence alters air flow may extend to twenty times the fence height (Castro, 1971; Phillips and Willetts, 1979). Thus, the inverse relationship applies for sediment transport, not wind velocity. The obstruction variable reflects the presence of a house or other structure upwind of the trap (boardwalks, roads) which would be expected to influence either air flow or sediment source. The negative equation coefficient for this variable implies that sediment trapped diminishes when an obstruction is present upwind.

7. MORPHOLOGIC CHANGES IN THE DUNE ZONE

A Mann–Whitney test of the dune elevation and slope data indicates significant differences between developed and undeveloped sites. The average foredune elevation at developed sites is 5.75 m above MSL; at undeveloped sites the elevation is 7.63 m. The average foreslope at developed sites is 11.6°; at undeveloped sites the average slope is 15.1°. These differences can be traced to the way in which the dunes are used along developed shorelines. The presence of sand fences at the base of the foreslope causes the accumulation of sediment just landward of the fence, creating a gentle foreslope. The overall elevation of the dune at developed sites is lowered because the dune zone often serves as a front yard for the house lining the shore and is heavily used for recreational activities. Pedestrian activities have been shown to produce a reduction in the dune crest elevation (Steiner and Leatherman, 1979; Baccus and Horton, 1980).

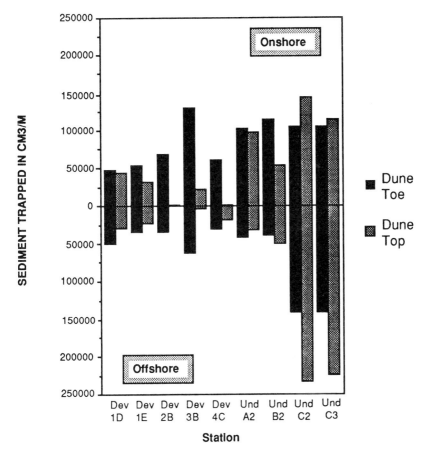

Fig. 5 Sediment trapped at sites at the dune toe and at the top of the dune foreslope
during onshore and offshore winds

Thus, it appears that dune morphology along developed and undeveloped
shorelines is different. An examination of sediment transport and elevation
change in specific morphologic zones emphasizes the differences.

7.1. Back-beach

Onshore sediment transport was highly variable at dune toe stations (Fig. 5).
During the study period, the undeveloped stations all recorded quantities of
sediment transport above $0.1 \text{ m}^3 \text{ m}^{-1}$, whereas all developed stations except 3A
were below $0.1 \text{ m}^3 \text{ m}^{-1}$. The presence of sand fences on the beach (Fig. 6) is
one reason why sediment transport was much less at developed sites than at
undeveloped sites. Fences were located less than 5 m upwind at station 4A and

Fig. 6   Developed site 4 in Lavallette, New Jersey. Note the fences on the beach which interrupted sediment transport in a landward direction. Station 4B is visible as 4 vertical white pipes just to the left of the boardwalk. Station 4A, not apparent, is located seaward of 4B, on the right side of the boardwalk, in its shadow

some 20 m upwind of station 5A. Given fence heights of 1 m, sediment transport would be limited within a distance of 20 m downwind of the fence except under very high wind velocity conditions (Castro, 1971; Phillips and Willetts, 1979).

Other potential reasons for the differences in landward sediment transport between the sites involve beach topography and the grain size of the beach sediment. Beaches fronting the developed sites averaged 47.4 m in width with an average slope of 4.1°. Beaches fronting undeveloped sites had an average width of 55.4 m and an average slope of 2.7°. It is difficult to determine whether these differences may be attributed to development or to simple variations in beach conditions alongshore. However, in a study of long-term changes at these sites, beach width was found to vary mainly with changes in the location of the dune toe rather than of the water line, and large changes in the location of the dune toe were attributed to the use of sand fences (Gares, 1987). Thus, differences in beach width between developed and undeveloped sites may be due to human interference. The average beach widths appear to be sufficiently different to explain partially why sediment transport at developed sites was so much lower than that at undeveloped sites. Wider beaches at undeveloped sites represent a larger source area for eolian sediment movement.

Sediment particle size also differed, averaging 1.09 ϕ at developed sites and 1.29 ϕ at undeveloped sites. More importantly, Sites 1 and 2 had significantly coarser sediment than all the rest, 0.80 ϕ and 0.91 ϕ respectively. These size differences cannot be attributed to development, but they may better explain variations in sediment transport between developed and undeveloped sites than do beach dimensions. On a steep beach, eolian sediment transport would more likely be limited due to the presence of coarse particles on the beach rather than due to the slope of the beach.

7.2. Dune foreslope

The dune foreslope was a zone where larger amounts of sediment were moved onshore than offshore, producing accumulation at both developed and undeveloped sites. The average accretion from November 1981 to April 1982 was slightly greater at undeveloped sites, and the range of elevation increase among undeveloped stations was considerably greater with the maximum reaching 0.422 m, a value nearly twice as large as that recorded at developed sites (Table 3). Only 8 of the 51 undeveloped stations recorded erosion, with an average of 0.125 m and an extreme of 0.377 m. Of the 19 developed foreslope stations, 6 showed erosion, but the average was only 0.022 m and the extreme was 0.065 m.

Much larger amounts of sediment were moved by wind at the undeveloped sites than at the developed sites (Fig. 3). A comparison of onshore sediment transport values from paired beach and dune stations shows the extent to which accumulation prevailed at each site. At undeveloped stations, the average amount

Table 3   Elevation change data for foredune zones at developed and undeveloped sites, from 29 October 1981 to 5 April 1982. (Values in meters)

	N	Mean	Max	Min
Accretion				
Foreslope—undeveloped	43	0.137	0.422	0.005
Foreslope—developed	13	0.112	0.215	0.015
Dune crest—undeveloped	32	0.085	0.330	0
Dune crest—developed	5	0.016	0.071	0.001
Backslope—undeveloped	8	0.017	0.049	0
Backslope—developed	4	0.005	0.014	0
Erosion				
Dune front—undeveloped	8	−0.125	−0.032	−0.377
Dune front—developed	6	−0.022	−0.007	−0.065
Dune crest—undeveloped	6	−0.063	0	−0.317
Dune crest—developed	5	−0.01	0	−0.018
Backslope—undeveloped	13	−0.024	0	−0.152
Backslope—developed	21	−0.025	0	−0.114

of sediment trapped at the top of the foreslope amounted to 95.9% of the average sediment trapped on the back-beach at the dune toe. At developed sites, the average amount of sand trapped at the top of the foreslope was only 27.7% of the average amount of sand trapped on the back-beach. Variation between sites ranges from 46.6% to 137.1% at undeveloped sites and 1.4% to 93.5% at developed sites. The lower proportion of sand transferred at developed sites is the consequence of the presence of sand fences along the dune toe at 4 of the 5 developed sites. Sediment deposition occurred in the immediate vicinity of the fences, reducing sediment transfer to the top of the foreslope.

In the case of offshore transfer at undeveloped sites, the amount of sand trapped at the dune toe is 67.4% of that trapped at the top of the foreslope (Fig. 3). This is attributed to the existence of a stagnation zone in the lee of a dune crest (Acrivos *et al.*, 1968; Taylor and Gent, 1974), where sediment deposition predominates. At developed sites, sediment transport increased from the top of the foreslope to the dune toe, with the change amounting to 279.8%. This larger sediment transfer is attributed to the gentler foreslopes at developed sites which would be less likely to produce wind flow separation (Taylor and Gent, 1974; Pearse, Lindley and Stevenson, 1981) and sediment deposition on the foreslope. A second factor which leads to increased sediment transport from dune to beach is the vegetation density on the dune crest. Vegetation cover at Sites 1 and 2 was quite dense, reducing the source area for offshore sediment transport. Vegetation cover on the dune crest at Sites 3 and 4 was sparse, but offshore sediment transport at these sites differed by an order of magnitude. The higher transport at Site 4 is explained by the lack of vegetation cover. The lower transport value at Site 3 is attributed to the presence of a house only a few meters landward of the dune crest which created a stagnation zone in its lee and resulted in low sediment transport seaward from the crest.

7.3. Dune crest

The amount of elevation change from November 1981 to April 1982 along the dune crest at undeveloped sites averaged only 0.085 m, but 32 of the 38 dune crest sample stations recorded deposition (Table 3). Only one site had accretion greater than 0.3 m, whereas 21 stations showed deposition less than 0.1 m. Erosion on the dune crest averaged 0.063 m, with a maximum of 0.317 m. However, 4 of the 6 dune crest stations with erosion lost less than 0.05 m in elevation.

The predominance of deposition along the dune crest at undeveloped sites is attributed to net onshore sediment transport (Table 4). The variation in sediment transport and elevation change at dune crest stations is related to variations in vegetation density along the dune crest. Stations with low vegetation density landward of the station showed higher offshore sediment transport. Onshore sediment transfer was affected similarly, but the amount moved inland

Table 4   Sediment transport, elevation change, and vegetation density on the dune crest at developed and undeveloped sites from 29 October, 1981 to 5 April, 1982

Station	Sediment onshore	Transport offshore	$(cm^3 m^{-1})$ net[1]	Elevation change (m)	Vegetation cover % Landward	Seaward
IBSP No. A3	16314	8489	7825	0.058	49	88
IBSP No. A5	19783	19230	553	0.08	26	15
Dev. site 1F	1122	1122	0	0.001	70	88
Dev. site 2C	433	637	− 204	− 0.018	88	20
Dev. site 3C	2779	4334	− 1555	0.005	12	1
Dev. site 4D	1046	969	77	0.001	14	66
Dev. site 4E	35081	88392	− 53311	NA	29	9

[1]Positive values indicate onshore transport; negative values indicate offshore transport; NA = not available.
IBSP = Island Beach State Park.

was generally larger because of lower vegetation densities along the dune front.

At developed sites, accretion occurred at 5 of the 10 dune crest stations (Table 3). The range of change was small and can be attributed to the low sediment transport values recorded at developed dune crest stations (Table 4). Explanations for the lower sediment transport rates vary from site to site. Dense vegetation cover limits sediment transport at Site 1. At Site 2, vegetation on the dune front prevents much sand from being transported to the crest, and a coarse sand and gravel parking area on the backslope limits offshore sediment transport. Vegetation density is low at Site 3, but sediment transport is limited. In the case of onshore transport this is due to the presence of rows of sand fences at the dune toe and at the top of the foreslope. The two-story house 10 m landward of the sample station creates a stagnation zone on the dune crest during offshore winds, limiting sediment transport. At Station 4D, a sand fence limits onshore sediment movement, and dense vegetation landward prevents sediment from being transported offshore. The lack of a fence seaward of Station 4E and low vegetation density both landward and seaward of the trap result in high sediment transport off the dune crest. A comparison of Sites 3 and 4 emphasizes the interference imparted by the house at Site 3 which sits on the dune and rises two stories above the dune surface. At Site 4, the house is located about 15 m landward of the sample stations, but it is nestled low behind the dune. This house has little effect on air flow and sediment transport is controlled by the vegetation densities along the dune crest.

7.4. Backslope

The majority of backslope stations at undeveloped sites have erosion which is attributed to net offshore movement of sediment in this zone. The elevation change from November 1981 to April 1982 ranges from 0.049 to − 0.152 m

(Table 3). Net offshore sediment transport is facilitated by the low vegetation density on the backslope and by the gentle slope of the dune. The backslope serves as a source for sediment that is moved offshore, as evidenced by the erosion which occurs there. When sand is moved up the backslope by the predominant offshore winds, it is deposited in the dense vegetation along the crest. The relationship between elevation change and net sediment transport is, thus, reversed on the landward side of the dune crest.

Erosion also predominates on the backslope at developed sites (Table 3). The range of elevation change from November 1981 to April 1982 is 0.014 to −0.114 m. Net sediment transport on the backslope at these sites is generally in the offshore direction, but the amounts of sand moved are small. There is little vegetation cover in this zone, which should create an ideal situation for sediment transport. Little sediment transport occurs, indicating that there are other factors which influence sediment transport at these stations. At Sites 1, 2, and 5, the offshore sediment transport is limited by the coarse sand and gravel used for parking areas. The surface surrounding the house at Site 3 consists of medium sand which should be conducive to sediment transport, especially because vegetation cover is lacking. The house is the primary cause for the small amount of sediment transport recorded at Station 3C (Table 4). The space between the houses at this site is very narrow, of the order of 7 m. The narrow passage might be expected to funnel wind so as to increase its velocity, producing greater sediment transport. The small amount of sediment transport in this zone suggests that the funneling does not generally take place. It seems likely that the flow separation which would occur on the upwind side of the house would reduce the velocity in the space between the houses. Thus, the house has a profound effect on sediment transport at this site.

8. DISCUSSION OF THE RESULTS OF MORPHOLOGIC ANALYSIS

Elevation changes in each morphologic zone examined are greater at undeveloped sites than at developed sites. The amount of elevation change is correlated to the sediment transport rate, which is statistically different for each type of shoreline. Three factors explain the differences in sediment transport and elevation change: (i) vegetation cover; (ii) sand fences; and (iii) houses. The slope of the seaward face of the dune also affects the amount of sand moved inland, particularly at undeveloped sites where steeper foreslopes existed.

Sediment transport was greatest when the sand surface was bare or when the vegetation cover was minimal. Vegetation effectively stabilizes the sand surface (Olson, 1958; Bressolier and Thomas, 1977), as seen along the foredune crest at the undeveloped sites and at Site 2. When vegetation cover was sparse, particularly where recreational use was heavy such as at Sites 3 and 4, the sediment had little long-term stability. Areas with little vegetation cover, such as Stations 1D or 4E, had high sediment transport rates and erosion prevailed.

Although sand fences do not significantly reduce the transport of sediment through the fence, they do cause deposition of sand into a mound on the downwind side of the fence. Unless this sand is stabilized, it can still be removed by wind from other directions so that overall dune growth is slow. This was the case on the incipient dune at Site 5. The use of sand fences also affects sediment transport and elevation change at locations farther downwind because the sediment deposited to their lee is not moved to those locations. This is particularly true when the network of fences is dense, as at Sites 3, 4 and 5. The fences on the beach at Sites 4 and 5 reduced the amount of sand transported to the dune toe and dune front. Multiple rows of fences at Site 3 prevented much sediment from being moved from the beach to the dune front and dune crest. Thus, fences primarily affect dune formation by concentrating sediment deposition, which prevents sediment transport to other locations in the dune zone.

The layout or configuration of houses also affects dune morphodynamics. Although the observations in the study are limited, it appears that the height of the house above the dune surface and the spacing between houses along the shoreline are important factors. The elevation above the dune of the house at Site 3 and its close proximity to neighboring houses seemed to produce reduced air flow, and therefore low sediment transport, in its lee. Houses on pilings, such as the one at Site 5, create a situation favorable to sediment transport due to altered air flow under the house and to the absence of vegetation beneath the house (Nordstrom and McCluskey, 1985). At Site 5, sediment transport was limited by the use of imported sand and gravel to create a parking area behind and under the house. Sediment transport at other sites was also affected by the use of sand and gravel. These bare areas represent potential sediment sources, but the coarseness of the particles prevents their movement.

The results of this study fit into general observations concerning long-term dune dynamics along developed shorelines. An analysis of several sets of aerial photographs taken between 1952 and 1982 was conducted for the same study sites to observe changes in the location of the dune toe and the dune crest and to measure changes in the width of the dune zone (Gares, 1987). The results indicate that dunes grew narrower at developed sites through time and that the dune crest at developed sites remained in a fixed position or moved seaward; whereas its position fluctuated considerably at undeveloped sites. The sediment transport study suggests that the dune crest remains stationary at developed sites, due to limited landward sediment transfers resulting from the use of sand fences to accumulate sediment along the dune toe. The presence of houses may also sufficiently affect air flow to modify sediment transport patterns. These same factors can be expected to contribute to the long-term stability of the dune system. In the long-term study, it was observed that the dune toe advanced seaward at developed sites over the course of several years. This is attributed to fences placed in more seaward positions, which advance the dune toe and widen the dune at the expense of the beach. The presence of fences at the dune toe concentrates sediment in that location and prevents the transfer of sediment farther inland, isolating the original crest and adding little sediment to that part of the dune system. The overall consequence of these activities is a stable dune system, which fluctuates only along the dune toe in response to erosion by storm waves and deposition of windblown sand along the fence line.

9. CONCLUSIONS

The results of this study support the hypothesis that different dune morphologies and dune dynamics exist at developed and undeveloped sites. The amount of sand moved by the wind and the amount of elevation change are considerably less at developed sites than at undeveloped sites. This leads to the conclusions that the net landward transfer of sediment along developed shorelines is restricted by human efforts to stabilize the dune and that the dune line becomes a barrier to sediment movement inland. Certain characteristics of developed sites are important factors affecting sediment transport. These include the presence of obstructions (houses, roads, boardwalks), the characteristics of the surface surrounding the house, and especially the use of sand fences in various locations and configurations. The primary interference is caused by sand fences. Their use on the beach severely reduces the amount of sand moved to the dune zone and their presence along the dune toe produces sediment deposition in that zone. Although the sediment deposited at the dune toe represents an addition to the sediment budget of the dune, the increase is limited because this newly-deposited sand is not stabilized and is subject to removal by offshore or alongshore winds, or by wave erosion.

One factor which affects sediment transport in the dune zone is the interference which buildings have on air flow. This interference needs to be quantified using detailed wind measurements. The effect of the buildings on the sediment transport patterns observed in this study was surmised on the basis of information from other studies, but most air flow studies have been done around high-rise structures, and single family homes, such as those found along much of the New Jersey shoreline, have not been examined in great detail. Information about the relationship between house height and stagnation zones downwind might, for instance, be used to determine setback requirements for new house construction which would permit better dune development. In addition, it was observed that dense housing patterns might account for lower sediment transport in the dune zone. Better air flow information about these conditions might help to establish housing density restrictions.

This study has shown that development is indeed a major factor contributing to the morphology of the dunes developed shorelines. However, more detailed process measurements are needed to understand better the degree to which development alters that morphology.

ACKNOWLEDGEMENTS

The research was made possible by a grant from the New Jersey Department of Environmental Protection, Division of Coastal Resources. The author would like to thank Nancy Humiston, Jim McCluskey, Jim Brosius and Mike Siegel for help with the field work.

REFERENCES

Acrivos, A., Leal, L. G., Snowden, D. D. and Pan, F. (1968) Further experiments on steady separated flows past bluff objects. *J. Fluid Mech.*, **34**, 25–48.

Ashley, G. M., Halsey, S. D. and Buteux, C. B., (1986) New Jersey's longshore current pattern. *J. Coast. Res.*, **2**, 453–463.

Baccus, J. T. and Horton, J. K. (1980) Pedestrian impacts: Padre Island. In Mayo, B. S. and Smith, L. B., Jr (eds), *Proc. Barrier Island Forum Workshop*. National Park Service, North Atlantic Region, Boston, pp. 89–97.

Bagnold, R. A. (1941) *The Physics of Blown Sand and Desert Dunes*. Chapman & Hall, London.

Bressolier, C. and Thomas, Y.-F. (1977) Studies on wind and plant interactions on French Atlantic coastal dunes. *J. Sediment. Petrol.*, **47**, 331–338.

Castro, J. P. (1971) Wake characteristics of two-dimensional plates normal to an air stream. *J. Fluid Mech.*, **46**, 599–609.

Gares, P. A. (1987) Eolian sediment transport and dune formation on undeveloped and developed shorelines. Unpublished PhD dissertation, Department of Geography, Rutgers, The State University of New Jersey.

Hauf, T. and Neumann-Hauf, G. (1982) The turbulent wind flow over an embankment. *Bound. Layer Meteorol.*, **24**, 357–369.

Hesp, P. (1983) Morphodynamics of incipient foredunes in New South Wales. In Brookfield, M. E. and Ahlbrandt, T. S. (eds), *Eolian Sediments and Processes*. Elsevier, Amsterdam, pp. 325–342.

Hoxey, R. P. and Moran, P. (1983) A full-scale study of the geometric parameters that influence wind loads on low-rise buildings. *J. Wind Eng. Indust. Aerodyn.*, **13**, 277–288.

Hsu, S. A. (1987) Structure of air flow over sand dunes and its effect on eolian sand transport in coastal regions. *Proc. Coastal Sed. '87, ASCE*, 188–201.

Leatherman, S. P. (1978) A new eolian sand trap design. *Sedimentol.*, **25**, 303–306.

Mather, J. R., Adams, H. A., III and Yoshioca, G. A., (1964) Coastal storms of the eastern United States. *J. Appl. Meteorol.*, **3**, 693–706.

Nagib, H. M. and Corke, T. C., (1984) Wind microclimate around buildings: characteristics and control. *J. Wind Eng. Indust. Aerodyn.*, **16**, 1–15.

Nordstrom, K. F. (1987) Shoreline changes on developed barriers. In Platt, R. H., Pelczarski, S. G. and Burbank, B. K. R. (eds), *Cities on the Beach*. University of Chicago Department of Geography Research Paper 224, University of Chicago Press, Chicago, pp. 65–79.

Nordstrom, K. F. and McCluskey, J. M. (1985) The effects of houses and sand fences on the eolian sediment budget at Fire Island, New York. *J. Coast. Res.*, **1**, 39–46.

Olson, J. S. (1958) Lake Michigan dune development—1: Wind velocity profiles. *J. Geol.*, **66**, 254–263.

Pearse, J. R., Lindley, D. and Stevenson, D. C. (1981) Wind flow over ridges in simulated boundary layers. *Bound. Layer Meteorol.*, **21**, 77–92.

Peoples, R. A., Jr and Gregg, W. P. (1984) Applications and limitations of science in the definition and delineation of coastal barriers to support formulation of government policy. *Coastal Zone '83 Post-Conference Proceedings*, California State Lands Commission, Sacramento, CA. pp. 1–45.

Phillips, R. A. and Willetts, B. B. (1979) Predicting sand deposition at porous fences. *J. Waterw. Port. Coast. Ocean Div. ASCE*, **105**, 15–31.

Steiner, D. B. and Leatherman, S. P., (1979) *A preliminary study of the environmental effects of recreational usage on dune and beach ecosystems of Assateague Island.* University of Massachusetts, National Park Service Cooperative Research Unit Report 44.

Svasek, J. N. and Terwindt, J. H. J. (1974) Measurement of sand transport by wind on a natural beach. *Sedimentol.*, **21**, 311–322.

Taylor, P. A. and Gent, P. R. (1974) A model of atmospheric boundary layer flow above an isolated two-dimensional 'hill': an example of flow above 'gentle topography'. *Bound. Layer Meteorol.*, **7**, 349–362.

US Army Corps of Engineers (1981) *Barnegat Inlet, New Jersey, Phase 1: General Design Memorandum.* Philadelphia District Office, US Army Corps of Engineers.

Chapter Seventeen

Directions for coastal dune research

K. F. NORDSTROM, R. W. G. CARTER AND N. P. PSUTY

One book on coastal dunes cannot cover all relevant topics or provide a definitive assessment of each of the more important subtopics. We have attempted to concentrate on eolian processes and landforms, while acknowledging the role played by waves, vegetation, and human action. The contributions in this book concentrate on characteristics of wind flow, sources of sediment available to the wind, rates and avenues of eolian transport, and erosional and depositional landforms created primarily by wind action.

Many of the chapters represent attempts to place dunes in the overall coastal system. The consensus is that wave processes play an important role in influencing the size, shape, evolution, and longevity of dunes and that dunes are an integral part of the coastal system, linked to the beach through onshore and offshore sediment exchanges by wave and wind processes.

Many of the chapters point to the variable as well as the persistent nature of coastal dunes. Dune characteristics are variable in that their dimensions and shapes may continually change through a variety of feedback mechanisms. The morphologic variability and mobility of dunes, often seen as undesirable by managers, is a natural characteristic that allows dunes to adapt to environmental change yet retain a long-term niche in the landscape. Management of dunes is often based on subjective assessments; epithets like 'fragile' and 'vulnerable' are still used to describe them, but such terms have no quantitative meaning and little conceptual validity. This type of value judgement stymies the formulation of scientific management. Furthermore, the lack of objectivity has pervaded the decision-making community, particularly where individuals with no training in dune research are involved. One message from the preceding chapters is that a more scientific approach to dune management is both possible and necessary.

Coastal Dunes: Form and process. Edited by K. F. Nordstrom, N. P. Psuty and R. W. G. Carter
© 1990 John Wiley & Sons Ltd

Several chapters call attention to the role played by dunes in coastal evolution across a variety of time scales (McLachlan, Chapter Ten; Orme, Chapter Fourteen; Pye, Chapter Fifteen). The widespread eolian sand deposits of the Holocene are a testimony to the past importance of dune deposition and to the potential value of the dunes as reservoirs of sediment. The predicted rise in sea-level, probably at a rate far greater than at present, will cause a remobilization of many coastal dune systems. It is not known where the release of shoreline sediments will lead to the reactivation of transgressive sand sheets or where sand will enter the littoral transport system to construct new dunes downdrift. The chapters in Section III are based on past events, but the results have application to the prediction of future changes. We require more of these discriminating studies of the structure of existing dunes, coupled with research in areas where sea-level rise is already accelerating. Rate of sediment supply will co-vary with rates of sea-level rise, and one goal would be to specify the dune alterations that would occur as the dimensions of each of these variables change.

The development of models of the physics of sand movement by wind, among the first geomorphological environments to be characterized in quantitative terms (Bagnold, 1941), has not advanced as fast as in other branches of geomorphology (rivers, slopes, glaciers). The superficially simple matter of estimating eolian transport has yet to be adequately described, and consequently, it is not possible to calculate precise sediment budgets, except at intensively monitored sites. Part of the problem may be the lack of a universally acceptable methodology for acquiring field data. Other factors include the complex nature of many source areas and the numerous environmental factors controlling eolian entrainment (Carter and Wilson, Chapter Seven; Carter, Hesp and Nordstrom, Chapter Eleven).

Several of the authors of chapters in this book use wind data gathered at established meteorological stations, remote from their study areas, whereas others use data derived from locally emplaced instruments. Hsu (1987) identifies some of the problems of assuming that remote wind conditions approximate local conditions. The use of remote wind data appears to be effective in predicting long-term potential sediment supply to the dunes or in predicting sand transport in locations where topography is uncomplicated (Davidson-Arnott and Law, Chapter Nine; Gares, Chapter Sixteen), but it appears that local wind data are required for effective process-response modeling of sediment transport rates and short-term changes in areas where there are buildings or complex topography (Gares, Chapter Sixteen). It must also be recognized that winds with the potential to move sand exceed sediment supply at many sites (Carter and Wilson, Chapter Seven). Recording instruments are preferable to visual observations in obtaining local wind data because subsequent evaluation of detailed records enables accurate determination of temporal trends in the characteristics of wind flow (gustiness, duration of flow at specific speeds above the threshold of entrainment). Bauer *et al.* (Chapter Three) provide an example of some of the advantages these instruments provide in calculating sediment transport, even as they caution us to think carefully about how these instruments should be employed.

The decreasing real cost of data recorders and the availability of reasonably priced anemometers and wind vanes provide geomorphologists the opportunity to collect the kind of process data needed to calculate eolian sediment transport rates. The major methodological problem in gathering reliable field data now may be the design of an effective way to measure sediment transport in the field—a problem that is shared by scientists working in other mediums. A variety of methods are illustrated in this book, including fluvial bed load samplers (Bauer *et al.*, Chapter Three), vertical cylindrical traps (Goldsmith, Rosen and Gertner, Chapter Five; Gares, Chapter Sixteen), natural topographic traps (Davidson-Arnott and Law, Chapter Nine), and sequential profiling (Carter and Wilson, Chapter Seven). Some of these methods replicate methods used in the past, and others represent attempts to find a more accurate or expedient method. Many of these and other methods have been reviewed and compared (Horikawa, 1988) but a thorough analysis of the advantages and accuracy of each methodology is still required to specify the direction that future sampling designs should take. We need these comparative studies of field methodologies to achieve greater consistency in study plans.

Most studies of storm effects on dune morphology have concentrated on the significance of wave erosion or overwash; eolian activity is examined in the context of subsequent dune rebuilding (Hosier and Cleary, 1977; Godfrey, Leatherman and Zaremba, 1979; Schroeder, Hayden and Dolan, 1979; Ritchie and Penland, Chapter Six; Gares, Chapter Sixteen). Goldsmith, Rosen and Gertner (Chapter Five) provide quantitative data indicating that storm events account for virtually all of the eolian transfers, although recovery may take several months or even years. Much of the morphology of the dune landward of the crestline is affected by storm winds blowing onshore over the crest (Psuty, Chapter Eight) or through gaps in the foredune (Rosen, 1979; Gares and Nordstrom, 1987; Carter, Hesp and Nordstrom, Chapter Eleven). More work is required to identify the significance of these winds, and field studies should be directed toward obtaining simultaneous wind and sand transport data under storm conditions at a variety of sites.

Many of the research needs identified by scientists working on eolian processes in the coastal zone are similar, conceptually, to those facing workers in beach processes. The establishment of a comprehensive research program similar in goals to the US National Sediment Transport Study or the Canadian Coastal Sediment Study is a worthwhile goal. The high value placed on coastal dunes as a form of shore protection or the ecological significance of dunes may provide the practical or applied rationale for funding this kind of programme, but the unresolved basic research questions must be scientifically coherent. The primary dune or foredune is the focus of interaction of wave erosion and deposition, eolian transport from the beach, and eolian transport to secondary dunes inland. A comprehensive quantitative model of foredune dynamics, including wind effects, sediment exchanges and topographic change using state of the art methodology would provide a viable long-term focus for a comprehensive international program. The specific goal would be the understanding of an important component of the dune system, but important by-products would be the achievement of consistency in study plans and reduction in the likelihood of repeating past mistakes.

There is a growing body of literature on human-induced changes to coastal dunes in a basic research context (Gares, 1983; Nordstrom and McCluskey, 1985), but our understanding of the relationship between waves, winds, and human structures in these environments is far from adequate. Human processes are rarely treated in an explicit process-response context. Several of the chapters in this book make use of data from human-altered systems without specifying the relationship of human effects to the landforms. Several authors refer to human alterations, using value-laden terms such as 'interference' and 'manipulatory' and the resulting landscape is described as 'disturbed' or 'degraded'. Human-induced sediment mobility (and stability) is considered undesirable by the authors of some of the chapters, although other authors, working in undeveloped areas, point to mobility and stability as characteristics of healthy dune systems. Proper assessment of human-altered systems requires a more objective approach, similar to that practiced in natural areas. The study by Gares (Chapter Sixteen) provides a blueprint for an objective approach, and a logical next step would be a definitive, highly instrumented, comparative study of natural and developed dunes. Undeveloped systems are not the only measure by which developed systems should be evaluated; developed systems have a different combination of processes, sediment sources, and geomorphic expression (Nordstrom, 1990). Pye (Chapter Fifteen) shows how human activities may be incorporated into models of dune change that are grounded in the natural system. The key to understanding more intensively developed shorelines may lie in the establishment of alternative models of dune dynamics that treat human-altered environments as the standard.

Several chapters provide quantitative data to help identify rates of transport on the beach (Bauer *et al.*, Chapter Three) and from beach to the dune (Goldsmith, Rosen and Gertner, Chapter Five; Davidson-Arnott and Law, Chapter Nine). Results of studies such as these provide the basis for equations of eolian sediment transport that are specifically tailored to beaches and coastal dunes. However, equations derived from non-coastal settings are still being used as the principal standards by which these field results are interpreted. There is an acute need for better understanding of eolian transport in coastal environments, more effective techniques for sand trapping, recognition of the variables that limit transport, such as scarps and width of source areas, and ultimately, more accurate definitions of windblown sediment budgets.

The state of knowledge of wind flow and sediment transport landward of the dune crest lags even further behind. Many of the results presented in this book are more conceptual or descriptive than the results of investigations on the beach or the seaward portion of the foredune. This is not surprising, considering the complexity of the flow of wind and sediment over the complex terrain (Robertson-Rintoul, Chapter Four). There is a need to quantify the linkage between rates and pathways of transfer between the foredune and secondary dune and mechanics of sand transport in dunes landward of the crest and in blowouts, especially the feedback between windflow and shape, as well as the role of wind erosion in subsequent deposition.

Many of the chapters stress the significance of the concept of the coastal sediment budget. Data are provided to quantify the role played by wave processes and eolian processes operating across the beach. There is a need to identify a standard budgetary period, extend sediment budgets through time, and quantify the rates of transfer farther inland. Previous studies have identified gross rates of transfer, with emphasis on management implications (Sless, 1957). The pathways that these transfers take and the resulting dune morphologies have also been described (Carter, Hesp and Nordstrom, Chaper Eleven; Hesp and Thom, Chapter Twelve). These rates should be linked to changes in the volume and morphology of the secondary dunes.

The chapters indicate that coastal dunes are distinguished from their desert counterparts by wave effects and the more active role played by vegetation and human processes. Some of the individual landforms may be similar (e.g. barchans migrating across deflation surfaces in dunefields—Hesp and Thom, Chapter Twelve; Borówka, Chapter Thirteen), and the results of some experiments conducted in secondary dune environments may be directly applicable (e.g. Robertson-Rintoul, Chapter Four; Borówka, Chapter Thirteen), but coastal dunes are subject to a greater variety of processes than desert dunes, and they appear to undergo more complex cycles of development (Psuty, Chapter Eight). The temporal variable for coastal dunes may extend across thousands of years and involve sequences of mobilization and relative stability (Orme and Tchakerian, 1986; Borówka, Chapter Thirteen; Orme, Chapter Fourteen). Thus, studies of regional or long-term development require a comprehensive research program to incorporate sand availability and limitation through the late Quaternary, sequences of sediment budget variation during the late Holocene, and spatially shifting natural processes, as well as modifications produced by humans in the modern time frame.

The greatest emphasis in this book on factors other than eolian processes is on the effects of waves, regarded as the basis for distinguishing coastal dune forms from other dune forms. This emphasis reflects not only the importance of wave processes but also the background and research interests of many of the authors. A few considerations of human processes are provided to give the reader a feeling for the significance of human inputs and the great need for further research in this area. The significance of vegetation is well documented in previous studies (Hosier and Cleary, 1977; Godfrey, Leatherman and Zaremba, 1979) and is perhaps under-represented in this volume. There is a need to establish more linkages between ecologists and geomorphologists to study vegetation dynamics and sand-deposition processes in order to build on these past works and to consider the regional variability of coastal dune development (Godfrey, 1977; Psuty and Millar, 1989).

Several other areas of inquiry, common in coastal dune research, have not been addressed in this volume, including sediment analyses (Greenwood, 1969; Williams, Evans and Leatherman, 1985), simulation models (Kriebel and Dean, 1985; van de Graaf, 1986), analyses of dunes in the rock record (Dott *et al.*, 1986; Eschner and Kocurek, 1986), and relationship to overwash (Leatherman, 1976). To these may be added several unresolved issues identified in Sherman and Hotta (Chapter Two), the promising avenues of research related to exchange of nutrients (McLachlan, Chapter Ten), and numerous management applications (Cullen and Bird, 1980; Doody, 1985; Tinley, 1985; Chapman, 1989; van der Meulen, Jungerius and Visser, 1989).

This book is not intended as a synthesis of results. It is presented as an example of the kinds of research now being conducted in coastal dunes, along with an indication of some of the potential avenues for future research. There is much work to be done. The large number of variables and the complexity of the interactions point to the need for new data to fill existing gaps, more comprehensive syntheses and models based on existing work, and cooperative efforts between specialists. Hopefully, this spirit of cooperation will take on world-wide scope, and organizations such as the European Union for Dune Conservation and Coastal Management (van der Meulen, Jungerius and Visser, 1989) can expand to reach a wider audience. It is hoped that this volume will help stimulate interest in these efforts and help establish a coordinated, large-scale research program.

REFERENCES

Bagnold, R. A. (1941) *The Physics of Blown Sand and Desert Dunes*. Methuen, London.
Chapman, D. M. (1989) *Coastal Dunes of New South Wales: Status and Management*. University of Sydney Tech. Rep. 89/3, Sydney.
Cullen, P. and Bird, E. C. F. (1980) *The Management of Coastal Sand Dunes in South Australia*. Geostudies, Black Rock, Victoria.
Dott, R. H. Jr., Byers, C. W., Fielder, G. W., Stenzel, S. R. and Winfree, K. E. (1986) Aeolian to marine transition in Cambro-Ordovician cratonic sheet sandstones of the northern Mississippi Valley, USA. *Sedimentol.*, **33**, 345–367.
Doody, P. (ed.) (1985) *Sand Dunes and their Management*. Nature Conservancy Council, Peterborough, UK.
Eschner, T. B. and Kocurek, G. (1986) Marine destruction of eolian sand seas: origin of mass flows. *J. sediment. Pet.*, **56**, 401–411.
Gares, P. A. (1983) Beach/dune changes on natural and developed coasts. *Proc. Coast. Zone '83, ASCE*, 1178–1191.

Gares, P. A., and Nordstrom, K. F. (1987) Dynamics of a coastal foredune blowout at Island Beach State Park, NJ. *Proc. Coast. Sed. '87, ASCE*, 213–221.

Godfrey, P. J. (1977) Climate, plant response, and development of dunes on barrier beaches along the US East Coast. *Internatl. J. Biometeorol.*, **21**, 203–215.

Godfrey, P. J., Leatherman, S. P. and Zaremba, R. (1979) A geobotanical approach to classification of barrier beach systems. In Leatherman, S. P. (ed.), *Barrier Islands*. Academic Press, New York, pp. 99–126.

Greenwood, B. (1969) Sediment parameters and environmental discrimination: an application of multivariate statistics. *Can. J. Earth Sci.*, **6**, 1347–1358.

Horikawa, K. (ed.) (1988) *Nearshore Dynamics and Coastal Processes: Theory, Measurement, and Predictive Models*. University of Tokyo Press, Tokyo.

Hosier, P. E., and Cleary, W. J. (1977) Cyclic geomorphic patterns of washover on a barrier island in southeastern North Carolina. *Environ. Geol.*, **2**, 23–31.

Hsu, S. A. (1987) Structure of airflow over sand dunes and its effect on eolian sand transport in coastal regions. *Proc. Coast. Sediments '87, ASCE*, 188–201.

Kriebel, D. L. and Dean, R. G. (1985) Numerical simulation of time-dependent beach and dune erosion. *Coast. Eng.*, **9**, 221–245.

Leatherman, S. P. (1976) Barrier island dynamics: overwash processes and eolian sediment transport. *Proc. 15th Conf. Coast. Eng.*, 1958–1974.

Nordstrom, K. F. (1990) The concept of intrinsic value and depositional coastal landforms. *Geog. Rev.*, **80**, 68–81.

Nordstrom, K. F. and McCluskey, J. M. (1985) The effects of houses and sand fences on the eolian sediment budget at Fire Island, New York. *J. Coast. Res.*, **1**, 39–46.

Orme, A. R. and Tchakerian, V. P. (1986) Quaternary dunes of the Pacific Coast of the Californias. In Nickling, W. G. (ed.), *Aeolian Geomorphology*. Allen & Unwin, Boston, pp. 149–175.

Psuty, N. P. and Millar, S. W. S. (1989) Coastal dunes: a question of zonality. *Essener Geogr. Arbeiten.*, **18**, 149–169.

Rosen, P. S. (1979) Aeolian dynamics of a barrier island. In Leatherman, S. P. (ed.), *Barrier Islands*. Academic Press, New York, pp. 81–98.

Schroeder, P. M., Hayden, B. and Dolan, R. (1979) Vegetation changes along the United States east coast following the Great Storm of March 1962. *Environ. Mgmnt.*, **3**, 331–338.

Sless, J. B. (1957) Coastal sand drift. *J. Soil Cons. NSW*, **13**, 146–158.

Tinley, K. L. (1985) *Coastal Dunes of South Africa*. Rep. 109 South African National Scientific Programs, Council for Scientific and Industrial Research, Pretoria.

van de Graaf, J. (1986) Probablistic design of dunes: an example from the Netherlands. *Coast. Eng.*, **9**, 479–500.

van der Meulen, F., Jungerius, P. D. and Visser, J. (eds) (1989) *Perspectives in Coastal Dune Management*. SPB Academic Publishing, The Hague.

Williams, A. T., Evans, N. H. and Leatherman, S. P. (1985) Genesis of Fire Island foredunes, New York. *Sediment. Geol.*, **42**, 201–216.

Index